HATHOR'S ALCHEMY

THE ANCIENT EGYPTIAN ROOTS OF THE HERMETIC ART

ALISON M. ROBERTS

NorthGate

First published in 2019 by
NorthGate Publishers
3 Court Ord Cottages
Meadow Close
Rottingdean
East Sussex
England BN2 7FT
www.northgatepublishers.co.uk

Design: James Lawrence

ISBN 978-0-9524233-3-1

Printed by KHL Printing Co Pte Ltd,
Singapore

ACKNOWLEDGEMENTS

This book has taken many years to write, the seeds being sown already in 1974 when I first visited Nefertari's temple at Abu Simbel, and was deeply moved by its mysterious beauty. Along the way I have valued the friendship of Warren Kenton, Joanna Lapage-Browne, Khojeste Mistree and John Ramsay. I have also greatly benefited from conversations with Jeremy Naydler, whose knowledge of Western alchemical treatises and *Books of Hours* has led me to manuscripts I would never have found otherwise. I would like to thank Pat Rae for taking me to a foundry in London to see molten copper being poured into moulds. It was an unforgettable experience, which brought home to me how a crucible filled with red-hot metal truly glows like a rising sun. I would also like to thank Charlotte Kelly, who represents my books, for her on-going encouragement.

Once again, I have been fortunate to have access to the unique library collection of the Warburg Institute in London, and thank the staff for their helpfulness at all times. I would like to thank Adam McLean for generously giving me a copy of *Aurora consurgens* and for help with alchemical queries. I am also grateful to Alan Williams for permission to quote from his translation of Rūmī's *Masnavi*. For help with photographs I would like to thank the following: Theodor Abt at the Living Human Heritage Project in Zurich; Christian Eckmann at the Römisch-Germanisches Zentralmuseum, Mainz; Giuseppe Fanfoni; Christine Green; Barbara Heller at the Werner Forman Archive; Erik Hornung; Christian Leitz at the Athribis-Projekt, Tübingen; Sister Philippa Rath OSB and all the Benedictine sisters at the Abbey of St Hildegard, Eibingen; Elisabeth Walczuk at Brepols Publishers; Arman Weidenmann at the St Gallen Library. My thanks are also due to the Institut français d'Archéologie orientale du Caire (IFAO) for kindly allowing me to reproduce line-drawings from their publications.

I am indebted to Elizabeth Henry for her careful editing, including her considerable help with all the detail in the notes.

For the integration of pictures and text, which is so important to the book's overall conception, I owe James Lawrence special thanks. Good-humoured throughout, not least when changes and additions needed to be made, and foreign scripts' diacritical dots and lines had a habit of going astray, his unerring attention to every detail has ensured the beautiful design. I have greatly appreciated being able to work with him and see the book emerging into its completed form.

The generosity of my brother, John Roberts, has also ensured it sees the light of day, and I am very grateful to him. Finally, I would like to thank Helen Maimaris and Sam Gladstone for their companionship on our journey to Egypt. In particular, I would like to thank Helen for our visit to the Qaitbay Mosque in Cairo and her helpful comments on the book, and Sam for his lovely photography. They both carefully prepared many photos for publication, and their help during the later stages has been invaluable. I offer them now my heartfelt thanks.

DEDICATION

This book is dedicated
to my son Sam

CONTENTS

PREFACE

Living in the modern world it is perhaps difficult to imagine how our souls circle with the sun each day, how we live attuned to the cycles of the moon and stars, and journey through a whole lifetime in the course of a day. Certainly, biologists now recognize the importance of circadian rhythms for our life on earth, and also know the 24-hour cycle of waking and sleeping, of day and night, is regulated by the eye and light. Yet these biorhythms were known to the ancient Egyptians millennia ago as sacred knowledge, a radiant science of illumination reverently expressed through image, myth and praise-giving. All earthly life grew and transformed in tune with these rhythms—and not just humans, but animals, plants and even the precious stones buried deep in mountains and rocks. And each day at dawn, these wonders of creation were birthed anew by the great solar Eye goddess, Hathor.

Such is the message in her magnificent Graeco-Roman temple at Dendara. It is also the message in Ramesses VI's stunning tomb in the Valley of the Kings at Thebes; and in the two temples at Abu Simbel where King Ramesses II and Queen Nefertari are everywhere to behold as cosmic beings, incarnations of the sun god, Re, and Hathor. Here the royal couple breathed into Egyptian culture a spirit of beauty and grandeur, excavating these vast chambers in the Nubian sandstone which have endured for more than 3,000 years, and where knowledge is preserved that eventually passed into alchemy.

That alchemists inherited this Egyptian knowledge might seem hardly credible. After all, were they not deluded gold-makers, whose futile quest to change base lead into gold ultimately led to modern chemistry? Moreover, according to most modern-day scholars, when alchemy first emerged in the diverse world of Hellenistic and early Roman Egypt, it originated from 'technical' craft recipes, its theoretical principles being provided by Greek philosophy and Gnosticism. In this Greek-Gnostic narrative, ancient Egypt is conspicuous by its absence—a fate tellingly encapsulated by the erasure of Hathor's features on temple columns at Dendara, as seen on the front cover of this book.

Yet alchemists insisted their knowledge came from the Egyptian temples. As guardian of its secrets, they also honoured Hermes Trismegistus, the 'thrice-great' Greek Hermes, whom the ancient Egyptians knew as Thoth, the god of wisdom, the divine scribe, and protector of the priestly arts and sciences. Hence, alchemy became known as the Hermetic art, though the re-expression in a new idiom unfortunately eclipsed its Egyptian origins.

This book seeks to restore alchemy to the healing sacred art it always was, a living and transformative wisdom known to the ancient Egyptians for centuries. It is also the fourth—and final—book in the series I have written on Hathor and the Divine Feminine, and, in a way, this Hathor quartet has journeyed through alchemy's paint-box, starting with the black and white publication of *Hathor Rising: The Serpent Power of Ancient Egypt* (1995), which remained firmly in ancient Egypt, to the present publication of *Hathor's Alchemy* in colour over 20 years later.

At the close of *My Heart My Mother: Death and Rebirth in Ancient Egypt* (2000), I had begun this alchemical exploration; then again, fleetingly, in *Golden Shrine, Goddess Queen: Egypt's Anointing Mysteries* (2008). I also knew such an exploration would require a much more detailed study, something that eluded me, and probably would have continued to do so, had I not discovered Ingolf Vereno's edition of the Arabic *Epistle of the Secret* in the Warburg Institute library. I soon excitedly realized that here was the missing link, enabling me to reconnect alchemy with ancient Egypt, though it has certainly not been one-way travel, since alchemy highlights aspects I might otherwise have overlooked in the Egyptian sources. While Vereno took a very different approach, interpreting the *Epistle of the Secret* from a Gnostic perspective, I would like to acknowledge here my indebtedness to his publication, without which my own work would not have been possible.

Parts 1 to 3 set out the Egyptian wisdom, including what can be gleaned about the highly-secret metallurgical beliefs presided over by Hathor as goddess of copper and gold. Parts 4 and 5 then turn to the transmission, and, inevitably, they cover a vast span of time, ranging from Hellenistic Egypt to medieval Europe via the Islamic world. However,

it should be noted the book is not intended as a comprehensive study of alchemy in these different eras (which is a subject in itself). Nor does it approach alchemy as a monolithic tradition, based solely on the Sulphur–Mercury theory. Instead, it explores those alchemical treatises which, if understood within the Egyptian tradition, cease to be a bewildering maze of 'incomprehensible' symbols, becoming instead a living way of transformation patterned on the old Egyptian 'work of a day'.

In order to keep the book within limits, I have needed (apart from brief references) to omit Renaissance and later alchemy, which anyway continued medieval knowledge. What has not been fully appreciated in studies of Renaissance Hermetism, though, is the extent to which the spirit of Hermes was already alive in the European Middle Ages through alchemy, long before Marsilio Ficino translated the newly-discovered *Corpus Hermeticum* into Latin in 1463 in Florence. Difficult though this influence may be to trace in detail, the great 12th-century Benedictine abbess, Hildegard of Bingen, seems to have been very familiar with alchemy's principles. Dante's absorption of the Hermetic art also becomes obvious once the Egyptian tradition on which he drew is understood. Further afield, Islamic medieval mystics similarly derived inspiration from its wisdom. As Ibn Umail observed, alchemy speaks to 'people of any religion'.

CHRONOLOGICAL OUTLINE

3000 BCE — **EARLY DYNASTIC PERIOD**
c. 3100–2686 BCE

2500 BCE — **OLD KINGDOM**
c. 2686–2181 BCE   DYNASTIES 3–6

FIRST INTERMEDIATE PERIOD
c. 2181–2040 BCE   DYNASTIES 7–10

2000 BCE — **MIDDLE KINGDOM**
c. 2040–1650 BCE   DYNASTIES 11–13

SECOND INTERMEDIATE PERIOD
c. 1750–1570 BCE   DYNASTIES 14–17

1500 BCE —

NEW KINGDOM
c. 1570–1070 BCE   DYNASTIES 18–20

1000 BCE —

THIRD INTERMEDIATE PERIOD
c. 1070–664 BCE   DYNASTIES 21–25

500 BCE — **LATE PERIOD**
664–343 BCE   DYNASTIES 26–30

PERSIAN KINGS 343–332 BCE
MACEDONIAN KINGS 332–305 BCE
PTOLEMAIC PERIOD 305–30 BCE

0 —

ROMAN PERIOD 30 BCE–395 CE

500 CE — **BYZANTINE PERIOD EGYPT**
395–641 CE

1000 CE — **ISLAMIC PERIOD EGYPT**
641 CE–PRESENT

HIGH MIDDLE AGES WESTERN EUROPE
c. 1000–1300 CE

LATE MIDDLE AGES WESTERN EUROPE
c. 1300–1500 CE

1500 CE —

2000 CE —

MAP OF EGYPT

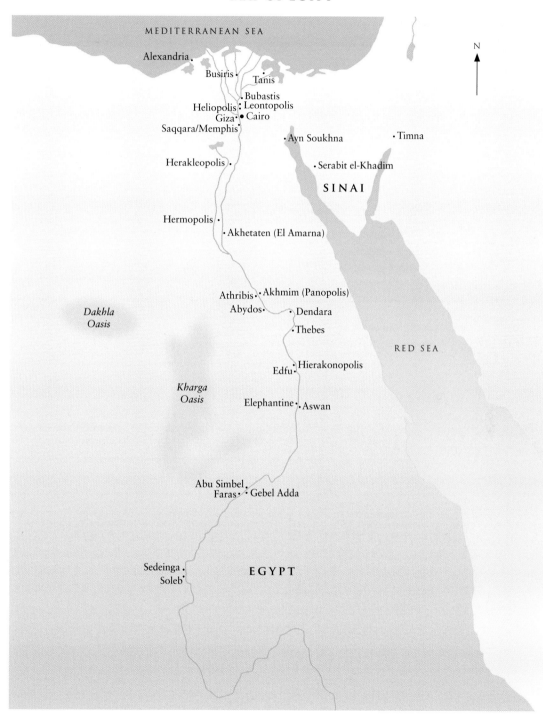

MEDITERRANEAN SEA

Alexandria.

Busiris.

Tanis.

Bubastis
Heliopolis. Leontopolis
Giza. Cairo
Saqqara/Memphis.

Ayn Soukhna.   .Timna

Herakleopolis.

.Serabit el-Khadim

SINAI

Hermopolis.

.Akhetaten (El Amarna)

Dakhla
Oasis

Athribis. .Akhmim (Panopolis)
Abydos. .Dendara
.Thebes

RED SEA

Edfu. .Hierakonopolis

Kharga
Oasis

Elephantine. .Aswan

Abu Simbel.
Faras. .Gebel Adda

Sedeinga.
Soleb.

EGYPT

N

INTRODUCTION

Deep in the south of Egypt, in the sandstone cliffs of Lower Nubia not far from the Sudanese border, lies a temple which, over 3,000 years after it was built, still draws people to its dark interior. It is the temple of Queen Nefertari at Abu Simbel *(fig.3)*, a mysterious cavern where visitors may sense a numinous presence, a deeply feminine presence seeming to speak across the ages.[1] Perhaps they hear the voice of its indwelling deity, the love goddess Hathor, or perhaps it is Nefertari herself who speaks, since this is her house, her holy place *(fig.3)*. According to a dedicatory inscription, Ramesses II built it for 'the Great Royal Wife Nefertari-beloved-of-Mut, a temple excavated in the pure mountain of Nubia, in beautiful shining sandstone as a work of eternity'. He also paired it with a larger temple nearby dedicated to the divine state trinity of the sun god, Re-Harakhti, the Theban god, Amun-Re, and the Memphite craft god, Ptah, whose divine figures, hewn from the solid rock, are shown enthroned with Ramesses within the innermost sanctuary of their temple.

Work on these majestic temples must have started early in the king's long reign (1279–1212 BCE), as Nefertari's temple was completed around year 21, when she had but a few years to live.[2] She was certainly alive when Ramesses signed his famous peace treaty with the Hittites that year, guaranteeing peace and stability after 20 years of inconclusive armed conflict, for a letter sent by Nefertari (or Naptera as her name was written in

LEFT fig.2 Wall-painting of Queen Nefertari wearing the vulture headdress and tall plumes characteristic of New Kingdom queens. (Tomb of Nefertari, Valley of the Queens, Western Thebes. 19th Dynasty.)

TOP RIGHT fig.3 View of Nefertari's temple at Abu Simbel showing colossal statues of the queen and Ramesses II carved in the recesses. (19th Dynasty.)

the cuneiform script) to her Hittite counterpart, Queen Pudukhepa, who had apparently written to her enquiring about her health, was discovered in the Hittite state archives in Boghazköy. In suitably flowery language, Nefertari expressed the wish that the Egyptian sun god and the Hittite storm god might bless the cordial relations between their husbands 'and make the brotherly relationship between the Great King, the King of Egypt, and his brother, the Great King, the King of Hatti, last forever.' [3] Around the time she was sending this greeting to the Hittite capital high on the Anatolian plateau, together with jewels and linen garments for Pudukhepa, her Nubian temple would have been nearing completion.

There had, of course, been other Hathor temples in Nubia, for this 'Gold Land' was indisputably the domain of the great goddess, whose precious metal was much sought after by the Egyptians, whether as dust, rings, nuggets or ingots. Just to the south of Abu Simbel at Faras towered a huge outcrop of rock known locally as 'the Tower of Gold'. Here several 18th-Dynasty pharaohs had honoured Hathor as 'Lady of Ibshek', in a terraced temple reminiscent of Hatshepsut's spectacular temple at Deir el-Bahri on the west bank at Thebes, with its Hathor cave excavated in the mountain. Further south, at Sedeinga in Upper Nubia, Queen Teye had her own Hathor temple, paired with Amenhotep III's temple at Soleb, whilst at Gebel Barkal below the fourth cataract, the later Sudanese rulers of Egypt dedicated a rock sanctuary to Hathor-Mut.

No other temple, however, provides such an intact picture of Egyptian queenship as Nefertari's, though that it remains at all is little short of a miracle. Teye's temple at Sedeinga is now in a ruinous condition, and the Faras site, which encompassed centuries of history through the Pharaonic, Christian and Islamic periods, was completely submerged by the rising waters formed when the Aswan Dam was built in the 1960s, and today lies beneath Lake Nasser. Nefertari's temple was originally located close to the water's edge (fig.6), but fortunately a massive international rescue operation, aided by UNESCO, ensured its removal to safety not far from its original location, thus preserving this 'work of eternity' for future generations through an extraordinary feat of modern engineering. To be sure, it had to be dismantled block by block and rebuilt against an artificial cliff on higher ground, yet still it continues to exude an air of profound sanctity.

To reach Abu Simbel, a modern-day visitor can travel by boat, by plane or by a road through the desert, though to fly there is the most magical. Glimpsed from the aircraft's windows, the rocks beneath form strange spirals in the yellow sand, whilst rivulets of the Nile meander finger-like into the desert, making firm purple ridges where water and sand meet. Everywhere, shapes seem to leap out from the rocks. Lizards, crocodiles and birds, as if sculpted by some unseen hand, impress themselves upon the imaginative eye. Then suddenly, as the plane circles above Lake Nasser's milky-green water, huge figures of Ramesses II

appear, seated in splendour fronting his temple in the western cliffs *(fig.50)*, and the magnitude of the building task the ancient Egyptians faced strikes home as the desert rock is silhouetted against the immeasurable silent space of the land.

Nubia is, of course, where Hathor flees as the raging Sun Eye, deserting the sun god, Re, who is ruling in Egypt, as told in the *Goddess in the Distance* myth inscribed in Graeco-Roman temples.[4] Manifesting as the savage lioness Sekhmet, the fleeing goddess prowls the Nubian deserts, eating the blood and flesh of her enemies as flames leap from her bloodshot eyes, fire blazes from her breath and her heart burns with anger.

Yet Re cannot rule without her—separation from his Eye leaves him unprotected against his enemies—and he desires her return. He sends the wise moon god, Thoth, and the air god, Shu, disguised as apes, to entice her back to Egypt with their music, dancing and humorous stories, promising her intoxicating festivities if she returns with them. When they finally arrive in Egypt, to cool her heat she is plunged into the waters of the

Abaton on the island of Biga and transformed into the joyful, beautiful Hathor, the 'green' goddess who returns 'in peace' from the South, the propitiated 'Lady of Favours, Mistress of the Dance, Great of Attraction', who once again bestows all her blessings on Egypt.

As queen of Egypt, Nefertari incarnates this loving-destructive goddess power. On her temple's façade she twice appears crowned with Hathor's horned headdress, a radiant epiphany of the beneficent Sun Eye and companion of the king *(fig.3)*, their towering figures sculpted in such a way that light and shade continually play upon their crowns and bodies throughout the day, whilst round their huge legs, at knee-height, peep their six children—two princesses grouped with their mother, four princes clustered around Ramesses.

Proclaimed on this façade, too, is that Re rises because of his love for Nefertari, just as it is said that he comes forth in order to see Hathor's loveliness at Dendara. Here it is the queen who compels him to shine. Hers is the power to move the sun in its heavenly circuit. Love, harmony and beauty radiate from her sacred dwelling, and how could it be otherwise when she accompanies a king ruling as 'son of Re'?

ABOVE fig.4 Hathor and Queen Nefertari. (Tomb of Nefertari, Valley of the Queens, Western Thebes. 19th Dynasty.)

RIGHT fig.5 Colossal statue of Queen Nefertari holding a sistrum on the façade of her temple at Abu Simbel. (19th Dynasty.)

FAR RIGHT fig.6 View of the two temples at Abu Simbel in their original waterside setting before they were moved to safety in the 1960's.

Twice a year, around the modern calendar dates of 22 October and 22 February, the great temple's full glory is revealed at sunrise, when the solar rays shine straight through the entrance doorway and creep slowly down the central aisle and through the huge chambers until finally they suffuse the sanctuary's gloom with solar warmth, illuminating the figures of Amun-Re, Ramesses and Re-Harakhti, who, along with Ptah, are enthroned impassively there *(fig.51)*, and revitalizing them with life-giving divine light. No wonder Thoth's baboons reverently greet the rising sun high up on the temple's outer façade *(fig.50)*, for to witness this solar miracle deep in the western mountain is a profoundly moving experience.

In contrast, unlike the great temple's orientation almost due east, the axis of Nefertari's temple lies in a southeast–northwest direction, so sunlight never reaches further than the entrance porch. Yet she is beloved of Re, and within her dark cavern, each relief, each architectural element and each royal gesture, through its placement and function, is designed to reveal her role in the mysteries of the solar circuit. Egyptian temples are built to mirror the eternal cosmic order—they are 'images of heaven', their rituals corresponding in every detail with the movements of the celestial bodies—and Nefertari's temple is no exception. It, too, is a cosmos in miniature.

Indeed, according to the classic Middle Kingdom *Story of Sinuhe*, the Egyptian queen is an 'image of heaven' ruling in a palace on earth. 'Your heaven is in the palace', says King Senwosret I in a letter to Sinuhe, an erstwhile courtier living in exile somewhere in Asia, as he urges him to return to the all-encompassing embrace of the palace queen he once served. She is 'a heaven on earth', an incarnation of the sky goddess for her people, whose head is 'adorned with the rulership of the land'—all of which is certainly not royal hyperbole on Senwosret's part, but rather an unusually explicit allusion to his queen's cosmic nature.[5]

Nobody would have entered the queen's temple at Abu Simbel without this knowledge, though how the cosmos is mirrored in its three chambers has gone curiously unstudied in the Egyptological literature. Yet there are clues, particularly in the cosmographic 'book' known as the *Book of Day*, which shows Re travelling from southeast to northwest, from dawn to dusk, along precisely the same directional axis as Nefertari's temple, a journey, perhaps not surprisingly, that is deeply feminine, being completely encompassed by the outstretched body of the sky goddess, Nut.

In fact, it is this course of the sun that unravels the meaning of Nefertari's dwelling, for day after day it lives and breathes in harmony with this heavenly rhythm—waking at dawn, being vivified through the day and then inexorably sleeping at night—and by living this cosmic rhythm on earth, the royal couple manifest their power as rulers of Egypt. Just as Re travels across Nut's body each day, so the royal couple journey through this feminine earthly palace, living their destiny attuned to the heavenly rhythms above and travelling the celestial goddess way. Therefore, finding the key to Nefertari's temple means delving further into the little-known *Book of Day*…

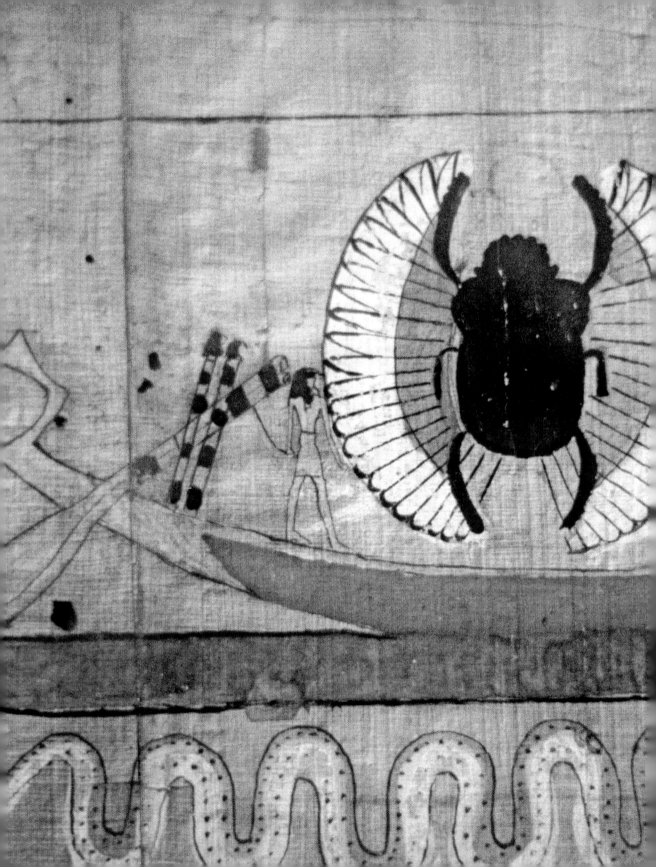

PART I

LIVING THE HOURS: THE BOOK OF LIFE

LEFT *fig.7* The power of 'Becoming' here symbolized by the winged scarab, Khepri, in the solar boat. Standing at the prow are Hathor and Maat, who vitalize and guide the sun god's journey through the hours of the day. (Vignette from the papyrus of Nesikhonsu, Egyptian Museum, Cairo, S.R.IV. 544/JE36465. Dynasty 21.)

ONE

RAISING BEAUTY: MYSTICISM OF LIGHT

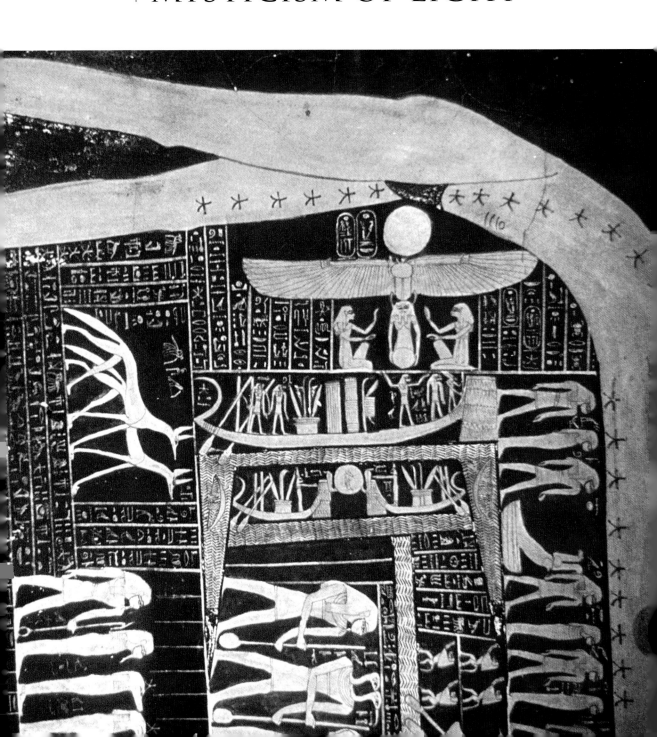

Nut's awesome way of life, death and rebirth is charted in great detail in Ramesses VI's tomb in the Valley of the Kings at Thebes, where, high above on the sarcophagus chamber's ceiling, two hugely elongated figures of the star-spangled goddess, lying back to back and floating in the heavenly waters, enclose the sun god's journey through the hours of the day and night.

The sheer scale of these two sequences is breathtaking, though their prosaic modern-day titles—the *Book of Day* and the *Book of Night*—scarcely convey their true meaning.[1] For their vivid images and texts are a 'celestial geography', mapping in great detail the mythic territory traversed by Re and his companions in the sun boat—the great life, death and regeneration cycle he lives with the celestial mother goddess in an alternating rhythm of 'waking and sleeping',

'living and dying', 'inhaling and exhaling'.[2] At night there is an inhalation as the sky goddess draws the ageing sun god towards the safety of her arms in the Northwest, towards the interior of her body, within which he travels during the 12 hours of the night. By day there is an exhalation as the sun comes forth, expressed in the expansive gesture of the air god, Shu, joyfully raising the sun boat aloft in the *Book of Day*'s first hour *(figs.8, 10)*.

FAR LEFT *fig.8* Detail of the *Book of Day*'s first hour. (Ceiling of Ramesses VI's sarcophagus chamber, Valley of the Kings, Western Thebes. 20th Dynasty.)

ABOVE *fig.9* View of the Valley of the Kings at Thebes. The doorway in the centre is the entrance to the tomb of Ramesses VI.

LEFT *fig.10* Enclosed by the legs and body of the sky goddess Nut, an unnamed goddess, surely Hathor, is about to give birth to the foetal child in her womb, assisted on either side by two attendants. Above her are a winged scarab and solar disk, and beneath is the air god Shu, his arms upraised to support the sun god's birth. (Line-drawing of the first hour of the day in Ramesses VI's tomb.)

This daytime sky encompassed by Nut is the sun god's 'light span' of life, beginning with his birth in the first glimmer of light in the Southeast and journeying through increasing brightness to noonday sovereignty, followed by fullness and maturity in the afternoon of life. Then, as the light wanes, he descends into the sunset of old age in the Northwest, where the great sky goddess swallows the sun in order for him to be reborn again at dawn (*fig.11*). Giving life and taking life, she forms the protective space in which the sun god lives, grows and changes during the daytime hours, and he is the fiery orb, the falcon-headed god, tracing his arc of light across her heavenly circuit, whose unfolding life brings light to the world from dawn to dusk, from sunrise to sunset.

His is a heavenly circuit, too, intimately related to life on earth.[3] For this passage of time influences all growth and development, blessing humans with the gift of life—and not just humans, since, according to the *Book of Day*'s introductory text, every animal, every worm even, receives life when Re rises in the eastern horizon. As present-day biologists are discovering through their research into circadian rhythms, this diurnal cycle is fundamental to all life on earth, they also know this 'body clock' is regulated by the eye and light. The difference is that in ancient Egypt this is a sacred science.

The *Book of Day* is best preserved in Ramesses VI's tomb, where in fact it is shown twice, and reappears subsequently in an abbreviated version in

ABOVE *fig.11* The king welcomes the sun boat in the Northwest at the close of day as the sky goddess Nut prepares to swallow the sun at her mouth. Among the deities in the bottom register is a serpent-headed figure named the 'Headless One'. (Detail from the *Book of Day*, Tomb of Ramesses VI, Western Thebes. 20th Dynasty.)

BELOW *fig.12* Here a vertical strip of zigzag lines, representing the river Nile, divides the *Book of Day* into its eastern and western phases. It also marks the ninth hour of the day when the sun boat meets the annual river axis. (Line-drawing of the abbreviated version of the *Book of Day*, tomb of Ramesses IX, Valley of the Kings, Western Thebes. 20th Dynasty.)

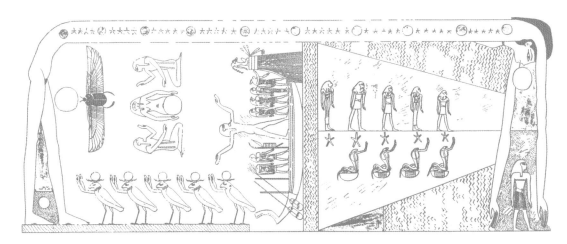

Ramesses IX's tomb *(fig.12)*. It is also depicted in Osorkon II's tomb at Tanis, and knowledge of it evidently continued right through into the Ptolemaic era, as witnessed by the inclusion of isolated scenes on private sarcophagi.[4] Yet its significance for Nefertari's temple has gone unnoticed, perhaps because at first sight its bewildering array of almost identical deities, all striding along with the sun boat beneath Nut's arched body, is utterly unlike anything at Abu Simbel. Indeed, it would be extremely difficult to understand its meaning were it not for the *Ritual of Hours*, 12 hymns praising the sun god's 12 transformations through the daytime hours and characterizing the different events of each one.[5] These secretly transmitted chants, first appearing in Hatshepsut's 18th-Dynasty temple at Deir el-Bahri and still inscribed in Graeco-Roman temples, contain the visionary wisdom inspiring the royal cult of the sun, and, together with the *Book of Day*'s heavenly 'map', form a powerful testimony to the Egyptian mystical understanding of daytime transformation. For not only does each hour mark the passage of time, but its mythic space also opens up a completely different dimension—the goddess 'life world' in which Re moves, grows and develops from birth to death.[6]

These hymns are not, though, sacred literature to be read silently on temple walls—they are for ritual use. True to the spirit of the Memphite demiurge Ptah, who forms creation through his heart and sacred word, they belong within a solar cult honouring the voice as a shaping power, an instrument bringing to birth a vibrant living world. Thus, whenever priests chant them in the sun temples, they attune themselves to Re's heavenly movements. Through their praise-giving, Nut's way becomes an animate reality filled with life and energy, a continuous circuit of light unfolding throughout the day. Like Thoth's baboons perched high at Abu Simbel, these priests participate in

creation, attuned to the moral order of the cosmos, the eternal source of light and life. Circling with the sun each day, their souls are illuminated by these solar mysteries of life and creativity.

RISING DAWN: GESTATING SUN

Each day begins at six o'clock, and each hour rises for a particular god or goddess, the keynote being struck in the very first hour, called 'She who raises the Beauty [*nfrw*] of her Lord'. This is the hour rising for the wisdom goddess, Maat *(fig.13)*, whom the king brings into the sun boat to guide the journey, proclaiming right at the outset his deep desire to establish her way.

All is veiled, however, in this secret region of the eastern lightland, where Re is still a gestating child,

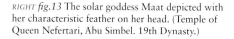

RIGHT fig.13 The solar goddess Maat depicted with her characteristic feather on her head. (Temple of Queen Nefertari, Abu Simbel. 19th Dynasty.)

not yet born and still embraced by his mother, a nourishing serpent mother no less, whose seven uraeus snakes gather together his *Ka*-power:

> Shine, Re, O become Khepri, the Self-Begetter…
> O you who appear in this your birth…
> May you come forth whole
> Within the embrace of your Mother
> Who raises your beauty daily.
> Open for you is the eastern lightland…
> May the king cause Maat to rise to the bull of Maat.
> May the king cause Maat to enter the boat of Re,
> So that Maat may unite with the throne.
> O hail to these
> Your seven uraeus snakes,
> Who provide your *Ka*-powers.
> Who cause havoc among your enemies
> In the Lake of Knives.
> O you who appear, O Horus-Re,
> On your seat at the front of the great barque,
> Giving light high above all lands.[7]

Correspondingly, the *Book of Day*'s first-hour icon, placed directly beneath Nut's vulva, graphically depicts the sun god within his mother's womb, which is shaped like a huge sun disk *(fig.14)*. Almost too small to see, located high on the tomb ceilings, this is the vision experienced at first light, a rare embryological portrayal of the gestating sun god, the 'little Re', as he is called on an ostracon discovered in Ramesses IX's tomb, though the goddess in her birth pangs, flanked by a kneeling midwife on each side, is apparently too numinous to be named during this rising dawn hour when all is flowing, all is vibrant, when baboons raise their arms in praise and great jackals, spirits of the East, stride towards the horizon gates, opening the ways of dawn.

No such reticence prevails in a later Demotic litany, however, which continually invokes Hathor as the mother of the sun, with each repetition of her name being followed by a different epithet praising her as the mysterious birthing goddess:

> Come, Goddess of Mystery, giving birth, Hathor.[8]

And here the word for mystery, *itn*, intentionally evokes the name of the sun disk, the Aten (*itn*), thus conjuring up an image of the great womb disk, the solar Eye, mirroring exactly the *Book of Day*'s birthing icon. Nameless though she may be in the *Book of Day*—perhaps because to name her would betray secrets known only to initiates in the mysterious region located below the visible horizon—this late litany unveils the parturient mother's true identity. She is Hathor, the great Sun Eye—not, as so often stated, the sky goddess, Nut.[9] Charged with the power of childbearing, this fiery serpent Eye goddess manifests with her seven snakes to vitalize and protect the 'little Re' she cradles within her. Born from the womb like all humans on earth, he is entirely dependent on her loving-destructive power for his rising. And what is desired, above all, as the first-hour chant concludes, is for 'greenness' to shine, for cool water to flow in the fire island's heat, and the 'green bird' to soar, that bird messenger of dawn sometimes depicted as a swallow perched on the prow of the day boat:

> May the king open the way of the green bird for you
> So that you drink water on the fire island's banks.[10]

All too easily this irascible solar Eye's fire can burn and blaze out of control; all too quickly Hathor, the life-giver, can become Sekhmet, the destroyer.

ABOVE *fig.14* The juxtaposition of the closing scene of the *Book of Night* (right) and the first hour of the *Book of Day* (left) in Ramesses VI's tomb.

Both rage and radiance emanate from this dawn mother, and necessarily so, since there are evidently hostile forces seeking to impede the sun god's birth in the 'Lake of Knives', but when this radiant mother shines, when the green bird truly flies and 'beauty is raised', she becomes the flame of life.

Soaring above her is the winged scarab, Khepri *(fig.14)*, the 'Becoming One' whose very name expresses everything desired for the sun god as he begins his life journey: the power to 'transform', to 'grow' and to 'change'.

In this heavenly circuit, where Nut provides the heavenly space, it is the 'leaving' and 'returning' solar Eye goddess who is the life-giving force, the source of vitality for the yet-to be-born sun god. She is the catalyst for his growth and development, hers is the rage and radiance, the volatile power roused for ascent, the life-giving magical force provoking movement and causing everything to stir, grow and live.

This power is also embodied by the goddess called 'She who causes to rise', shown standing at the sun boat's prow in the *Book of Day*, incarnating fiery serpent energy, her name derived from a verb meaning 'to rise' (*i'r*), and resonating with a name of the uraeus: 'Iaret'.[11]

What is powerfully experienced in this first hour, too, is a synchronicity between night and day, between the Osirian ancestors in the *Dwat* and the living on earth, encapsulated in the *Book of Day*'s carefully worded text placed near the birthing goddess, which expressly states that the ninth night hour (named 'She who creates Harmony') is also the first hour of the day (named 'She who raises the Beauty of her Lord').[12] Correspondingly, the *Book of Night*'s ninth hour honours the ancestors as they come forth to be nurtured with food offerings.

Thus, as darkness gives way to dawn light, there is a confluence of worlds in this borderland region, a mystical communion between those waking by day and those travelling in the *Dwat* during the ninth hour.[13] A powerful night stream of ancestry merges with the dawn of incoming life, giving rise to a mysterious transmission between departing and incarnating souls, a dynamic transmission of vital *Ka*-energy passing from life to death and from death to life. It is perhaps difficult for us to grasp this complex fusion, but it is absolutely intrinsic to the ancient Egyptian rising dawn experience.

EMPOWERING YOUTH: SEEING EYES

On the sun boat travels into the second and third hours, its occupants still hidden from earth-dwellers, though striding among the companions in the top register is a figure called 'the Youth', embodying the sun god's adolescent phase as he secretly transforms from foetal child to 'beautiful youth of attraction' within the eastern horizon.[14]

All-important in this mysticism of light are the sun god's 'Eyes', his instrumental organs of vision radiating light in the world *(fig.16)*. These need to open now during the second and third hours, since it is only when both shine that earth truly brightens, and, correspondingly, living creatures open their own eyes to behold the manifold beauty of creation coming forth at dawn. Hence, as the horizon realm begins to shine radiantly white and

the contrast between light and darkness becomes
more sharply defined, the second-hour chant
celebrates Re's left Eye, the victorious Moon Eye
shining in the hour called 'She who drives away
Darkness' and rising for the god Hu, 'divine
utterance':

> O you rising in his uraeus snakes,
> Coming forth with his wings,
> Phoenix in lightland.
> Rise, Re, shine, Re,
> Be radiant, Re, be radiant, Re,
> Ascend, Re, in your White Crown,
> Ascend, Re, in your Red Crown,
> Through which you are mightier than the
> other gods…
> O Bull of his uraeus snakes,
> Who comes forth with his scorching breath,
> Towering high above, his power centred in his Eye.
> May the king drive the storm clouds away from
> the path of Re…[15]

This is no weak, effete god celebrated here in the
Ritual of Hours, but rather a ruler who has
banished conflicting loyalties from his realm and
dispelled storm clouds from the heavenly ways, a
victorious, youthful ruler wearing the Red and
White Crowns of Lower Egypt and Upper Egypt,
now set in motion and beginning to travel.

Of course, what lies behind the sun god's
manifestation here is the archetypal conflict
between Horus and his uncle, Seth, instigated by

ABOVE *fig.15* Aerial view of dawn over Western Thebes
with the mortuary temple of Ramesses II clearly visible
in the foreground.

ABOVE RIGHT *fig.16* Accompanied by a baboon, the
chantress of Amun, Henuttawy, kneels in adoration
at the glorious rising of the sacred Eye in the eastern
horizon. (Papyrus of Henuttawy, British Museum,
London, EA 10018. Third Intermediate Period.)

Seth's driving will to inherit the throne of Osiris
after brutally murdering his brother. This great
spreader of confusion thrives on division in the dry
and arid desert land he rules.[16] He also damages
Horus's Moon Eye during their conflict, and it is
this potential instability, with its storm clouds and
atmospheric disturbances, that imperils life in the
second hour.

Yet this conflict is intrinsic to the Egyptian
concept of life, for in Seth's fire is great strength,
and the initially weak Horus needs to wrestle with
his burning sexual energy if he is to grow strong.
It is said that the Egyptian king rules as Horus and
is mighty as Seth, and underlying their struggle is
'the deeply rooted Egyptian tendency to understand
the world in dualistic terms as a series of pairs
of contrasts', opposites needing to be brought
into balance and represented by the Red and
White Crowns.[17]

Thus, when the contrasts between light and
darkness are particularly keenly experienced in the
dawning day, finding this equilibrium is key in the

second-hour hymn, with its praise of Re rising in harmony with the waxing moon, his lunar Eye shining luminously white in the horizon lightland, able to heal all divisions and provide a stable foundation for life to flourish. Correspondingly, such a vision of unity is also unveiled to night travellers in the *Dwat* in the *Book of Night*'s tenth hour, when 'the son becomes the nurturer of his father' and all contrasts are subsumed, all opposites dissolved, as the powerful Horus ruler, shining for the living by day, simultaneously serves his Osirian ancestors, remembering their names through the power of his creative voice. As new life longs to come forth, so all the memories and experiences of the generations are reawakened in this transfigured gateway to the world, held together by the life-bestowing strength of the radiant horizon ruler.[18]

Through much struggle and conflict, the 'little Re' of the first hour has become a powerful second-hour 'reconciler', belonging to the potent lifeline of the generations and able to form a vast living link with Osiris and the ancestors. Serving simultaneously in different realms, possessing the power to move between worlds, he has awesome influence, for in him all opposites are held in equilibrium—the Red and the White, North and South, the fertile Black Land and the arid Red Land, right and left, night and day—and through the stable foundation he provides, warring factions no longer impede Maat's way.

Gradually the sky warms and the colours glow in the rising flush of life, as the sun boat sails into the third hour, that mystical time just before sunrise when a sudden green flash is seen momentarily shining on the horizon. This hour, rising for Sia, 'perception', is called 'She who praises the *Bas* of the gods, who sees millions', and now 'the gods come in praise' before Re as 'the beautiful youth in his right Eye', the 'beloved' as he is sometimes called:

> Awake in peace, O Primordial One,
> Re, going forth from the field of the Double Lion,
> Ascending encircled by his uraeus snakes,
> Rising in lightland…
> Circling round Egypt,
> Travelling round the deserts,
> Who plans for time and cares for everlastingness,
> The beautiful Youth in his right Eye, manifesting
> as the Falcon…
> Coming forth as the Lord of Night, Ruler of Hours,
> The far-striding Lord of Terror.[19]

When his right Eye shines *(fig.16)*, the youthful horizon ruler radiates the power of the loving-destructive Eye goddess Hathor-Sekhmet, as he must, since Hathor's attraction empowers him with the necessary charisma to attract followers and captivate hearts, to bind together all living creatures in this glorious rising creation, magnetically uniting the Two Lands through the dynamism of love.[20]

It is Sekhmet, too, who transforms him into a fierce warrior, a 'far-striding Lord of Terror', capable of protecting Egypt's boundaries and repelling enemies in the desert regions, though if her circulating energy is to become a source of positive transformation—if this volatile Eye goddess is not to create disturbance in vision—this dawn god must 'awake in peace' during the third hour.[21]

Thus this youthful adolescence phase requires healing the two Eyes, and until they are balanced and harmonious, neither further movement nor the flight of the green bird is possible. What is needed is to contain and transform their fiery sexual energies rather than to be consumed by them, and it is only when this crisis of 'separation' has been overcome, when both Eyes have been propitiated and healed, that Re is truly in possession of 'seeing Eyes'.

Now his mystical glance, which awakens all creatures to new life, awaits the revelation of the diadem adorning his beautiful face—the awesome all-powerful uraeus serpent issuing from his brow, ready to unleash her striking power in the fourth hour, ready to be seen blazing in the eastern horizon by those on earth.

In this climactic fourth hour, day and night, heaven and earth, Nut and her partner, the earth god Geb, finally separate, and Re, the 'Majestic One', is visible at last in all his victorious glory to those on earth, coming forth in his boat in the East. That 'instantaneous flash' of the risen sun reveals a dazzlingly powerful realm, as the eternal enemy, the Apophis snake, falls back, scorched by the heat of the sun god's four fiery snakes, powerless to prevent his coming forth.

Finally, the sun boat is visible to earth-dwellers and night is truly at an end.

TESTING ASCENT: BLISSFUL UNION

However, the struggle with Apophis is not over yet, as the sun god needs to move beyond the bloom of youth towards maturity at the zenith. This is undoubtedly also a challenging time. For as the sun rises higher in the sky, intensifying in heat, Apophis once again seeks to impede the sun boat, trying to strand it on his dreaded sandbank of catastrophe and famine *(fig.17)*. Filling his dreadful body with water, he seeks to sweep it away from the natural

flow of life, and his activity continues through the middle of the day from the fifth hour (10–11 a.m.) to the eighth (1–2 p.m.), that is, the two hours before and after midday, when the heat is greatest.

The attacks of this incredibly inflated demon, both at the hottest time of day and in the middle of life, pose an enormous threat to fruitful growth, as the psalmist in the Hebrew Bible well knew, asking, in Psalm 91, for protection under the shadow of God's wings from the 'scourge that lays waste at noon'. Members of Christian monastic communities are repeatedly warned of the kind of struggles they face when the noonday demon of *acedia* attacks, trying to draw them away from their chosen path of life *(see page 234)*.

This phase of ascent is critical, carrying the very real possibility that creative life will dry up, that growth will cease abruptly in the very prime of life, preventing everything from reaching maturity. Significantly, the Egyptian word for 'midday' means literally 'standstill', conveying the sense of suspension prevailing in the midday hour, which chapter 144 of the *Book of the Dead* is specifically designed to overcome, being a secret book, according to its rubric, for use from the fifth hour of the day onwards, helping 'to protect a person from standstill in the sky'.

Defending Re from this monster during the *Book of Day*'s airless hours is Seth himself, whose unbounded brute instincts so threatened the sun boat in the second hour. Interestingly, he has neither been banished nor demonized; on the contrary, as the 'son of Re' in the solar barque, he has become the sun god's anger and aggression incarnate, and his turbulent power, along with the spellbinding power of his sister Isis, now provides the strength needed to confront Apophis.

This is no time for vacillation, no time for conflict over the right of Horus to rule. Brother and sister stand forward in the sun boat to keep it on course, fighting together to pin down the monstrous threat to life. Seth's rebellious instincts have become the sulphuric strength needed to protect Re from his enemy, whilst Isis bewitches Apophis with her incantations. Both Seth's iron will and the magic of Isis are needed to cut the serpent's vertebrae and pin him to the ground.

And though their struggle with Apophis is here measured in hours, the Egyptians know it can take years to vanquish this foe, who poses an ever-constant threat to stability. After all, they experienced the long rule of the Hyksos, the earthly incarnations of Apophis who overturned Maat's way and ruled from their kingdom in the Delta, sowing discord and dividing the land into warring factions, before King Ahmose finally drove them out at the beginning of the 18th Dynasty and united Egypt once again.

The whole cosmos waits breathless in the heat whilst the crew struggle to overcome the dreaded demon in order to secure Re's glorious union with the wisdom goddess, Maat, during the seventh hour. This is the hour rising for Horus, the hour called 'She who expands the heart', the time when the whole of nature is fully open to the influx of cosmic life:

> Re appears, the Powerful One of Heaven…
> Lord of Manifestations within the shrine,
> driving away darkness,
> Who sits on the lap of Maat,
> Beloved of the heavenly Goddess in the
> midst of lightland…
> Gold of the People, Electrum of the Gods,
> Appear, O Bull…
> Appear, appear, Powerful One of the Sky,
> Chief of the Gods in the sky…[22]

ABOVE fig.17 Detail of fettering the Apophis snake during solar ascent in the *Book of Day*. (Tomb of Ramesses VI, Western Thebes. 20th Dynasty.)

The shining sun god now appears with his daughter Maat in the noonday shrine, guaranteeing creative order in the heavenly heights.[23] He is a 'far-striding' sun of justice, 'perfect of colours, with beautiful eyes', a potent bull, full of creative energy, identified elsewhere with Atum, and sometimes shown as a monkey holding a bow and arrow, a virile archer 'shooting rays' as the Heliopolitan creator god. The 'little child' nurtured by the first-hour snake mothers has become a towering ruler capable of 'driving away darkness', borne aloft by fiery serpent power to bring an 'expansion of the heart' in this heavenly realm. Shining 'within the shrine' like a golden image, he rules Egypt's cult world with Maat, whose joyful wisdom way has prevailed, and their union marks the climax of ascension, the culminating moment in the zenithal South, when the sun reaches maximum height in the daytime sky.

TWO

MIDHEAVEN MOUNTAIN: MAKING THE RED

Like the phases of the moon, the solar lifetime through the day has a waxing and a waning phase, but in between comes the crucial ninth hour, when Re triumphantly crosses the midheaven sky, a profound transition in the journey bringing everything to fruition. During ascent, the challenges have come from desiccative forces seeking to prevent creative growth, but now, in the afternoon of life, the sun boat takes a different tack, a northerly direction ultimately leading to Nut's embracing arms in the Northwest *(fig.11)*.[1]

With this change of direction comes a distinct shift of location and mood, starting already in the eighth hour, rising for Khons, whose name means 'crosser' or 'wanderer'. As the son of Amun-Re and Mut in the divine triad ruling at Thebes, Khons embodies the fertile ever-renewing moon, waxing and waning through three distinct phases and ensuring the whole of nature swells to fullness.[2] His are the moon forces now setting limits to expansion in the solar circuit, drawing the sun boat to cross the sky in the glorious ninth hour— the hour 'rising for Isis', called 'Mistress of Life', when a vision of great fruitfulness awaits those able to reach these heavenly heights.

NURTURING FIELDS: SECOND BIRTH

During the ninth hour, when the full extent of the victory over Apophis becomes a reality, the sun boat travels through the blissful 'Field of Reeds', described in the *Book of Day* as a heavily fortified place with metal walls, where emmer and barley grow to incredible height. Here is the Egyptian paradise, the 'field of life'. It is difficult to reach, but Sothis, the star of Isis, guides travellers on its 'beautiful ways'.[3]

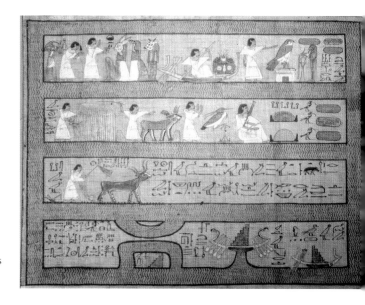

LEFT fig.18 Sennedjem and his wife Iineferti reap grain in the Field of Reeds. (Wall-painting in the tomb of Sennedjem at Deir el-Medina, Western Thebes. 19th Dynasty.)

TOP RIGHT fig.19 Vignette from chapter 110 of the *Book of the Dead* showing the scribe Ani undertaking various agricultural activities in the Field of Reeds and adoring the heron of plenty. In the bottom section are two boats carrying stepped thrones; in the top register Ani worships various deities. (Papyrus of Ani, British Museum, London, EA 10470. 19th Dynasty.)

According to chapter 110 of the *Book of the Dead*, this is also the realm of the 'Mistress of Winds', the divinizing 'Birthplace of the God' and place of 'enthronement' encapsulated by the thrones of Osiris and 'Food' depicted in vignettes illustrating chapter 110 *(fig.19)*. The struggle in the stifling noonday heat has been left far behind, the air is fresh again, and everything is surrounded by pools and islands intersected by canals filled with running water. In a sense it is a nowhere land, yet it is also a wonderfully prosperous one. The landscape, though familiar, is also strangely unfamiliar, for grain grows to an extraordinary height, unlike anything seen on earth, promising bread and beer in abundance *(fig.18)*. Everywhere there is perpetual transformation, as seeds are sown, lengthen into stalks and ripen in the warm sun, ready to be harvested and sown again in an ever-recurring cycle bringing food and fertility. Everything has an awe-inspiring magnificence in this land of fruitful grain. And here the blessed ones reap and sow, revelling in everything that belongs to their Ka-existence *(figs.18,19)*. Every fibre of their being has been purified; everything is the very opposite of the dreaded Apophis sandbank of the previous hours, as fruitful life unfolds in an amazing wealth of colours and burgeoning vegetation.

In the *Book of Day* this region has been simplified, condensed into a scene of the Egyptian king and 18 deities, shown with grain growing beneath them and standing within what seems to be a hieroglyphic mountain sign *(fig.20)*. When this sign encloses a sun disk, it is used in the word for 'horizon', though here it is split into two halves, as if to indicate a juncture or crossing-point in the journey, thus not only symbolically marking the eastern and western sides, but also dividing the 18 deities into two groups, including three forms of Osiris: Osiris the 'Eternal One', Osiris 'Gold' and, facing them on the other side of the mountain, a White-crowned figure called 'Powerful of Face'.

'All gods are three', declares chapter 300 of *Papyrus Leiden I 350*, and here, in the 'Birthplace of the God', Osiris's three figures illuminate his divinized state.[4] As 'Gold', too, his divinity is proclaimed, for gold is the flesh of the gods. It vivifies, showing Osiris to be resurrected and alive, stable and firm, a life-giving source of nourishment whose presence symbolizes the permanence and renewal at the heart of this daytime solar circuit.[5]

Also among the deities is a figure called Henu, surely a personification of the boat sacred to Sokar, the Memphite earth god so important in the Osirian resurrection mysteries *(see Part 2)*. In addition there is a deity named 'He who is on the Divine Throne', encapsulating this realm as a divinizing place of heavenly royalty. But the most startling inclusion is the astral figure named Meskhetyu, standing beside Osiris and representing the Great Bear in the northern sky, the constellation of Seth. Clearly, the death-bringer has his place alongside his erstwhile victim, and it is not Osiris, as might be expected, who is inert and mummiform, but Seth, a reversal suggesting the containment of his destructive impulses. For this ninth-hour field is the realm of 'peace' *(hetep)* in every sense, where all stand contentedly together, entirely centred on Horus's rule, which is reinforced by the presence of the 'Praiser of Horus' and the 'Just One' standing alongside Meskhetyu. Reconciled now with Seth, it is Horus who protects this realm from intrusive forces, assisted by the knife-wielding deities shown above the group.[6]

And as the sun, the gold and the grain all mature together, growing to fullness in the heavenly heights, so the corresponding ninth-hour chant in the *Ritual of Hours* praises Re's glorious second birth, bursting forth from 'the egg' containing the mysterious germ of life. This chant is a wonderful opening of the heart, a celebration of the primordial egg from which hatches the great god of light:

> Rise, rise,
> Shine, shine,
> O ascend, he who goes forth from his egg,
> Lord of Sunrises, Primordial God of the Two Lands,
> Bull of Heliopolis,
> Who crosses the islands in the Field of Reeds,
> Whom the Turquoise Deities praise,
> Their oars dipping in the river,
> You sail your two skies, you are arrived in peace...[7]

Clearly, this is not simply the sun god's first-hour birth repeated, but rather his rebirth into a completely new dimension of life, his liberation into an expanded realm of immense beauty and illumination, where he manifests like the rampant Mnevis bull of Heliopolis, rising and setting in his birthplace as the eternal source of light and renewal:

> You have crossed your two skies, Re, in peace,
> The Western-dwellers rise up for you,
> Your enemy, now driven back before you,
> lies fallen…
> Grant that she [Hatshepsut] breathes the sweet
> air of the north wind,
> Install her as a follower among the just,
> Who are in your following forever.[8]

Rising up towards the boat are the Western-dwellers. All come to praise the sun god, who is animated by the 'sweet air of the north wind' and crossing his 'two skies' in peace. This is the refreshing wind that cools the torrid dry land and heralds the return of the annual floodwaters each year, a wind associated with the rebirth of plant life, with Osiris and with everything that makes life green and verdant. Judging by Ramesses IX's highly abbreviated version of the *Book of Day (fig.12)*, which shows a vertical strip of water dividing its eastern and western phases, here in the midheaven the sun's journey directly intersects with the course of the river Nile. So the rhythm of the day meets the annual seasonal cycle, hence the inclusion of the inundation god, Khnum, in the ninth-hour group of deities in Ramesses VI's tomb.[9] Coasting along on the north wind's sweet breezes, travelling northwards, away from the zenithal South, the triumphant sun boat flows in the same direction the Nile inundation takes when returning each year to bring new life to Egypt.

LEFT fig.20 Detail from *fig.17* showing some of the deities in the Field of Reeds in the *Book of Day*, including three forms of Osiris. (Tomb of Ramesses VI, Western Thebes. 20th Dynasty.)

ABOVE fig.21 Re's second birth in the Field of Reeds. Here Herytuben, a musician priestess of Amun and grand-daughter of the Theban high priest, Menkheperre, praises the young sun child enclosed in a disk supported by the lions of 'yesterday' and 'tomorrow' facing in different directions. To the left she worships the earth god Geb in the form of a crocodile, and sows and reaps in the Field of Reeds. (Papyrus of Herytuben, Egyptian Museum Cairo, no. S.R.VII.10254/J.31986. Late 21st Dynasty.)

As the whole vista of his lifetime from sunrise to the waning of the day is glimpsed in its entirety, Re is born as the 'sun of the year' *(fig.21)*, able to ascend in the East and descend in the West and span the totality of the solar circuit *(fig.12)*, the glorious flow of life sustained by the three elements of sun, air and water.

This power is wonderfully captured in chapter 77 of the *Book of the Dead*, with its praise of the mighty falcon bursting forth from the egg:

> I am a great falcon who goes forth from his egg.
> I fly up and alight as a falcon whose back is four
> cubits, his wings green stone. I come forth from the
> shrine of the night boat. I have brought my heart
> from the eastern mountain. I have alighted in the
> day boat. The primordial gods are brought to me,
> bowing down. They worship me when I appear,
> having been reassembled as the falcon of gold upon
> the pointed stone.

Alighting on the 'pointed stone', with his wingtips made of 'green stone', this golden falcon spans the complete solar round, the circuit that is celebrated in the ninth hour, when everything is experienced in an unbroken cycle of life on the annual river axis ruled by the great bird of light. Time moves now not in linear progression, but rather in a full circle, like the ouroboros-snake biting its tail. So the living, breathing, developing creation brought forth by this sunrise perpetually transforms in a ceaseless cycle of fertile renewal.

A spirit of justice also reigns, for the ninth-hour chant specifically requests Re to place the pharaoh among 'the just ones' in his following—all those companions who have cultivated wisdom, matured and developed according to Maat's way. They have not been blown off-course, their victory over Apophis in itself being a 'justification',[10] and now these people who 'reap the field' have reached the afternoon of life and become Re's companions in the sun boat. With the sun strengthening their hearts, they are free to travel across the whole of the heavenly circuit, their consciousness expanded to know both the future and the past, to bear fruit and generate abundant life.

Hence the presence of a double-headed bird figure called 'He who encircles the Heart' above the ninth-hour group of deities in the *Book of Day*, for this realm is reached through heart wisdom. Blessed with his 'two-headed' expanded vision, his capacity to look both forwards and backwards, the travellers in the sun boat are able now to carry ninth-hour fruitfulness into the last three hours of the day.

GOLD OF REGENERATION: THE REDDENING HOUR

During these closing hours, when the sun boat sails inexorably towards Nut's protective arms in the Northwest, towards her beautiful face floating in the waters of Nun *(fig.11)*, nothing but the prospect of old age and decline might seem to loom. Yet this time is particularly radiant, a deep illumination beginning in the tenth hour called 'She who illuminates the heavenly water, who cools the steering oar', and rising for the god 'Magic' (Heka).

According to the tenth-hour hymn, this is when Re enters the night boat for crossing into the West, and curiously, immediately after the departure from the Field of Reeds, three torchbearers, shown in the register above the boat, light the way, including a figure named 'He who makes Redness',[11] the colour

associated with the North and its fiery tutelary deity, the serpent goddess Wadjet. As the boat sets sail upon the western waters, everything now shines in the redness he creates.

This has already been hinted at in the first-hour's introductory text, with its praise of the sun god's 'four faces' named '*Ba* of Shu', '*Ba* of Khepri', 'Green *Ba*' and 'Red *Ba*'. Like the serpent Eye goddesses ruling this solar circuit, whose oscillating green and red colours characterize their life-giving and death-wielding nature, the sun god manifests these same astral colours.[12] And just as 'greenness' is longed for in the rising dawn *(see page 20)*, redness is all-important here in the sunset realm.

The tenth-hour hymn in the *Ritual of Hours* praises Re's golden head, completely encircled by his fiery uraeus snakes:

> Hail to you, Gold of the Stars
> whom the crew of the night boat praise,
> Hail to these your uraeus snakes
> who bring you everything heaven and earth contain,
> who burn your enemies
> with the great flames that come from their jaws.
> You are Re, Overlord of the Powers,
> You distribute to those in your following,
> You have repelled those in the turbulence of Shu,
> You sail south and are given praise,
> You sail north and receive homage,
> The Immortal Powers announce truth to you,
> The king has invoked you,
> He has propitiated the face of your cobras…[13]

Shooting forth flames against enemies, the sun god's all-consuming female cobras here bring him 'everything heaven and earth contain'. He is the cosmic ruler of heaven and earth, the 'Gold of the Stars' shining as a celestial body in this reddening time, his divine countenance gleaming with the metal that forms the flesh of deities.

Evidently, too, as 'Overlord of the Powers' *(sekhemu)* he is perceived to be like a divine cult image, since the word *sekhem*, often linked with the fiery, leonine goddess Sekhmet as the 'Powerful One', means both 'cult image' and 'power', especially power manifesting through the face. Hence the frontal aspect is rarely shown in

ABOVE *fig.22* Sunset at Aswan.

Egyptian art, though the *sekhemu*-inhabitants of the netherworld are specifically portrayed facing directly towards the viewer, as if to convey that their tremendous power is unleashed whenever their faces are turned frontally towards a person.[14] Like Sekhmet's knife-wielding demons, their glance has the power to cut off heads, for, as one demon boasts in an Old Kingdom *Pyramid Text*, 'He on whom my face falls, his head shall not stay in place.'[15] Thus to invoke Re as 'Overlord of the *sekhemu*' means to encounter directly the awesome power of his uraeus-encircled face, with all its potential for decapitating destruction, and, not surprisingly, the tenth-hour hymn gives the reassurance that the king has propitiated 'the face of your cobras'. Yet *sekhem*-power is also the source of eternal regeneration, as travellers passing through Sekhmet's fiery gateways in the *Book of Night* well know[16] and, according to Isis in a scene in Petosiris's Ptolemaic tomb at Tuna el-Gebel, it is gold that ensures that Khepri, the 'Becoming One', is regenerated. 'You renew life through the gold which goes forth from your limbs', she tells the transforming sun god.[17]

It is as 'Gold', too, that Hathor is invoked by ecstatic dancers and musicians during Amenhotep III's first Sed Festival at Thebes, when they 'make jubilation for her at twilight', imploring the fiery serpent goddess to rise, be propitiated and take the

king safely through the night to be reborn at dawn.[18] Likewise, here in the tenth hour, all is gold and flames as the hissing female snakes coil around the sun god, bringing everything to fiery fulfilment in the above and below.

Shining forth in this dangerous fire-heat, Re manifests his eternal starry nature, his exalted divine image gleaming with the promise of eternal regeneration, flashes of serpent flame surrounding his celestial head in the reddening sunset land.[19] But nourishment is also offered here, for if the pharaoh propitiates these fierce cobras, he will become 'a follower among the peaceful ones', on whom great abundance is bestowed, 'joy in the marshes of the wild birds', and 'water to drink on the riverbanks', and will be able to 'enter and go forth with Re, being called [to eat] with the ancestors.'

MATRIX OF SILENCE: THE THIRD SUNRISE

Onwards the sun boat travels, towards the beckoning arms of Nut, the encompassing maternal embrace to which all safely return at the close of day. Now, in the 11th hour, called 'Beautiful to behold', and rising for the 'Adjuster of the Tow-rope', the shapeshifting Re becomes the ageing god Atum, praised as 'Lord of the Festivals of the sixth day and the 15th day'. Clearly he is descending into the West with the moon, attuned to its monthly restoration to fullness, which culminates on the 15th day, when Thoth completes his healing of the lunar Eye.

What is celebrated in this hour, too, is Re's union with the mother goddess, a self-regenerating act he performs as the virile Bull-of-his-Mother (Kamutef), the self-begotten god ever-renewing himself through the mother goddess, paradoxically both 'father' and 'son', the source of his own rebirth as the 'two in one':

> Hail, Re ... the journey is achieved for him
> who emerges in his perfection,
> who unites the primordial waters
> with his Mother at dawn.
> Lord of the Festivals of the sixth day and
> the 15th day.
> The Upper Egyptian king of the Western gods,
> The Lower Egyptian king of the Eastern gods,
> 'Welcome in peace' say the gods of the
> western horizon…
> The king greets you, uniting with Re in life,
> Like your union with your Mother,
> her arms around you.
> May you join the king in life, that he be
> justified before Atum.[20]

Just as the Bull-of-his-Mother manifests at the harvest festival celebrated each year at Thebes, when the first sheaf of grain is cut and offered to the regenerative bull god, so here in the *Book of Day* the sun god's regenerative power streams forth, emphasized by the names of two figures standing close to Nut's arms, one called 'Great Bull', the other 'Begetting Soul of Re'.[21]

To obtain these seeds of renewal from the precious ripe 'head', the ninth-hour grain must be cut, the harvest reaped and the death endured—a 'separating' experience encapsulated by the group of Heliopolitan deities striding towards the West in the *Book of Day*'s top register. At each end are Nut and Geb, here enclosing their children Osiris, Isis, Nephthys and a deity named Wedja, whose very name means to 'separate', being written here with

the hieroglyph that regularly replaces the name of Seth.[22] His dismembering murder of Osiris sows division in the world, and certainly the nature of Osiris's death is conveyed by the deities in the lower registers—a serpent-headed god called 'the Headless One' *(fig.11)* and, in the register below him, the 'Chief of the Westerners (Osiris)', flanked by 'She who is in the midst of Abydos' and 'Sekhmet the Great'. According to tradition, it is at Abydos that Osiris's head, the most powerful symbol of his dismemberment, is preserved, and this needs to be restored if he is to rise up and manifest his resurrected power, thus redeeming the catastrophe of his slaying *(see page 127)*.

Egyptian sources do no more than hint at these death-dealing events, but evidently this awesome sunset region guards the 'headless' god and his rebirth mysteries, a positioning made particularly poignant by the juxtaposition with the tenth-hour's glorious celebration of Re's golden head.

Undoubtedly this 'red time' can be a bewildering experience, generating great fear and trembling as the 'red' powers wield their fiery flames, making it imperative for travellers not to be overwhelmed by their drying forces. Indeed, the *Book of Day*'s closing text specifically refers to the stars of the Great Bear in the northern sky, the constellation of Seth, affirming that they are tied with a golden chain to the dual mooring-posts guarded by Isis.[23] During this passage into darkness, Osiris's feared enemy must be restrained, anchored to a stable point in the heavenly firmament, and what is needed when the forces of life begin to wane and the fiery decapitating deities seem to hold sway is the strength to pass through the tumult without suffering complete separation from the life-giving deity—in short, the power to unify.

In the second hour the youthful sun god, as ruler of the terrestrial land of Egypt, united the binary opposites of Red and White, North and South *(see page 22)*, but now this unification is on a cosmic scale and is accomplished by divinity as the 'two in one', uniting above and below, heaven and earth, the living and the dead, to ensure eternal regeneration. It explains the unusual inversion in the 11th hymn in the *Ritual of Hours*, which praises Re as the Upper Egyptian king ruling the

ABOVE *fig.23* View of the Great Peak towering over the Valley of the Kings.

Western deities and the Lower Egyptian king ruling the Eastern deities. Normally in this solar circuit, South (the realm of the White Crown) would be equated with East, and North (the realm of the Red Crown) with West, but here they are reversed, an indication of complete unification, the binding together of the Two Lands.

Such a cosmic unification is encapsulated, too, in Ramesses VI's corridor version of the *Book of Day*, where, after the 11th-hour text, the cartouche enclosing the king's name is paired with that of the 18th-Dynasty pharaoh Amenhotep I, thus connecting Ramesses to his illustrious predecessor who had ruled around 400 years before him. Early 19th-Dynasty kings, notably Seti I and his son, Ramesses II, had particularly revered Amenhotep I, enacting a special ritual celebrating their union with him and honouring the vast chain of past rulers, the lineage of the royal *Kas* who sustained the reigning king's rule.[24] Ramesses VI is continuing this tradition, and evidently the tenth-hour chant's request that the king might 'come and go with Re' and be called to eat 'with the ancestors' has been heard.

However, this is not just about 'father' and 'son', since the name of Meretseger, 'She who loves

Silence', the snake goddess of El Qurn, the holy mountain towering above the Valley of the Kings at Thebes *(fig.23)*, is inscribed between their cartouches. Within her eternal 'silent' womb the New Kingdom rulers lie buried, hidden beneath a huge natural pyramid enclosing the secrets of renewal; and here in the *Book of Day*, this great Theban mountain goddess unites Amenhotep I and Ramesses VI at sunset, bringing together two rulers of Egypt in her cavernous abode, whose immortal royal 'seed' never dies, but renews the generations in eternal succession.[25]

Centuries later, Hermes Trismegistus (whom the Greeks identified with Thoth) taught his pupil Tat about the 'silent womb' and regeneration in Book 13 of the *Corpus Hermeticum*, the collection of Greek Hermetic writings that originated in Egypt sometime during the early Roman period.

In this revelatory teaching, entitled *On Rebirth and the Promise to Be Silent*, Tat asks Hermes to initiate him into the secrets of rebirth, the sowing of the seed that 'is the true good', though, to his initial confusion, he is told the womb 'is the wisdom of understanding in silence' and that this mystery of rebirth is a lineage that cannot be taught but rather 'god reminds you of it when he wishes'.[26] Even so, Hermes then proceeds to initiate Tat into this regenerative mystery, in which the begotten is 'a god and a child of god, the all in all, composed entirely of the powers', though if his pupil is to experience this birth of divinity, he must first learn to let his senses sleep, no longer 'picturing things in three bodily dimensions'. After purging him of negative passions and restoring the positive virtues, Hermes then reveals the deathless state, fulfilled in a mystical union between teacher and pupil, in which Tat experiences himself as 'son', completely united with Hermes, who becomes for him 'father' and embodiment of the 'great divine spirit'. The culminating moment comes when, having been instructed by Hermes to stand in the open air facing the south wind, Tat reverently bows down in adoration first to the sun setting in the West and then again when it returns in the East, hearing the Ogdoadic powers singing an ecstatic hymn of praise to the universal creator. Clearly, his initiation is attuned to the mysteries of the solar circuit. He also experiences himself both as 'son' and the source of his own rebirth, a self-created being who knows the secrets of an 'undying harvest' sown in the womb of silence, for, as Hermes says in his closing words: 'Truth has borne good fruit for you, my child, an undying crop.'

Whatever controversies may surround Egyptian influence on the *Corpus Hermeticum*, Hermes's 'regeneration' teaching here distinctly recalls the sun god's regenerative experience in the *Book of Day*, albeit shorn of its mythic associations and re-expressed in a Greek idiom to make it understandable to his pupil.[27] And it comes to completion in the 12th hour, called 'She who

illuminates the islands, uniting with the Living One', rising for the 'Protector in the Twilight', the hour when the ageing sun god, Re-Atum, enters his 'sacred fields in the western horizon', encircled by his rising cobras.

From the standpoint of those dwelling in the *Dwat*, the 12th hour heralds Re's arrival as the great redeemer who comes 'to give food, to tend needs, to speak truth, and provide nourishment for the Westerners', bringing them renewed light and life when his *Ba* enters the West. From the perspective of the daytime cycle, however, it marks the sun god's concealment within the coils of his Eye, his return to his primal origins within the fiery flame of the 'Eye of Atum'. For the Eye is the pre-eminent force and bearer of the sun god's life through the hours, the instrument of his cyclical movement by day and night, whose protection in the twilight 'makes you well for life'. Coiled back into his snake origins, she will return him to new life at dawn, ensuring his regeneration in this ceaseless cycle of 'Life'.[28] Such a return is beautifully visualized in the *Book of Day* in Ramesses VI's tomb, where as the sun boat progresses towards Nut's mouth and sheltering arms in the West, it is greeted by the kneeling figure of the king, his arms raised in praise and his face turned not only towards the approaching boat, but also towards the rising sun in the East, towards rebirth in the rising dawn *(fig.11)*.

Thus, in addition to the sun god's first birth in the early daytime hours and his second birth from 'the egg' in the ninth hour, here on the western threshold of death, as he returns to the arms of the mother, there is the promise of a third sunrise on this journey, a third 'coming forth' by day in the rising dawn, and this profound experience is the supreme vision granted to travellers in the 'silent' realm.

HALL OF AWAKENING:
HEAVEN IN THE PALACE

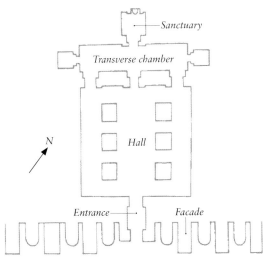

Sanctuary

Transverse chamber

N

Hall

Entrance

Facade

Nefertari's temple is ready now to be opened, its porch threshold crossed (fig.24), with its scenes of Ramesses offering flowers to Hathor (left) and Nefertari to Isis (right).[1] Moving further into this dark female cave, the bright sunlight is left behind and a large chamber opens, miraculously hewn from the sandstone mountain, its central aisle flanked on either side by three pillars supporting the ceiling above. On each pillar, a mask-like face of Hathor stares watchfully across the aisle, her curled striated wig, tucked behind the familiar cow ears, still streaked with traces of black and red pigment. The faces seem beneficently disposed, although, as ever, the rearing uraeus snake in the naos superstructure

PREVIOUS PAGE fig.24 View into Queen Nefertari's temple at Abu Simbel. (19th Dynasty.)

ABOVE fig.25 Ground-plan of Queen Nefertari's temple at Abu Simbel.

TOP LEFT fig.26 Hathor transmits solar attraction to Ramesses II through the beads of her menit-necklace, initiating his ascent to the

zenith of the sky. (First relief on the south wall of the hall, Nefertari's temple.)

TOP RIGHT fig.27 Seth and Horus bless the Red and White Crowns of Lower and Upper Egypt worn by Ramesses II. (Second relief on the south wall of the hall, Nefertari's temple.)

OPPOSITE LEFT fig.28 Nefertari offers flowers to Anukis, a member of the divine triad worshipped at Elephantine, and shakes her naos sistrum, the musical instrument played by mediating daughter goddesses at the noonday threshold. Here Anukis is identified with this daughter role. (Third relief on the south wall of the hall, Nefertari's temple.)

OPPOSITE RIGHT fig.29 Ramesses II offers Maat to Amun-Re, ensuring their union at the zenith. (Fourth relief on the south wall of the hall, Nefertari's temple.)

above warns of lurking danger, and, in fact, each face belongs to a huge naos sistrum, Hathor's cult instrument which is used to propitiate wrathful deities.[2] Instantly, this cave is transformed into a resounding hall of the daughter goddess, a musical chamber of intercession shaking the very depths of the earth – though why here in the queen's temple?

These pillars perfectly frame a sequence of four reliefs on each side wall, with the culminating scene on the south (or left) side showing the supreme union between the Theban ruling god, Amun-Re, and his beloved daughter Maat, symbolized by Ramesses offering a tiny figure of the wisdom goddess to the deity *(fig.29)*. South takes precedence over north in Egyptian temple decoration and, as in the *Ritual of Hours (see page 25)*, here on the south wall the goddess who directs and guides the order of Egypt has reached the zenith, the midday goal, as male and female unite to rule this musical realm. Manifestly, Maat is no abstract justice goddess when united with Amun-Re, but one who is a living part of the regular course of Egyptian life, as in the following hymn from the daily ritual for Amun at Thebes:

You live with Maat,
You join your limbs to Maat,
You make Maat rest upon your head
That she may take her place upon your forehead.
You become young again
Through seeing your daughter Maat,
You live from the perfume of her dew.
You wear Maat like an amulet at your throat.
She rests on your breast.
The gods pay tribute to you with Maat,
For they know her wisdom…
Your food is Maat, your drink is Maat,
Your bread is Maat, your beer is Maat…
You exist because Maat exists,
Maat exists because you exist,
She unites with your head,
She manifests before you for eternity.[3]

Proclaimed here is Amun-Re's complete and utter dependence upon the truth goddess. His body is entirely encompassed by her, renewed by her and adorned with her. Reabsorbed into the fragrance of her life-giving power and surrounded by the wonderful pleasures of union, he is able to dispense great blessings with her. She is his food, his very life, and only when they are united is there wise rule. But Maat without Hathor, order without intoxication? This journey skyward has more to tell.

ROUSING HATHOR:
MANIFESTING MAAT

Initiating the sequence on the south wall is the Nubian Hathor, 'Lady of *Ibshek*', holding out her *menit*-necklace for Ramesses to touch *(fig.26)* and thereby placing his hands upon 'beauty' *(nfrw)*, 'the adornments of the Lady of the Two Lands', according to the brief inscription between them:

> [Your] hands upon beauty
> O enduring King,
> The adornments of the Lady of the Two Lands.

If Ramesses is to rise like Re in the dawn, he needs the loving-destructive Eye goddess to create his 'beauty'; and indeed, Hathor's gesture, the bestowal of her shining jewels, evokes the name of the *Book of Day*'s first hour: 'She who raises the Beauty of her Lord' *(see page 19)*. Strung with rows of tiny turquoise-green beads, the necklace transmits her life-giving attraction, and here she empowers Ramesses with all the solar 'greening' vitality that is hers to give in the eastern horizon.[4]

Next the falcon-headed Horus and his erstwhile rival, Seth, are shown blessing the Red and White Crowns adorning Ramesses as ruler of the Two Lands *(fig.27)*, corresponding with Re's manifestation in the daytime second hour wearing these same crowns *(see page 22)*. Having withstood Seth's unbounded fiery energy, his brutal attempts to usurp the throne, Ramesses has triumphed, and on the pillar nearest this scene is the ibis-headed god Thoth, who played such a crucial role in healing the Moon Eye during the conflict. Interestingly, though, Ramesses faces not towards the victorious Horus, as might be expected, but towards Seth. Indeed, the Ramessid kings in general, who originated from the eastern Delta, seem to have been particularly devoted to Seth, and here at Abu Simbel the warrior king, renowned for his great prowess on the battlefield, implicitly honours the turbulent god.

As already noted *(see page 24)*, Seth is not cast out after being forced to concede the throne to his rival, nor is he left to fester in some remote corner of Egypt. The Egyptians were well aware of the need to integrate him into the solar order, and hence at the close of the conflict between the two gods, when 'right' (not 'might') has finally triumphed, Re takes him into the sun boat, where he becomes his protector against Apophis. For when Seth's great strength is joined to Horus's light, it becomes a positive force working to maintain a united Egypt, as Ramesses here acknowledges by turning to face Seth at the blessing of his crowns.

Thus the first two scenes on the south wall correspond exactly with Re's 'eastern horizon' transformations in the *Book of Day*, opening the way for Ramesses to move further towards the zenith.

Now Nefertari enters the sequence in the third scene, a graceful Hathorian figure crowned with her plumes and horned headdress, shown offering fresh flowers to Anukis, daughter of Satis and Khnum in the divine triad guarding the inundation waters at Elephantine *(fig.28)*. She also rhythmically shakes her naos sistrum, the musical instrument a daughter goddess plays to propitiate zenithal deities, turn away their rage and make them beneficently disposed towards those seeking entry into their presence. Mid-life power, in all its magnificence, carries with it the potential both to blast and bless, striking to the ground anyone unprepared for an encounter with such tremendous energy, and just as Apophis threatens to blow the sun boat off-course in the *Book of Day*, stranding it on his dreaded sandbank *(see page 24)*, so here in this royal realm, as the heat intensifies, a different but no less dangerous challenge is faced during ascent—the difficulties of coming before the zenithal father god.

Hence, here in Nefertari's earthly temple, all eyes are focused on the queen as musical enchantress. Her offering to Anukis holds out the promise of cool water flowing in the noonday heat and, with her plumed feathered crown ready to fan overheated solar deities, the Elephantine daughter goddess is herself the image of coolness, as she beneficently manifests with Nefertari at this difficult threshold giving access to Amun-Re. Where others might fear to tread, the musical skills of these daughter goddesses create harmony, their charm procures the god's mercy and compassion, and Ramesses needs his sistrum-shaking queen, otherwise he might remain forever caught and

RIGHT fig.30 Faience statuette of Amun embracing Amenirdis I, daughter of the Kushite ruler, Kashta. During the reign of his successor Piye, she was formally installed as the 'God's Wife of Amun', the title of the high priestess in the temple of Amun-Re at Thebes, thereby helping to consolidate Kushite control of Egypt. (Egyptian Museum, Cairo, CG 42199. 25th Dynasty.)

crystallized in the youthful zest of dawn leadership, which, powerful though this may be, nevertheless stops short of midday rule and maturity.

Drawn on, however, by the ever-transforming female, he reaches the goal of ascension in the end scene, when he brings Maat before Amun-Re *(fig.29)*, thus fulfilling his sacred function as the representative of the sun god on earth and perfectly mirroring the sun god's heavenly union with Maat in the *Book of Day*'s seventh hour *(see page 25)*.

The similarities with the much-later Indian Tantric way of transformation, in which the goddess is the 'power-holder', can hardly have escaped the notice of some readers. Just as the female Kundalini snake lying dormant at the base of the spine has to be roused through specific rituals in order to make her journey through the chakras, so here at Abu Simbel the Hathorian energy of four females must be awakened and rise up towards the zenithal crown.[5] Represented by Hathor, Nefertari, Anukis and Maat, this fourfold Divine Feminine energizes the king's ascension and provides an early hint of Hathor's all-important fourfold epiphany in her Dendara temple *(see chapter 6)*.

However, unlike Tantra's bold depictions of sexual union, little in Egyptian temple decoration explicitly conveys the deep desire, the sexuality and the fire coursing through this skyward journey, apart from the fact that the particular deities shown are renowned for their volatility, so there is always the threat that their power might run amok, out of control. Great insight and knowledge are needed if such power is to be channelled fruitfully, but, discreet as ever, and mindful of the power latent in pictorial images (to the extent that the heads of dangerous animals were sometimes severed from their bodies in Old Kingdom hieroglyphic inscriptions to nullify their power),

the Egyptians shy away from any overt expression, preferring to veil what is, after all, a highly sensual path, and one easily misunderstood.

An exception to this secrecy is a small statue *(fig.30)* showing Amun-Re locked in a tender embrace with Amenirdis I, daughter of Kashta, the Kushite king of Napata in the Sudan who invaded Egypt at the end of the eighth century BCE. Amenirdis became the 'God's Wife of Amun' in Amun-Re's temple at Thebes, an office which included playing the sistrum to propitiate the god, and whilst such a portrayal of a deity and priestess together is extremely rare, this late statue, reminiscent of certain sculpture from Akhenaten's reign, provides an unusual example of erotic intimacy.[6]

RITUAL AND COSMOGRAPHY: SINUHE'S STORY

These four scenes in Nefertari's hall, unique in Egyptian temple decoration, are not simply sacred art but have ritual significance, coming alive in a palace ritual performed by the Egyptian queen and her daughters as narrated in the Middle Kingdom *Story of Sinuhe*. Even though it was composed nearly 600 years before Nefertari's lifetime, the 'wisdom of the queen' inspires this much-loved story and there is the same intent to reveal Egypt's sovereign lady as the companion of the solar king ruling in the palace on earth.[7]

Central to the plot is Queen Neferu, a 'heaven on earth' in the palace, according to her husband, Senwosret I *(see page 13)*. The story's hero, Sinuhe, is her favoured courtier, who is on a military campaign with the king when the story opens. Senwosret is ruling with his father, Amenemhet I, as co-regent, but Sinuhe accidentally overhears news concerning Amenemhet's sudden death in the royal residence. Fearing there may be turmoil arising from the king's untimely death, or perhaps for some other, unexplained reason, he then feels impelled to flee into exile somewhere in Syria-Palestine.

Years pass, during which he rises to prominence as a successful tribal ruler in his adopted homeland, but, as the prospect of old age in a foreign country looms, he is filled with the longing to return to Egypt to live out his days and be buried according to his native customs. Alerted to his condition, Senwosret writes to him, inviting him to return home and reminding him of his need to prepare for burial, protected in a coffin by the 'Lady of All' (Nut) outstretched above him.

We join the story at the moment when Sinuhe, looking more like a desert nomad than a clean-shaven Egyptian, is about to enter the Egyptian palace and face Senwosret for the first time since his flight. Fearfully, he bows to the ground between the sphinxes guarding the palace entrance, then passes through the forecourt, being sent on his way by courtiers, until finally he is ushered into the throne room, where Senwosret sits, regally robed, like a resplendent cult image 'upon the great throne in a kiosk of gold'. Can Sinuhe, an Egyptian who, for some inexplicable reason, has preferred to live like a nomad rather than serve as a loyal member of Egyptian society, expect anything but death?

Suddenly he feels himself jerked to the ground, gripped by the power emanating from this awesome presence able to dispense both life and death. Fear seizes his heart, pressing down upon him like a great weight and compelling him to prostrate himself before his sovereign. He says of his ordeal, 'I was like a man seized by darkness. My *Ba* was

LEFT fig.31 Statue of Queen Tuya, mother of Ramesses II, who was the daughter of a high-ranking soldier before marrying Seti I. This statue, from Tanis in the Delta, originally represented a 12th-Dynasty queen, but the features have been reworked in the 19th-Dynasty style, indicative of Middle Kingdom influence in the Ramessid ideal of queenship. (Egyptian Museum, Cairo, JE 37484.)

gone, my lips trembled and my heart was not in my body. I did not know life from death.' Like a condemned man, he swoons at Senwosret's feet, plunged into a limbo-like state reminiscent of Re's old enemy, the snake Apophis, during his dramatic encounter with the sun god, for Apophis, too, when overcome by the power of those in the sun boat, falls fainting to the ground.

It is precisely at this critical moment, when Sinuhe's fate hangs in the balance, that Hathor's erotic spirit enters the room, embodied by the queen and her daughters adorned with their sacred *menit*-necklaces and sistra. Initially they shriek out in horror at Sinuhe's foreign appearance, yet his plight is evident, and immediately the princesses intervene on behalf of the stricken man. Shaking

their sistra, they sing a chant of intercession in which they beseech Senwosret to put aside his arrows of wrath and look with mercy on Sinuhe.

The purpose of their rhythmic magic is connection; its aim is to raise Senwosret to the very zenith of the sky, uniting him with his queen and opening him to the same joyful participation in the cosmic way of the world that Re experiences when united with Maat in the *Book of Day*. First they urge the king to touch the adornments of the 'Beautiful One', and then, just as at Abu Simbel, his journey skywards involves four successive transformations until finally he is acclaimed together with the 'Lady of All'.

Surrounding the king with Hathor's rhythmic music, this is what they sing:

SINUHE CHANT

Your hands to the Beautiful One,
O Eternal King,
to the adornments
of heaven's lady.
May Gold give life to your nose,
May the Lady of Stars
embrace you.

The White Crown fares north,
The Red Crown fares south,
Joined and united
by your Majesty's word.

May the uraeus
be placed on your brow,
You who have kept
the poor far from disorder.
May Re, Lord of the
Two Lands, be merciful
to you.

Hail to you and to
the Lady of All [Maat].

ABU SIMBEL

Hathor Gold offers her
necklace to Ramesses.

Horus and Seth bless the
Red and White Crowns.

Nefertari as daughter holds
a sistrum before Anukis.

Ramesses offers Maat
to Amun-Re.

BOOK OF DAY

The first hour called 'She who
raises the Beauty of her Lord'.

Re manifests with the Red and
White Crowns in the second hour.

The fifth hour is called 'Protectress
of Re, the Flaming One [uraeus]'.
The solar boat's crew battle
against Apophis.

In the seventh hour called
'Expansion of the Heart',
Re and Maat unite at the zenith.

Through his daughters' rhythmic music, Senwosret is ritually raised to the zenith of heaven, brought into union with his heavenly queen, and, with her now beside him, appears like Re, united with Maat, ready to enact their solar sovereignty here in the palace on earth.[8]

'Loosen your bow, lay down your arrow, give breath to the one who is breathless', the princesses implore him, 'give us our good gift on this good day, our reward being the son of the north wind, the bowman born in Egypt.'

Like the Jewish Queen Esther, who risked her life to intercede before her husband, King Ahasuerus, on behalf of Mordecai and the stricken Jews, an action which is joyously commemorated each year in the Jewish festival of Purim, so here at the Egyptian court, the queen and her daughters persuade a potentially wrathful and stubborn ruler to be concerned with the plight of others, thus connecting him with his people. Indeed, they change this stern king into a merciful ruler who welcomes Sinuhe back to the court and to the Egyptian way of life.

Sinuhe is washed and shaved, new linen robes are given to him and his body is anointed. 'Years were removed from my body', he says, as all his impurities 'return to earth'. He is given a prince's house, food in abundance and a garden to delight in—everything his *Ka* needs to enjoy a peaceful lifetime on earth. His tomb is also prepared in the necropolis. Despite his failure to serve his monarch for many years, he is rewarded with all the favours usually due to a loyal courtier.

So, like the 'just ones' sailing with Re in the daytime ninth hour, Sinuhe is reborn as a companion of Egypt's solar ruler and given to the princesses as the 'son of the north wind', the refreshing wind blowing in the ninth hour when the sun boat enters the nourishing Field of Reeds *(see page 29)*.

The Story of Sinuhe is no storyteller's fantasy, even if cast in the form of a folktale, but a deep teaching about the royal couple's role as Egypt's solar rulers. It also reflects the huge changes in perception of the pharaoh that occurred during the Middle Kingdom, who now functions within a solar circuit intimately connected with human life

on earth.[9] In short, what is being enacted here is a palace ritual which is completely in tune with the earth-related lifetime Re lives in the *Book of Day*, and which will be set out much later in Nefertari's temple at Abu Simbel. In essence it is not a historical individual who is the story's hero, but an archetypal person whose life ultimately derives its meaning from his relationship with the Egyptian king and queen ruling as representatives of Re and Hathor-Maat on earth.[10]

Sometimes it is difficult for a modern reader to appreciate just how much ancient storytelling is rooted in ritual and myth, so that it becomes all too easy to dismiss the narratives as simplistic, poorly constructed and illogical, or even to see them as isolated fragments strung together without any coherent form or structure. Nothing could be further from the truth, for they are a powerful mode of transmitting sacred knowledge.

TRAVERSING EGYPT: THE IMAGE STATE

Clearly, the roots of Nefertari's temple decoration reach deep into the Middle Kingdom, but before proceeding further into her temple palace, the four reliefs on the north wall need to be considered, since each directly complements the one across on the south side. Hence the hall can be 'read' not only lengthways, as ascension to the zenith, but also breadthways, creating an intricate maze of cross-correspondences within the temple.

The first scene on the north wall shows Ramesses wielding a *sekhem*-sceptre of power as he administers offerings before Ptah, 'Lord of Maat', typically seated like a cult image within his shrine *(fig.32)*. As the Memphite craftsman responsible for creating images, Ptah's presence here, across from Hathor, serves to ground the south wall's solar ascent firmly in his metallurgical cult world. He also, according to the *Decree of Ptah-Tatenen* for Ramesses II in the great temple nearby, fashions the king's body in electrum, his bones in copper and his flesh in iron, creating him in the likeness of the sun god, a creation arousing such deep joy in him that he is impelled to take Ramesses in his arms in an 'embrace of gold'.[11]

Significantly, when Ramesses touches Hathor's necklace in *fig.26*, he wears the distinctive crown of Ptah-Tatenen. Indeed, at Memphis it is Hathor who infuses Ptah's manifold forms with life, bestowing on them all her solar beauty and all her fiery warmth and desire.[12] Thus not only does she initiate ascent on the south wall, she also vitalizes the 'image' king fashioned by Ptah, and what could be more fitting, therefore, than to show her across from Ptah, holding out her beautiful necklace to Ramesses, seeking to enfold his divine body in her passionate embrace and bestowing on him all her radiant life?

Manifestly, 'creating beauty' is a metallurgical art in Egypt's royal cult of the sun *(see Part 3)*.

Next, opposite the scene of Horus and Seth blessing his Red and White Crowns, Ramesses offers flowers to Harsaphes, the ram-headed god of Herakleopolis in Middle Egypt, here wearing the *Atef* crown *(fig.33)*. According to some versions of chapter 175 of the *Book of the Dead*, it is in Herakleopolis that Seth is forced to bow low before the enthroned Osiris, resplendent in his *Atef* crown, and begrudgingly acknowledge him as the heir of Re. To add injury to insult, Seth receives a bloody snout in the process. His redness mingles with the earth when Re buries the blood, which, the chapter says, is how the agricultural rite of 'hacking the earth' began.[13]

Thus the presence of Harsaphes across from Horus and Seth is highly appropriate, though there is another possible layer of meaning, too, since the ram, like the bull, is revered in ancient Egypt for its food-giving power and sexual vigour, being particularly associated with semen and other liquids. In fact, the same chapter 175 of the *Book of the Dead* praises Harsaphes as 'the *Ka*-power of sweet copulation, which brings him all food offerings', and, in light of the emphasis on semen and masculine strength in the struggle between

TOP RIGHT fig.32 Ramesses II wields a *sekhem*-sceptre before Ptah. (First relief on the north wall of the hall, Nefertari's temple.)

RIGHT fig.33 Ramesses II offers flowers to Harsaphes, the ram-headed god of Herakleopolis. (Second relief on the north wall of the hall, Nefertari's temple.)

ABOVE LEFT fig.34 Nefertari shakes two naos sistra before Hathor, 'Lady of Dendara'. (Third relief on the north wall of the hall, Nefertari's temple.)

ABOVE fig.35 Ramesses II offers wine to Re-Harakhti, the solar 'Lord of Nubia'. (Fourth relief on the north wall of the hall, Nefertari's temple.)

ABOVE RIGHT fig.36 Line-drawing of the central entrance leading from the hall into the transverse chamber (vestibule). On the left of the doorway, the queen shakes her naos sistrum and offers flowers to Hathor: on the right she offers to Mut, the Theban mother goddess.

Horus and Seth, it is easy to see why Harsaphes is directly across from them in Nefertari's temple.[14]

Moving further southwards from Herakleopolis in terms of Egypt's geography, the next scene shows Nefertari shaking two sistra before Hathor, 'Lady of Dendara' *(fig.34)*, an obvious pairing with her sistrum-playing role before Anukis on the south wall *(fig.28)*.[15] At the same time it creates another forerunner of Hathor quadrifrons, whose 'four faces' so powerfully adorn the columns in her temple at Dendara *(fig.67)* and are ubiquitously praised in late inscriptions.[16]

Finally, leaving not the slightest shadow of doubt that intoxication belongs with the truth, the justice and the order of Egypt, Ramesses offers wine to Re-Harakhti, the solar 'Lord of Nubia' *(fig.35)*, across from the scene of him offering Maat to Amun-Re *(fig.29)*. Channelled and grounded in the cult forms shaped by the skilful hands of Ptah, 'Lord of Maat', the ritual practice of king and queen is shown to be that of a bacchanal love which does not abandon social and moral values. Order and intoxication, Maat and Hathor, though seemingly such disparate daughters, have need of each other. They both belong to the wholeness of this solar, though by no means sober, order, and everything in the hall converges to disclose this truth.[17]

Progressing along the right wall of the temple, revering the cult deities of the land, the king and queen symbolically travel southwards through Egypt's 'image' realm, starting implicitly in Ptah's cult centre in Memphis in the North, then moving to Herakleopolis and Dendara before finally arriving in Nubia. As they make their way southwards, on the opposite wall they also ascend skywards to the zenithal South, thus creating a radiant web of solar power circulating between the two walls, empowering and vitalizing the 'images' in this Hathorian royal realm.[18]

This royal journey also leads to the hall's central doorway, giving access to two rooms carved further into the rock. A hint of what lies beyond can be gleaned from the reliefs flanking the doorway *(fig.36)*. On the left (south) side, Nefertari offers flowers and shakes her sistrum as 'daughter' before Hathor, whereas on the right, she simply holds out flowers to Mut, the 'mother' goddess of Thebes, so breaking the symmetry between the two scenes— a small detail perhaps, but one preparing for the change of mood in the transverse chamber, with its focus on goddess birthing power. For on the other side of the wall from Mut is the enthroned goddess Taweret, 'who gives birth to all the gods', and Nefertari's homage to Mut in the hall paves the way for an encounter with this great mother goddess.[19]

FOUR

CHAMBER OF THE MOTHERS: ETERNAL CREATION

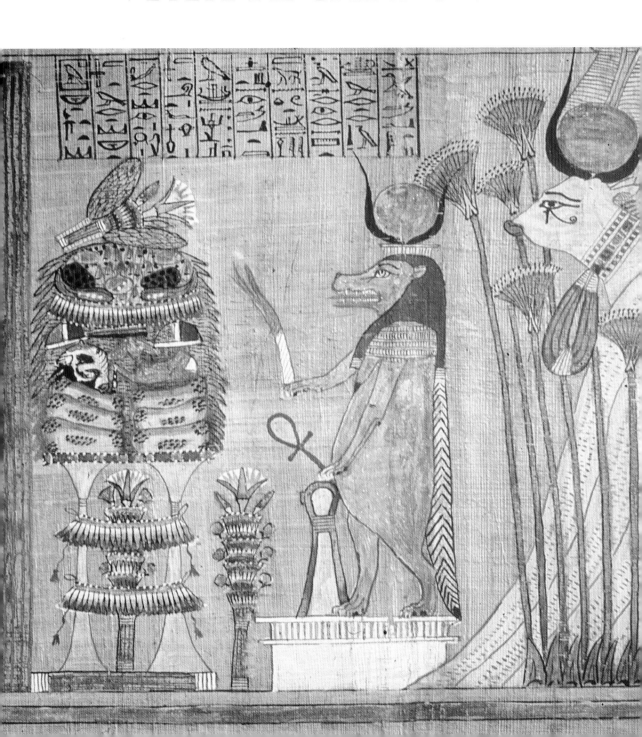

On entering this much smaller chamber, the reduction in scenes is immediately noticeable. At first glance there seems little to help unravel their meaning, for the inscriptions contain only the names and brief epithets of the various deities. Again, the ancient temple designers have given little away, though this room is clearly intoxicating and solar, as is evident from two large reliefs paired on either side of the doorway leading to the innermost sanctuary. On the right, the king offers wine yet again to Re-Harakhti *(fig.39)*; on the left he makes a similar offering to Amun-Re *(fig.38)*.

What is initially striking, however, is that everywhere the law of 'three' seems to rule. Thus, on the west wall a small relief next to Amun-Re depicts the three Horus rulers of three major fortresses in Lower Nubia: Horus of Aniba, Horus of Kuban and Horus of Buhen. Across on the other side of the doorway, next to Re-Harakhti, the corresponding small scene shows the Elephantine triad: the potter god, Khnum, his consort, Satis,

LEFT fig.37 The cow of Hathor emerges from the Theban western mountain and a papyrus thicket. She is accompanied by the birth-goddess Taweret in hippopotamus form, who holds a flaming torch and an *Ankh*-sign, both deities here ensuring the renewal of life. (Papyrus of Ani, British Museum, London, EA10470. 19th Dynasty.)

ABOVE LEFT fig.38 Ramesses II offers wine to Amun-Re. (Relief on the west wall [south side] of the transverse chamber, Nefertari's temple.)

ABOVE fig.39 Ramesses II offers wine to Re-Harakhti. (Relief on the west wall [north side] of the transverse chamber, Nefertari's temple.)

and their daughter, Anukis *(fig.43)*. On the east wall is the graceful feminine triad of Hathor, Nefertari and Isis, 'Mother of the God' *(fig.41)*. Across from them, on the other side of the doorway, Ramesses and Nefertari worship the enthroned Taweret *(fig.44)*. Clearly, the number three is all-important here, and Taweret's epithet makes it clear why it should be so.

BIRTHING DIVINITY: THE RIVER AXIS

Sometimes called the 'pure water', Taweret is the birth goddess par excellence, presiding over all the manifold functions of childbearing and usually depicted as an upright hippopotamus, her belly heavily distended and swollen like that of a pregnant woman *(fig.37)*. Here she is in human guise, however, a slender Hathorian figure, resplendent in her solar-horned headdress, the goddess who, according to her epithet, 'gives birth to all the gods'.

Remembering the *Leiden Papyrus*'s statement that 'all gods are three' *(see page 28)*, instantly the predominance of triads here makes perfect sense, encapsulating this chamber as a divinizing place.[1] Importantly, when Re is born again in the *Book of Day*'s ninth hour, coming forth from 'the egg' during this blissful hour 'rising for Isis' *(see page 27)*, he is in the Field of Reeds which is known as the 'Birthplace of the God' *(see page 27)*. This is the hour, too, when three divine forms of Osiris manifest in the midheaven lightland *(see page 28)*.

'As above, so below', and, making their way to this chamber, the royal couple reach their own divinizing 'birthplace'. In fact, the scene above the niche at either end of the room perfectly conveys a terrestrial 'reed' landscape: on the north wall Ramesses offers plants to Hathor as a cow, sailing in her boat amidst the northern papyrus marshes, whilst across on the south wall Nefertari makes a similar offering *(fig.40)*.

Called *wȝḏ* in Egyptian, papyrus is the heraldic plant of Lower Egypt; it is also used as a hieroglyph in words meaning 'to flourish', 'be healthy', 'fresh', 'happy', 'vigorous'. A 'green' plant symbolizing new vegetation, growing crops and fertility, its colour evokes 'peace' in every way. To sail in the papyrus marshes, amidst all the scenic delights of burgeoning nature, is 'to follow the heart' and 'bring contentment' to the *Ka*. Hence the fiery Eye goddess, Sekhmet, holds a papyrus stalk, symbolic of her propitiated green state. So, too, in Hatshepsut's temple at Deir el-Bahri, Hathor is shown as a cow sailing in her boat in the verdant papyrus marshes, where, just as she suckles Horus in the marshes of Chemmis, she nourishes Hatshepsut and her *Ka*.[2] It is the *Ka*, a word closely associated with the word 'bull' *(ka)*, that infuses a person's physical body with energy, life and vigour, and here Hathor enlivens Hatshepsut's *Ka* with all the solar life streaming from her divine milk.[3] Hence the boat scene above each niche creates the atmosphere of Nefertari's central chamber, transforming it into a nurturing royal birthplace filled with abundant life, and conveying also a profound sense of the

Egyptian word *hetep*, with all its various meanings of 'peace', 'rest', 'contentment' and 'nourishment'.

It is instructive, too, that entry into this room is also possible through a side door at each end *(fig.36)* —an important architectural feature contributing to its meaning. For whoever enters through the southern side door (south having precedence over north during the solar ascent) must then turn northwards, hence traversing an axis symbolically aligned with the annual course of the river Nile flowing northwards each year to bring much-needed water to Egypt's parched ground.

Returning in harmony with these waters is the pacified 'green' Eye goddess, Hathor-Sekhmet, whose return from Nubia is joyfully celebrated in great festivals of drunkenness each year. Indeed, it is towards the north that Nefertari herself faces when Hathor and Isis bless her crown *(fig.41)*, as do the royal couple when they make offerings to Taweret *(fig.44)*, thus aligning themselves with the inundation's life-giving flow. As Christiane Desroches-Noblecourt observed, a powerful current of water flows through this central chamber; hence, once again, there is an exact correspondence with the *Book of Day*'s ninth hour, when the sun boat meets the annual river axis ruled by Sothis, the star of Isis *(see page 29)*.[4]

GREENING RENEWAL: THE JUSTIFIED QUEEN

During the ninth hour the 'just ones' sail in the sun boat *(see page 29)*, and here in Nefertari's temple a unique scene on the east wall proclaims the queen's own 'justification' as she stands between Hathor and Isis, 'Mother of the God' *(fig.41)*, three female bearers of life, symbolized by the *Ankh*-sign of life each holds. Each goddess also raises an arm to create a beautiful *Ka*-like gesture enclosing the queen and thus showing her to be united with her *Ka* in life and truth, nourished by Maat in every way, since the very same gesture is also made by 'justified ones' when they come forth vindicated from the Judgement Hall after the weighing of the heart in the afterlife. Such a 'judgement' is also deeply maternal, and perhaps nowhere more movingly expressed than in the following lines from

---

BELOW *fig.41* A unique scene showing the divinization of Nefertari as Egypt's 'justified' queen. Here Hathor (left) and Isis, 'Mother of the God' (right) bless her crown, and Hathor also touches the rearing uraeus on the queen's brow, ensuring this fiery serpent 'awakes in peace' in the ninth hour of the day. (Relief on the east wall [south side] of the transverse chamber, Nefertari's temple.)

chapter 30B of the *Book of the Dead*, which is often inscribed on heart scarabs:

> O my heart of my mother,
> O my heart of my mother,
> O my heart of my transformations *[kheperu]*,
> Do not rise up against me as a witness,
> Do not oppose me in the tribunal,
> Do not be hostile to me before the
> guardian of the scales.
> For you are my *Ka*
> That dwells in my body,
> The protector *[Khnum]* prospering my limbs.

In this character-revealing realm where a lifetime's deeds are exposed to view, the heart weighed against Maat's tiny feather derives from the mother. She it is who sustains the power to transform *(kheper)*, develop and grow continually throughout a lifetime *(fig.42)*.

This 'maternal' aspect of judgement comes to the fore also in the Luxor temple when Amenhotep III 'becomes young again' in his 'place of justification' *(st-m3't)*, from which he 'goes forth in joy … his transformations being visible to all'.[5] He is the 'excellent egg of Amun who is suckled in the Magistrates' Chamber', and here the determinative for 'suckle' depicts the reborn child-king standing before the enthroned goddess, being nourished by drinking her divine milk.[6]

Similarly, as 'the actualized feminine aspect of the kingship',[7] Nefertari shines in her own maternal 'place of justification', and, just as Horus and Seth bless the crowns of Ramesses in the hall *(fig.27)*, here Hathor and Isis make a similar gesture for Nefertari—though in a quite different realm, a watery heart chamber of *Ka*-life, a blissful uterine space of second birth. Furthermore, by touching the rearing cobra on the queen's brow, Hathor acknowledges the powerful uraeus 'daughter of Re', the supreme crown goddess now marking Nefertari with the indubitable sign of solar sovereignty.[8] Perhaps her gesture conveys an anointing of the queen's brow with sacred oil. It is surely intended to ensure that this heated serpent, now fully roused to the crown, 'awakens in peace' around the queen's divine head—that coolness assuages the heat generated during the serpentine ascent.[9]

No other scene of its kind has survived so explicitly showing the deification of Egypt's queen, the glorification of her eternal *Ka* and her transformation into a divine being, a crowning perfectly encapsulated by the word *khenem (ḥnm)* which means not only 'to wear a crown' but also 'to join, unite with', 'enrich', 'keep safe', 'embrace' and 'be provided with'. From the *ḥnm* root are also derived the words for 'nurturer' *(ḥnmt)* and 'well' *(ḥnmt)*, and three water-filled vessels sometimes occur as the determinative in the name of the northern marshes of Chemmis *(Pḥww)*, where Hathor sails in her boat to nourish the king *(see page 48)*.

The same root is intrinsic to the name of the god Khnum, who appears in this room, the ram-headed god worshipped at Elephantine, guarding the deep cavernous source of the inundation waters, closing and opening this flowing fount of life that encircles the land each year, turning fields green and bringing food to Egypt. He is also responsible for opening the womb, letting forth the rush of encircling waters preceding every birth, and in this maternal *Ka*-chamber of second birth, his activity is surely needed.[10]

Perhaps, too, the presence of Isis at Nefertari's crowning evokes her role as Isis-Sothis, the star that makes its heliacal rising just before dawn in mid-July, heralding the start of the New Year, around the time when the annual inundation waters are also returning.[11] Yet here she is specifically named 'Mother of the God', as she is when she accompanies the ithyphallic Min-Amun and

Ramesses II in the great temple nearby. This is also her characteristic epithet when partnering the regenerative Theban bull god, Amun-Bull-of-his-Mother, the ageing god who returns to the mother at the close of day for renewal (see page 32).[12]

Probably this chamber works on different levels, and the meaning is intentionally ambivalent, for whilst on the one hand it is aligned with the annual river axis, it also corresponds with the midheaven realm where the solar circuit is glimpsed in its glorious entirety from East to West during the ninth hour (see page 29). Paired with Hathor as goddess of the East, Isis is the mother in the West, and hence, even though Nefertari faces towards Hathor at her crowning, this chamber is no less hers, since both goddesses are needed to encompass the solar circuit's totality.

Across from this crowning scene, on the other side of the doorway, Ramesses worships the enthroned Taweret, 'who gives birth to the gods' (fig.44), offering fresh plants (rnpyt) to her and thus honouring her power to make everything 'fresh' again (rnpj), everything new (rnpi), like the year (rnpwt) itself. His gesture almost exactly mirrors that of Nefertari offering flowers to Mut in the scene on the other side of the wall in the hall (see page 45).

TOP LEFT fig.42 Vignette illustrating chapter 30B of the Book of the Dead. Here Astweret clasps a heart amulet to her breast as she kneels before an offering table and a scarab, making manifest the purity of her heart and her eternal power of transformation. (Papyrus of Astweret, British Museum, London, EA 10039. Late Period.)

ABOVE fig.43 Nefertari offers to Khnum, Satis and Anukis, the divine triad guarding the inundation waters at Elephantine. (Relief on the west wall [north side] of the transverse chamber, Nefertari's temple.)

Behind the king stands Nefertari shaking her naos sistrum as 'daughter', so that, as with the pairing of Mut and Hathor in the hall, there is a duality of solar 'mother' (Taweret) and 'daughter' (Nefertari) here. Now, however, the king overtly stands between them, thus he is interwoven with two successive female generations, represented also by Satis (mother) and Anukis (daughter) across on the opposite wall, a prototypal mother–daughter pairing which is particularly important in the New Kingdom ideology of queenship.[13] Just as in the ninth-hour Field of Reeds seeds are sown in one cycle, then take root, flourish and are harvested in order to be sown again in the next cycle, 'mother' and 'daughter' sustain the pharaoh's ever-renewing

begotten by the male in this chamber, but comes forth from the maternal womb and returns there. No wonder Ramesses feels impelled to offer intoxicating wine in such a heart chamber.

The whole of nature hails his 'becoming young again'—every plant, every tree, every crop grows green again with him, and the river flows freely, as all are renewed together in one great annual cycle of second birth. And it is his deified solar queen, the very incarnation of the loving-destructive Sun Eye, who is the source of this inexhaustible fecundity. For to be queen of Egypt is to be beautiful, shining and on fire with Hathor's heavenly flame; it is to be a queen who manifests the radiance of Hathor's liquid light and life for her people on earth, and in this watery birth-chamber, where royal life reaches maturity and fulfilment, she guarantees Egypt's freshness and prosperity.[15]

These royal cultivators have found their way to that deep well of life whence all forms rise and whither all return, gloriously 'made young' again like the ninth-hour sun god, and it is by drawing

ABOVE *fig.44* Ramesses II offers flowers to Taweret, 'who gives birth to the gods', accompanied by Nefertari shaking her naos sistrum as the mediating daughter. Here the divinized king celebrates his eternal power of renewal in the solar cycle of life, guarded by Taweret as 'mother' and Nefertari as 'daughter'. (Relief on the east wall [north side] of the transverse chamber, Nefertari's temple.)

RIGHT *fig.45* View into the sanctuary of Nefertari's temple at Abu Simbel. Carved on the back wall is a damaged statue of Hathor in cow form, emerging from the mountain in the Northwest with Ramesses II between her forelegs.

life, ensuring he 'becomes young' again forever. Holding the thread that unites West and East, the beginning and end of the cycle, the harvest and the inundation, they unite his *Ka* with the eternal rhythm of 'greening life', forever replenishing and renewing him.

Thus, after all the exertion of ascent to the zenith ruled by Amun-Re and Maat in the hall, a very different vista opens up in this threefold central chamber, connecting the royal couple with the rhythm of the beating heart. Here Ramesses lives within a feminine lifecycle that is renewing him as Egypt's ruler year by year *(rnpwt)*, 'becoming young again' *(rnpi)* through the endless generations of mothers and daughters.[14] Life is not from this eternal source of life that they continually renew Egypt's vitality. No matter how many times a new pharaoh might appear on the throne, this circulating *Ka*-life never diminishes and always creates him 'fresh again'.

Neither Seth's arid treeless desert nor the sandbank of Apophis has triumphed in this healing life realm. It is an abundant green land, where bubbling springs of living water flow and a spirit of 'peace' reigns, presided over by Horus. For, according to chapter 29A of the *Book of the Dead*, which guards against the loss of the heart, whenever a justified person 'lives in truth [Maat]', Horus dwells in the heart at the very centre of the body—as he does here in Nefertari's temple, represented by the three falcon-headed Horus gods wearing the Red and White Crowns, the divinized protectors of three Nubian fortified cities. They bring a peace resulting not simply from the cessation of the conflict with Seth, but from the reunion of all the conflicting elements within a higher threefold totality, a peace stemming from the genuine reconciliation and co-operation of all opposites.[16]

Nevertheless, the royal couple are not yet at the end of their journey. Still another doorway beckons *(fig.45)*, one taking them deeper into this mountainous cavern, its entrance guarded by Ramesses on either side, warning those who step across its threshold: 'Be pure'. Probably never sealed off from the central chamber, this pure gateway gives access to the innermost sanctuary, where the sun's descending course brings the royal couple to the place of 'mooring' and the Northwest, the realm of the setting sun and the abode of generations.

ESSENTIAL REGENERATION: TRANSMUTING POWER

Inside the sanctuary, on either side of the doorway, are two fat-bellied fecundity figures. Both wear a papyrus crown, symbolic of the North, and, interestingly, both face towards the doorway as they bear aloft food offerings, taking food out into the world rather than into the sanctuary as a gift for the temple goddess. Manifestly, this most holy place is a source of abundant life for Egypt, though

food is not the only theme in the three main scenes adorning this small dark chamber.

On the south (left) wall, a single scene shows Nefertari standing behind an offering table, burning fragrant incense and shaking her naos sistrum before two seated goddesses—Mut, 'Serpent Lady of the Sky, Mistress of the Gods', and Hathor, 'Lady of Dendara, Mistress of the Two Lands' *(fig.46)*, divine epithets associating them with 'heaven' and 'earth' respectively. Indeed, according to a Theban hymn to Mut, to burn incense and make offerings is the appropriate cult service for a shining goddess ruling sky and earth:

> When she shines, sky and earth are like gold.
> Gods and people rejoice
> when they see her Majesty as Lady of Sun-rays.
> Divine offerings and
> burnt incense belong to the Lady of Sky and Earth.[17]

Here Mut's shining rays turn heaven and earth 'gold', the very same realms presided over by the sun god's fiery uraeus snakes in the tenth-hour hymn in the *Ritual of Hours* when they bring him 'everything heaven and earth contain', coiling around his divine image to reveal his regenerative power as 'Gold of the Stars' *(see page 30)*.

Correspondingly, Nefertari shakes her sistrum to propitiate the serpent goddesses of 'above' and 'below'. Previously, in the hall of ascent, she has offered to Hathor and Mut on either side of the central doorway *(fig.36)*, so there is, in a way, a repetition of themes here in the innermost sanctuary. Now, however, the goddesses sit side by side, though strangely their brief epithets allocate the Theban mother goddess to the sky and Hathor to the earth, an unexpected reversal given Hathor's status as the solar goddess par excellence, yet an interchange utterly in tune with the reversal of realms in the 11th-hour hymn in the *Ritual of Hours* expressing complete cosmic unity *(see page 32)*.[18]

Worshipping these two particular goddesses also holds special meaning for Nefertari, since by doing so she is simultaneously honouring her own royal name, Nefertari-beloved-of-Mut, which is written in an oval cartouche, or *shen*-ring, symbolizing 'all the sun encircles'. Whilst the second element defines her loving relationship with Mut, the first

element, Nefertari, means 'She who belongs to the Beautiful One', incorporating the word *nefer*, so characteristic of Hathor, with its root meaning of 'beauty', 'vitality' and 'perfection'.

To the Egyptians, names reveal an individual's essence, and knowing a particular name means being able to release its creative potential in the world. It is also by remembering names that earth-dwellers establish a spiritual connection with the Osirian ancestors, ensuring their permanent survival in the *Dwat*.[19] Hence, as Nefertari's lifetime nears completion, her supreme name, the power of the eternal word, resounds, with the divine queen of Egypt shining in all her fiery glory as the living image of Hathor and Mut, her indestructible golden essence gleaming in her enduring temple. Reinforcing her divinity is the incense rite she performs, since the word *sntr*, meaning 'to make godlike', resonates with the word for 'fragrant incense', *sntr*, used in the temple cult to transform statues into a divine state. In short, by burning incense for these two goddesses, Nefertari is simultaneously divinizing herself.

Across on the north (right) wall, the single scene shows Ramesses, protected by Nekhbet's vulture wings outstretched overhead, pouring a libation and burning incense for two enthroned figures seated before a table laden with offerings *(fig.47)*.

ABOVE fig.46 Nefertari burns incense and shakes her naos sistrum before Mut and Hathor, the female powers of earth and sky. She honours the two serpent goddesses with whom she is identified in her royal name 'Nefertari-beloved-of Mut', which encapsulates her divine 'essence'. (Relief on the south wall of the sanctuary, Nefertari's temple.)

One is Nefertari; the other, curiously, is Ramesses himself, wearing the typical crown of the Memphite creator god Ptah-Tatenen, or 'Ptah-of-the-Risen-Earth', the very same crown that he wears when he touches Hathor's *menit*-necklace in the hall *(fig.26)*. Hence, just as there is a repetition and deepening of the hall's experience when Nefertari burns incense for Hathor and Mut, so there is a similar recapitulation here, though, once again, the accompanying inscriptions offer little to unravel this strange scene's meaning. Is this offering to himself another instance of Ramesses II's self-aggrandisement—something of which he has so often been accused? It hardly seems plausible in such a hidden, inaccessible place, especially since Seti I, in his temple at Abydos, was shown offering to Osiris and his deified self, both recipients being adorned with the White Crown, and Seti being named as '*Menmaatre*, the Great God'.[20]

Clearly, when receiving offerings, Seti is here a divinized king, as is his son Ramesses at Abu Simbel, his divinity reinforced by the incense rite he performs (see above), as well as the ram's horn curled around his ear, often an attribute of deified kings.[21] He also sits with his beloved Nefertari, who, with her solar plumed headdress and fly whisk over one shoulder, seems to be the very incarnation of her famous namesake, Queen Ahmose-Nefertari, who reigned some 300 years before, and was particularly revered in the Ramessid era, together with her son, Amenhotep I. Indeed, such an intentional identification with their

ABOVE *fig.47* Ramesses II manifests as the 'two in one', consecrating offerings for himself and Nefertari seated at an offering table. Here in the Northwest of the temple (as in the *Book of Day*) the unification of the ancestor king and the reigning Horus king is celebrated at the close of day. Nefertari's iconography identifies her with the revered Queen Ahmose-Nefertari from the 18th Dynasty. (Relief on the north wall of the sanctuary, Nefertari's temple.)

BELOW *fig.48* Inherkhau, accompanied by his wife, Wab, burns incense for Queen Ahmose-Nefertari (lower register) and her son Amenhotep I (upper register), who head two rows of male and female royal ancestors dating back to the Middle Kingdom. (Line-drawing of a scene in the tomb of Inherkhau at Deir el-Medina, Western Thebes. 20th Dynasty.)

illustrious 18th-Dynasty predecessors would be entirely consistent with the *Book of Day*'s inclusion of the cartouches of Amenhotep I and Ramesses VI after the 11th hour, inscribed on either side of Meretseger's name, two kings united in the 'silent matrix' at the close of day *(see page 33)*.

As beneficent mediators, able to intervene directly in the affairs of the living, Ahmose-Nefertari and her son were not remote predecessors living in the far distant past, but ever-present rulers from whose hidden realm abundant new life continually poured forth, blessing the reigning king and the whole of Egypt. Nor was their influence restricted solely to royalty. At Thebes there were special shrines where the populace could place votive objects dedicated to them, stelae were inscribed for them, and their oracular voices could be heard speaking whenever their sacred boats were carried in procession during festivals.

Interestingly, too, some posthumous portrayals of Ahmose Nefertari show her skin painted black *(fig.48)*.[22] For, like the black Osiris, she was a source of fertility hidden deep in the earth, rising year by year with the royal ancestors to bring back the buried seed as food, ever-living, ever-reborn, sustaining the eternal regenerative cycle, which, at Thebes, was also closely linked with Amun-Re as Bull-of-his-Mother (Kamutef), who was both 'father' and 'son' and procured his regeneration by fertilizing his own mother.[23]

The sun god possesses the same self-regenerative power at the close of the *Book of Day (see page 32)*. So too, as the ageing sun god, Atum affirms this merged unity, telling the sun's enemies in the *Book of Gates*:

> As my Father Re triumphs over you, so I triumph over you. I am the Son who proceeds from his Father, and I am the Father who proceeds from his Son.[24]

Thus, just as the ageing sun god enters the mother goddess in the West, seeding the maternal womb to be reborn again at dawn *(see page 34)*, so Ramesses and Nefertari must ensure the continuity of the royal line, born of the essence that never dies, a mysterious renewal expressible only through the paradox of the 'two in one'.[25]

Surely then, this scene is to be understood as Ramesses, the living priest king of Egypt, burning incense before himself as Ramesses, the buried earth king adorned with Ptah-Tatenen's crown *(fig.47)*, the divinized 'ancestor' king, participating in the divine *Ka*-life he shares with the bull god in a royal lineage stemming from Amenhotep I and his mother, Ahmose-Nefertari.[26] Holding the crook of rulership and an *Ankh*-sign of life, he appears as a powerful lord of latent life, a potent source of renewal like all the other past rulers of Egypt who have 'become earth' at Memphis.[27]

And, just as her husband becomes an ancestor, so Nefertari appears as a transfigured evening queen united with her illustrious predecessor Ahmose-Nefertari, a great mother of the generations, whose womb encloses the eternal 'seed', a queen whose power rises ever and again to bring forth new life, eternally alive and the source of cosmic renewal. Perhaps Nefertari had indeed died and been buried in her glorious tomb in the Valley of the Queens by the time her temple was nearing completion *(figs.2,4)*. If so, it would have made this scene particularly poignant, though its theme also mirrors the *Book of Day*'s union between Amenhotep I and Ramesses VI *(see page 33)*, so that, right from the start, it must have belonged to the temple's overall conception.

Crucially, too, the careful recapitulation of themes from the hall *(see page 54)* points to something fundamental in the Egyptian experience of transformation, for the royal couple are not simply progressing in a linear direction through the temple, with each phase inexorably superseding the previous one, leaving behind everything that has happened before. Nor is the divinized Ramesses wearing Ptah-Tatenen's crown in the innermost sanctuary *(fig.47)* separate from his 'image' adorned with the very same crown in the scene with Hathor in the hall *(fig.26)*. Rather, the hall forms the counterpart to the sanctuary, the two chambers being linked together by the crucial 'unification of opposites' occurring in both places. However, whereas the hall's unifying work requires the king to balance the earthly dual 'opposites' forming the totality of Egypt's cult land, in the sanctuary, accessed through the divinizing chamber of the heart, all 'opposites' are merged in one continuous creation—heaven and earth, life and death, father and son—united by the same king, to be sure, but one who manifests now in a qualitatively different dimension of time-space as the divinized 'two in one'. In the hall of ascent, the principle of 'four' has ruled; in the transverse chamber, that of the 'three'; and now here in the holy of holies, whilst triads still adorn its walls, the mysterious paradox of the 'two in one' is key.

A transmutation is happening here, a change of modality in which the scale is cosmic, the dimension vast. It involves participation in the Memphite creative process itself, at the heart of which is the transmutation of an earthly body into a radiant, glorious image filled with divine life, eternally regenerated through the 'naming' power of the Word.[28]

The creative power to become the 'two in one' opens up a completely different vista of death and life, for whereas these might seem 'opposites' in the hall's three-dimensional earthly realm, in the sanctuary the veil is lifted to reveal their utter unity in the radiant flow of creative life that, in its very essence, is indivisible. This is a unification accomplished also by the Red and White-crowned sun god at the close of day *(see pages 32–3)*. And it is in this flow of life that Nefertari and Ramesses ultimately live and have their being as Egypt's rulers, channelling great blessings and abundance for their people on earth.

HATHOR RETURNS: THE JOYFUL EYE

The royal couple's marriage has brought them to the completion of their lifetime here in the Northwest, but there is also a third triad carved in the rock on the chamber's back wall, looking down the central aisle towards dawn in the Southeast. Here the Nubian Hathor steps forth in her cow form, entering through a doorway framed by a naos sistrum on each side and guarded by fiery snakes above *(fig.49)*. Between her forelegs stands Ramesses, securely moored to the goddess who ensures his rebirth, and although the figures are damaged now, the markings on the cow's neck indicate she would originally have encircled the king with her sacred *menit*-necklace.

ABOVE *fig.49* Return of the goddess. Ramesses II, adorned with the White Crown, offers to Hathor in cow form as she brings him forth from the northwest mountain between her sistrum pillars. In contrast to *fig.47* his 'two in one' aspects are here reversed, since, as the recipient of offerings, he now comes forth with Hathor as her regenerated 'son'. (Back wall of the sanctuary, Nefertari's temple.)

Once again the 'two in one' is honoured here, since to the left of them stands Ramesses, offering papyrus plants and wearing the Upper Egyptian White Crown, the same crown worn by Seti I as an Osirian 'ancestor' king at Abydos *(see page 54)*.[29] Whereas, however, on the right wall Ramesses serves as 'son' for himself as divinized 'father' seated with Nefertari *(fig.47)*, here on the back wall he offers as 'father' to himself as son coming forth with Hathor.

Encountering an image such as this, one can almost hear that great chant to the primordial Hathor in the *Coffin Texts*, celebrating her return with Horus, whom she has sought at the very limits

of existence and now brings back adorned with all
her radiance and beauty:

> I am Hathor who brings her Horus,
> who smites her Horus.
> My heart is like a lion…
> there is no limit to my vision…
> I am she who raises his beauty
> and unites his brilliance…
> I came into being
> before the sky was fashioned…
> Before the earth was released…
> I have searched out,
> And see I have brought.
> Come with my horns
> and raise my beauty,
> Come with my face
> and I will cause you to be exalted…
> I have given my tears…
> I make warmth for them
> In this my name of Shesemtet…
> Such am I.
> I am the cobra Wadjet.
> I am indeed
> the Lady of the Two Lands.[30]

This fiery Hathor, incarnating here as the Lower
Egyptian cobra goddess Wadjet, brings with her
a Horus king from the outermost regions of the
world, returning him to new life at dawn. 'What
does it matter if the *Wedjat*-Eye enters the mountain
of the West … She returns, she returns, the Eye of
Horus in peace' is the great invocation of trust and
hope chanted for the returning Eye in the Ramessid
ritual performed for the ancestor king Amenhotep I.[31]

Hathor returns, she returns, and such a return is
also promised to the solar king as he turns to face
eastwards at the close of the *Book of Day*, looking
towards the approaching sun boat but also to a
third sunrise in his journey through the hours
(see page 34). It is promised to Ramesses, too, here
at Abu Simbel. The candle has been rekindled and
the threshold of death has become the doorway of
life, as the warm-hearted goddess of love and desire
returns and, with overwhelming pleasure, brings
him forth from her holy mountain into a new
incarnation, birthed anew between her musical
sistrum pillars.[32]

And how else could that ninth hour of the night
called 'She who creates Harmony' be rendered in
stone, that hour that is also the first hour of the
day, known as 'She who raises the Beauty of her
Lord', that hour when the parturient Eye goddess
forever bears new life in the mysteries of the rising
dawn? She it is who makes all things grow, expand
and celebrate, she it is who joyfully returns,
bringing with her from the utmost reaches of sky
and earth a vital new Horus, whose flesh is adorned
with all the fragrance and freshness of her desire.

And as goddess and king stand together on the
threshold of life, between her musical sistrum
pillars, a new waxing light, a sun child growing
strong, beholds a living temple of beauty before
his eyes as he awakens to new life in this most
holy dwelling of the queen. Here at Abu Simbel
the returning goddess manifests her great impulse
to give birth to new life, forever ascending to
'the ramparts of the sky' to turn earth into a
revelation of beauty, rekindling life's flame on
the 'way of the green bird'.

Everything in this glorious temple moves
towards this central image of king and goddess,
the divine cow bringing forth her hidden treasure
from deep within the sacred mountain, sustaining
the continual flow of life, and she does so within
a temple dedicated to the queen. It is not hard to
imagine Nefertari once standing before them, her
'missing' presence in fact completing the three
triads on the walls to turn this chamber into a
mystery of 'the ten'. For whether the king journeys
towards the Northwest or comes forth from the
Southeast, he feels himself to be completely
surrounded by his divine beloved, sheltered by her
house. She is the royal mother, consort, daughter,
growing and changing with him, his 'heaven on
earth', his nurturing temple, eternally reaching out
her life-giving arms to guide his journey of life
through the power of her cosmic love. Together
they are the perfect images of Hathor and Re on
earth, complementary aspects of the solar life-force
that underlies all growth and renewal, not only in
this temple but in the whole of the daytime solar
circuit, and it is by manifesting this love wisdom
that the royal couple truly become divine children
of the sun.

FIVE

KINDRED TEMPLES:
SOLSTICE REVERSAL

Journeying through Nefertari's temple feels complete in itself, the fulfilment of a day. Yet close by is Ramesses II's majestic temple (figs.6, 50), oriented almost due east, towards those two magical sunrises occurring around the modern calendar dates of 22 October and 22 February, when sunlight shines through the entrance doorway and glides along the central aisle until it illuminates the divine images within the innermost sanctuary in a brief moment of indescribable glory (fig.51). Was there a plan linking these two temples together? And why did Ramesses choose to build here in Nubia in the extreme south of Egypt?

The October sunrise occurs when the sun is moving towards its furthest southerly declination, where it appears to stop on the horizon before changing its southerly course and turning back to the northern skies and to Egypt. The February sunrise occurs when the sun is returning northwards. So these sunrises move in rhythmic regularity along the river axis in an oscillating pattern from north to south and south to north.[1] What were they marking?

It was not the winter solstice, or 'standstill', as generally understood nowadays in more northerly lands. Whilst the October sunrise coincided with the sun's weakest phase, it occurred not in winter, but in the middle of the springtime season. By then, the Nile flood would have long passed its peak—according to travellers' accounts in the days before the Aswan Dam, the river attained its greatest height towards the end of August in Nubia—and would have dwindled noticeably. As these waters ebbed, in harmony with the sun's withdrawal southwards, the 'Black Land' was clothed anew in a mantle of verdant greenness. The next critical phase was fast approaching: the food-bearing season, with its ploughing and sowing in fields rich with the fertile deposits left behind by the retreating floodwaters. The 'Inundation' season, *Akhet*, was over and the eagerly awaited season of *Peret*, 'Coming Forth', was here, a time when the earth came alive again.

Much has been made of the midsummer period's importance in ancient Egypt, and rightly so, since this was when the sun, rising in the

Northeast, reached its greatest height. It was also the New Year time when the Nile flood returned, bringing welcome water to the thirsty dry earth, heralded by the heliacal rising of Sothis/Sirius, the star of Isis, on the eastern horizon just before dawn around 19 July (Julian calendar), after a disappearance of 70 days. Less well known, however, is the huge significance attached to the sun's southern 'turning-point', when it was furthest from Egypt, reaching the nadir of its annual changes in declination. Crucially, at that point its burning summer heat had diminished, yet it was still working powerfully to bring everything to fruition in the season of sprouting grain and blossoming vegetation, the time of year most favourable for food production, indeed for the growth and development of life on earth.

This southern turning-point was feared, too, as a perilous moment when time might stop, when the sun might not return, a 'standstill' which could interrupt life's continuity. Hence, in their way, these October and February sunrises at Abu Simbel bear witness to the sun's unimpeded power of

movement, its uninterrupted rhythm of descent and return around this southern turning-point.

Moreover, oriented towards this point, at the end of the north terrace outside the great temple, Ramesses built a solar shrine *(fig.52)*, which originally contained a naos housing two stone statues, one depicting a baboon crowned with a lunar disk, the other a scarab, symbols respectively of the moon god, Thoth, and the sun god, Re-Harakhti.[2] Also inside the shrine were four praising baboons perched on an altar, which was flanked by

---

PREVIOUS PAGE *fig.50* View of the great temple at Abu Simbel fronted by four colossal figures of the deified Ramesses II.

ABOVE LEFT *fig.51* Ptah, Amun-Re, Ramesses II and Re-Harakhti enthroned in the great temple's inner sanctuary. Each year in October and February they are illuminated by the rising sun, though the figure of Ptah on the far left remains unlit.

ABOVE *fig.52* The solar shrine at the end of the north terrace outside the great temple at Abu Simbel. The shrine is oriented towards the southern solstice.

I Akhet Day 1   *III Peret Day 1*

two obelisks (now on display in the Nubian Museum, Aswan). Other famous New Kingdom monuments were likewise oriented towards this southern 'turning', including the 'high room of the sun' in the temple of Amun-Re at Karnak, Hatshepsut's temple at Deir el-Bahri *(fig.65)*, and the colossi of Memnon fronting Amenhotep III's mortuary temple on the west bank at Thebes.[3]

Whether the two sunrises at Abu Simbel marked significant ritual moments in the agricultural cycle is more difficult to determine, for whilst they must approximately have spanned the four-month season of 'Coming Forth', the temple's inscriptions are silent.

Interestingly, though, archaeo-astronomers have shown that not only did the much later Inca astronomer priests in Peru orient their sacred buildings towards the solstices and equinoxes, they also recognized two 'zenithal' sunrises, occurring on 29/30 October and 12/13 February, as the sun moved to, and away from, its 'standstill', in this case the summer solstice in the southern hemisphere. Furthermore, these sunrises coincided with key agricultural ceremonies, the February sunrise being linked to the corn and potato rituals, and the October sunrise heralding the return of the rainy season, when the gods were invoked to bring rain.[4]

Given the approximate correspondence with the Inca dates, it is tempting to surmise Abu Simbel's sunrises likewise had agricultural significance. Certainly, during the fourth month of the year, in the days leading up to the 'Coming Forth' season, the lengthy Khoiak Festival was celebrated, when seed was sown in specially prepared troughs dedicated to Osiris, the god whose 'rising forth' from the earth brought fruitful life and abundant

grain *(see Part 2)*. Then, in the ninth month of the year (the first month of the *Shemou* season), came the annual harvest festival dedicated to Min-Amun-Bull-of-his-Mother, who is shown with Isis and Ramesses in the great temple at Abu Simbel, the ithyphallic god forever renewing himself by sowing and resowing his seed in the womb of the 'Mother of the God' *(see page 32)*.[5] Around the time of Abu Simbel's two sunrises, these must have been the kind of sowing and reaping agricultural activities anticipated in Egypt,[6] though this still leaves unexplained the great temple's relationship with Nefertari's 'heaven on earth'.

HINGE OF THE YEAR: THE JANUS FACES

Solstices are usually treated from a purely astronomical perspective, but astronomy alone, at least how it is understood nowadays, cannot account for everything this southern turning-point meant in ancient Egypt. For during the New Kingdom it also had mythical significance in the civil calendar, being particularly associated with a half-year transitional period between the second and third months of *Peret* (the sixth and seventh months). This 'mobile', or 'wandering', civil calendar, which fell short of a solar year by a quarter of a day, was formed of 12 months of 30 days divided into three seasons, plus five epagomenal days. It was also divided into two halves consisting of 180 days, as counted from the start of the 'Inundation' season, delimited by the 'New Year' ceremonies on 1 *Akhet* day 1 (which, in an ideal calendar, would have coincided with the heliacal rising of Sothis), and by the festival of 'Raising the Sky' at the beginning of the seventh month (III *Peret* day 1).

This half-yearly 'Raising the Sky' festival, marking the completion of a 180-day period and the beginning of a new cycle, honoured the Memphite god Ptah, and also became a festival of Amun at Thebes.[7] On the same day, Osiris's burial was celebrated, along with the victory of his son, Horus, the powerful falcon of light whose heavenly home was recreated when time changed direction.[8]

Much later, the significance of these two dates is very clearly shown in a Roman-period frieze at Dendara depicting the deities governing the days of the year (*fig.53*). At their head, facing towards each other on either side of two cartouches, are the representatives of 1 *Akhet* day 1 and III *Peret* day 1, the two ends of the year, who are placed directly above an arbour where Hathor suckles her newborn son, symbolically encompassing the north–south limits between which the sun oscillates in this 'counting' of the year.

Because the 12 months were tied to the 12 night hours (*see below*), this annual change of direction between the sixth and seventh months also created a significant solar 'turning' during the sun god's nocturnal voyage through the hours in the *Dwat*, influencing both Osiris's regeneration and the fate of all humans journeying with the sun.

Nowhere perhaps is the sun's relationship with this reversal more explicitly marked than on Amenhotep III's beautiful 18th-Dynasty clepsydrya, or water clock, discovered at Karnak (*fig.54*). The clock's interior has 12 graduated scales, each measuring the hours of the night during a particular month, with the months themselves being named on the rim of the vessel. The exterior is divided into three registers, and in the bottom register the king is portrayed offering to the representative month deities, beginning with the 'Inundation' season (*Akhet*). Representing the sixth month, called 'Great Flame' (*rkḥ wr*), and the seventh month, 'Little Flame' (*rkḥ nḏs*), in other words the second and third months of *Peret*, are two hippopotami, each placed on a standard and symmetrically ranged on either side of paired *Was*-sceptres (*fig.54*). Like the double-faced Roman god Janus, the guardian of thresholds, doorways and beginnings, who looks both backwards and forwards at the same time, these Egyptian month deities face in different directions, symbolically marking 'reversal', and also the passage from one cycle to another after a period of 180 days, as counted from the first day of the 'Inundation' season.[9] Furthermore, when the registers of this water clock are 'rolled out' on Ramesses II's rectangular astronomical ceiling in the Ramesseum at Thebes, these month deities (represented here by a jackal rather than a hippopotamus) appear at each end of the bottom register, defining the extremities.[10]

ABOVE LEFT fig.53 Representative figures marking the two 'ends' of the year at Dendara, depicted directly above Hathor suckling her child in an arbour. (Line-drawing of a frieze in the Roman birth-house at Dendara.)

RIGHT fig.54 Line-drawing of the section of the Karnak clepsydra marking the sixth and seven months. Here the paired 'month' animals face in opposite directions to indicate a division of the year after 180 days. Above them Amenhotep III, accompanied by the god 'Moon', offers to Re-Harakhti and receives 'life' from him. (Egyptian Museum, Cairo, JE. 37525 18th Dynasty.)

Clearly, annual 'reversal' also involves the sun, since above the clepsydra's sixth and seventh-month animals is a large scene of Re-Harakhti receiving offerings from the king, who is accompanied by an ibis-headed god called Iah, or 'Moon' *(fig.54)*. The same combination of 'moon' and 'sun' reappears in Ramesses II's solar shrine at Abu Simbel *(fig.52)*, which is oriented towards the southern 'turning-point' and originally contained a small naos housing a lunar baboon and a solar scarab *(see page 61)*, the link with this transitional period being reinforced by the paired scenes on the naos's side walls showing Ramesses offering to Re-Harakhti and the moon god, Thoth.[11] Interestingly, in a later scene at Dendara depicting 'filling the Moon Eye', as celebrated at the end of the sixth month *(see chapter 10)*, a baboon perches on the stern of the boat containing the Eye, while a scarab is located behind Thoth *(fig.55)*. Similarly, a Roman-period astronomical ceiling, which once adorned the sanctuary in the temple at Deir el-Haggar, shows these same creatures waiting to greet the six hour goddesses towing the night boat of the sun, their number indicating they have reached the crucial sixth-hour turning-point. Depicted in the register beneath the sun boat are the 12 month deities, thus tying this nocturnal journey to the yearly cycle.[12] In the grand scheme of the Abu Simbel great temple, the baboon and scarab within the solar shrine might seem a minor detail, yet they belong to very precise 'reversal' iconography which lives on right through into the Roman period.

In short, this southern 'turning' acquired 'mythic' meaning extending far beyond what is usually understood as astronomy nowadays and, whilst not astronomically exact in terms of the solstitial solar year, it was given official existence, as it were, within the 360-day civil calendar (plus five epagomenal days), a relationship with the months and 'hours' which played into the conception of the twin temples at Abu Simbel *(see pages 72–4)*.[13]

What is also important to remember, though, is that in Ramesses II's reign this civil 'wandering' calendar was once again in step with the seasons and the heliacal rising of Sothis after a Sothic cycle of 1,461 years, making it a particularly important

time. Certainly, the great temple's precise orientations suggest something very significant was at work in Ramesses's reign connected with this southern turning-point;[14] and it is known he particularly favoured the *Peret* season, choosing to begin at least two of his Sed Festivals on its auspicious first day.[15]

EGYPTIAN RESURRECTION: PLATO'S MYTH

Curiously, Plato's account of the 'age of Kronos' and the 'age of Zeus' in the *Statesman*, as told by the mysterious character called the Eleatic Stranger, inadvertently illuminates Egyptian 'reversal' beliefs, especially when he describes how, in the peaceful golden age of Kronos, the period of 'cosmic change' and 'reverse direction' was intimately associated with the resurrection of the 'earth-born' dead. There was, he says, a reversal of the ageing process, as the old returned to youth and the dead lying in the earth were reformed and revived:

> Such resurrection of the dead was in keeping with the cosmic change, all creation being now turned in the reverse direction.[16]

This rebirth of the dead in the earth, Plato says, was 'divinely regulated'. Moreover, when the counter-revolution was put in motion, the 'age of Zeus' began, inaugurating the processes of ageing forwards and sexual procreation, the way of generation favourable to human development in Plato's own 'contemporary' era. According to him, it was the ancestors living at the beginning of the 'contemporary' rotation who remembered the previous epoch's mode of rebirth and passed on their memories of this 'earth-born' resurrection to those living in the 'age of Zeus'.[17]

To be sure, Plato presented these alternating cycles in terms of successive historical epochs ruled by Kronos and Zeus, but his notion of cosmic reversal evidently drew on astronomical 'solstice' terminology, which he applied to his own ideas of death and rebirth.[18] Or were they his own? According to Strabo, Plato spent 13 years with the Egyptian priests in Heliopolis, learning from them about the heavenly bodies, including the length of the solar year.[19] He also says they hid most of their

knowledge from him, perhaps somewhat implausibly in light of the remarkable similarities between the *Statesman*'s account and Egyptian 'resurrection' beliefs tied to cosmic 'reversal' between the sixth and seventh months.

So, for example, the famous 'declaration of innocence' in chapter 125 of the *Book of the Dead* explicitly links the end of the sixth month with resurrection. 'I am pure', a person confirms before the tribunal in the Judgement Hall:

> My purity is the purity of the great *Benu*-bird which is in Herakleopolis; I am the nose of the Lord of the Wind, who ensures all people live, on that day when the *Wedjat*-Eye is full in Heliopolis on the last day of the second month of *Peret* before the lord of this land. I have seen the completion of the *Wedjat*-Eye in Heliopolis and no misfortune can befall me.[20]

No wonder the *Benu*-bird is invoked, for this is the Egyptian phoenix, the symbol of rebirth, closely associated with the course of the sun and the rule of Maat. This mythical bird presides over the recurrent cycles of time, rising from primordial darkness in the remotest regions of the world as the herald of renewed light and life. When the Eye

ABOVE *fig.55* Scene of 'Filling the Moon Eye' at the end of the sixth month, which is precisely located on the west side of Dendara temple's roof terace above the doorway giving access to the central Osiris room. Here Thoth (left) and Shu (right) support a large disk enclosing the Eye. A scarab is behind Thoth, and a baboon perches on the end of the boat behind Shu, the same creatures originally housed in the solar shrine at Abu Simbel *(see fig.52)*. Thoth and Shu are also the deities who entice the fleeing Sun Eye back to Egypt from Nubia in the *Goddess in the Distance* myth associated with the southern solstice. (Late first century BCE.)

is 'filled' in Heliopolis on the last day of the sixth month (the second month of *Peret*), it takes flight in Herakleopolis, the city in Middle Egypt where the ram-headed god Harsaphes is worshipped *(fig.33)*, and where, according to chapter 175 of the *Book of the Dead*, Osiris is crowned as the sun god's triumphant heir, grudgingly acknowledged by Seth *(see page 43)*.[21]

The *Benu*-bird specifically manifests at the end of the sixth month, the critical 'hinge of the year', the time when Seth's rampaging forces seek to damage the Eye and disrupt renewal—the period of year, too, when a person's 'lifetime' is judged in accordance with Maat's way.[22]

LEFT *fig.* 56 The weighing of the heart. (Papyrus of Anhai, British Museum, London, EA 10472. 20th Dynasty.)

RIGHT *fig.* 57 Detail of the *Aker*-lions in the *Book of the Earth*'s bottom register. (Right wall of Ramesses VI's sarcophagus chamber, Valley of the Kings, Western Thebes. 20th Dynasty.)

According to a mythological late text, it was Seth who 'damaged the Eye in Heliopolis during the Festival of the Faces, on the last day of the second month of *Peret*'.[23] The date is repeated in chapter 140 of the *Book of the Dead* for 'raising up' a person in the *Dwat* and entering the sun boat, which has to be recited when 'the *Wedjat*-Eye becomes full on the last day of the second month of *Peret*'. The vignette for this chapter shows the initiate praising a reclining jackal on a shrine, behind which is a kneeling god with arms upraised and bearing a large Eye on his head—the jackal not only being the traditional animal of Anubis but also representative of the sixth and seventh months *(see page 63)*.[24]

Clearly, this annual change of direction is perilous, threatening the integrity of travellers with Re and their capacity to 'live in truth', a betwixt and between time, when the teetering sun boat faces standstill at the cusp of the world. Yet when the two Eyes are healed and full (the rubric here instructs that two sacred Eyes are to be made, one of lapis lazuli, the other of red stone), the cosmic disaster is averted, the sun god's enemies are driven away and the 'shape of Khepri comes into being'.

Chapter 140 also mentions the fourth night hour, the Eye and the vitality of the land, thus linking this reversal at the end of the sixth month with both the sun god's nocturnal journey and Egypt's fruitfulness:

> His Eye rests in its place upon his person
> [the sun god] during the fourth hour of the night,
> and the land is vital in the second month of *Peret*,
> last day.[25]

And just as the ageing process is reversed in Plato's account of the 'age of Kronos', the same occurs during the sun god's journey in the *Dwat* in the *Book of Night*. Thus, during the third night hour, the Egyptian king manifests as the ageing sun god, declaring himself to be divine and powerful on the throne of Atum:

> I am divine, I am an *Akh*, I am powerful and I am
> seated on the throne of Atum.[26]

Moreover, beneath the sun boat in this third hour, various persons are shown in different phases of renewal, culminating in the birth of a child during the seventh night hour. Here, in short, between the third and seventh hours, when the 'old sun' is transformed into the new infant light, the natural ageing process is reversed—all of which is distinctly reminiscent of Plato's account of the reversal of the ageing process and the resurrection of the dead at the time of cosmic reversal.[27]

FILLING THE EYE: RAISING THE SKY

All this is not simply 'astronomy' as it is understood nowadays, it is 'sacred science', a southern reversal of time expressed completely mythically. It is also visualized in extraordinary scenes from the *Book of the Earth*, as depicted on the right wall of Ramesses VI's sarcophagus chamber in the Valley of the Kings, directly beneath the *Books of the Day and Night* on the ceiling above,[28] though without the background of 'Filling the Eye' and reversal at the end of the sixth month,

it is extremely difficult to understand what is going on here, and in fact the *Book of the Earth*'s relationship with this annual 'reversal' has gone completely unnoticed, as has its relationship with Plato's myth.[29]

Yet this right wall is a template of time in transition after a period of 180 days, time at the end of the sixth month and beginning of the seventh, time deep in the earth, time in renewal, encapsulated by the *Aker*-lions depicted in the centre of the bottom register *(fig.57)*.[30] Called 'Yesterday' and 'Tomorrow' in chapter 17 of the *Book of the Dead*, these *Aker*-lions face in different directions.[31] They also reappear in the centre of the top register, supporting the sun boat *(fig.62)*, thus creating a vertical axis along the centre of the right wall.[32] Whereas in the top register they are human-headed and unified, in the bottom register they are divided in half by the 'two arms of Nun', a separating act reinforcing the sense of 'time divided' as the sun sinks into the primeval ocean, here called the 'Place of Destruction' (*ḥtmyt*),[33] for renewal.

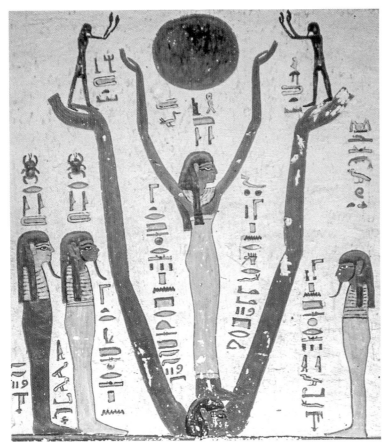

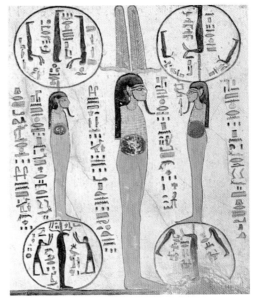

LEFT *fig.58* A goddess called 'She Who Vanquishes' raises the sun aloft to re-establish the solar circuit at the beginning of the seventh month. She stands on the head of a god rising from earth, whose hands support the praising figures of 'East' (left) and 'West' (right). (Detail from the *Book of the Earth*'s middle section, Ramesses VI's sarcophagus chamber. 20th Dynasty.)

BELOW LEFT *fig.59* A weeping eye enclosed in a mound, which is surmounted by a recumbent mummy. (Detail from the *Book of the Earth*'s bottom register, Ramesses VI's sarcophagus chamber. 20th Dynasty.)

BELOW *fig.60* The year's reversal between the sixth and seventh months symbolized by a bearded mummiform figure flanked by a mummy on each side. In the disk crowning the mummy on the right are three inverted figures, each with the head of a shrew-mouse, a creature associated with Horus as the blind god of Letopolis. Here their inverted posture highlights 'reversal' in the annual solar circuit. (Detail from the *Book of the Earth*'s middle section, Ramesses VI's sarcophagus chamber. 20th Dynasty.)

Clearly renewal is needed, conveyed by the four 'flesh' signs and a tiny eye dripping with tears shown further along the bottom register at the far right-hand corner of the wall *(fig.59)*.[34] Just as the Eye is damaged by Seth at the end of the sixth month *(see page 66)*, and the Eye of Horus drips in his conflict with Seth, so this eye, enclosed in a mound encircled by the 'arms of Geb', needs to be healed and 'filled' if the recumbent mummy above is to become a living body.

Above this group, moreover, a long vignette shows seven human-headed *Ba*-birds towing the sun boat and its occupants—the ram-headed sun god and the scarab symbol of Khepri, 'old sun' and 'renewed sun'—here being steered by Horus towards the 'cavern of Nun' *(fig.59)*. Also standing in the boat, on the far left, are the goddesses 'She who judges' and 'She who reckons', their names reminiscent of the 'judgement' at the end of the sixth month when the *Wedjat*-Eye is 'filled' *(see page 65)*.[35]

Facing towards this group is an *Aker*-lion with a sun boat shown above him *(fig.57)*. A young child is perched on the prow as it tilts precariously downwards towards the waiting arms of the Memphite earth god, Tatenen, its occupants, a ram-headed scarab being praised by two *Ba*-birds named '*Ba* of Atum' and '*Ba* of Khepri', all lurching towards Nun's arms.

Across on the left, above the *Aker*-lion facing in the opposite direction, a sun boat is being raised upwards, with a swallow, the 'herald of dawn', on its prow *(fig.57)*. Again a ram-headed scarab is in the boat, this time being towed eastwards by 14 human-headed uraeus serpents in order to shine 'in this eastern mountain'. Transforming from old man (Atum) to 'vital' child (Khepri), this 'descending' and 'ascending' sun has reversed the ageing process.

But it is the end tableau on the extreme right of the central section, located above the 'weeping eye' in the bottom register, that truly synchronizes this solar journey with the 'reversal' between the sixth and seventh months. Here a bearded mummy, crowned with ram's horns and plumes, stands between two shorter mummies, both placed on a disk and both crowned with another disk *(fig.60)*.[36] Within each of these disks are three more figures,

and, curiously, in the upper disk on the far right, these all have the head of a shrew-mouse. The Egyptian worldview extended to the smallest insect, but a shrew-mouse here seems incomprehensible— that is, until it is remembered that this creature is sacred to Horus as the 'blind god' of Letopolis, in other words the Horus whose Moon Eye is injured during his conflict with Seth.[37] Even more startling is the upside-down posture of the three shrew-mouse figures in the upper disk, which is surely intended to convey 'reversal'. Just as instructive are the brief inscriptions, telling how the head and feet of the right-hand mummy are 'in the Lower Region', whilst the head of the mummy on the far left is 'in the Upper Region' and his feet 'in the Lower Region'. This time of reversal has a vertiginous effect requiring an urgent cosmic reorientation, a return to equilibrium, a restoration of the heads of these strange 'upside-down' beings to an upright position, correctly poised once again between above and below, with their vision unimpaired.[38]

In fact, on the ceiling above, in the *Book of Night*'s seventh hour, the victorious Horus appears in his 'Letopolis' guise as 'Foremost of the Two Eyes' (Khenty-Irty), guarding enemies who have sought to fetter Osiris but are now fettered by this all-seeing god of Letopolis, beside whom stands a male figure raising aloft a child in the gesture of 'regeneration'.[39]

Correspondingly, in the *Book of the Earth* beneath, everything is similarly progressing from this pivotal 'reversal' to 'regeneration' in the seventh month. Thus, further along the register, to the left of the 'upside-down' figures in disks, a small male head is seen rising from the earth, and not only supporting a goddess with upraised arms, but also extending his own elongated arms to raise aloft the personified figures of 'East' and 'West', who are praising the sun disk between the goddess *(fig.58)*—a 'raising of the sky' gesture not only perfectly in tune with the festival of 'Raising the Sky' inaugurating the seventh month *(see page 63)*, but also creating the sky anew for the 'reversed' sun now to sail across. 'Seeing this god in his entirety after the two arms come forth' is the message beside the goddess, affirming everything is now reoriented anew and the sun's 'descent' has been successfully reversed, as it has reached the limits marked by the transition between the sixth and seventh months.[40]

What this reversal brings is depicted in the extraordinary tableau to the left, showing a procreative god called 'He who conceals the Hours'. A lower sun disk is behind him, an upper sun in front of him, and he stands in a stylized snake-encircled clepsydra with the 12 night-hour goddesses ranged along its sloping sides, each one holding a sun disk *(fig.61)*.[41] And, just as a child comes forth in the *Book of Night*'s seventh hour on the ceiling above, so this 'hour-concealing' god engenders a tiny child from the fiery seed falling from his erect phallus when the year reverses.[42] In fact, when he is depicted on a 28th-Dynasty cartonnage pectoral, in a much abbreviated version, seven stars shine above him, clearly locating him in the seventh hour, the crucial turning-point of the night and year.[43]

In the register above the clepsydra stride 14 ram-headed figures, all with their heads symbolically 'turned back' towards the approaching sun boat, here being greeted by a deity reaching out to touch the prow as if pulling it towards him *(fig.61)*. As already mentioned, the Egyptians worshipped the ram, like the bull, as the embodiment of male virility, a vigorous bringer of food and fertility *(see page 43)*, hence these 14 ram-headed deities reinforce the god's virile power in the clepsydra. Their number also evokes the totality of Osiris's integrated 14 bodily members. It also corresponds with the time when the luminous moon is full, on the 14th day of the month. Indeed, when the year 'turns back' at the critical end of the sixth month, on the 15th day a lunar ritual of 'Filling the Eye' is carried out.[44]

All this mythology is certainly relevant to the weeping Eye in the right-hand bottom corner of the wall *(fig.59)*, for when the moon reaches fullness and seed is channelled into life-giving fertility, this injured Eye becomes a sound Eye, a healed Eye, a revelation of Osiris's heir and Seth's defeat, an Eye deriving its significance from the fertile seed life it brings.[45]

Clearly, the cosmos has not fallen into chaos, and the clepsydra's flow of the hours and months continues unabated, though this is not just about the moon and the 'filled' Eye', since the ram-headed gods turn their heads towards the figure of Khepri, praising the ram-headed god in the sun boat *(fig.61)*. Within this fiery boat also sails a uraeus. It is a secure boat, not a boat adrift, having the word 'moored' inscribed above it, and its life-giving solar light is certainly needed by 'He who conceals the Hours' in his clepsydra beneath, for if his seed is to be fertile and develop, it needs the influence of sunlight in these dark chthonic depths.[46] And just as sun and moon deities mark the reversal of time between the sixth and seventh months on Amenhotep III's clepsydra, so they function together here in the *Book of the Earth*, precisely when the 180-day revolution is complete and a new cycle of time begins.

All this carefully chosen imagery—combining fertile rams, the night hours, the mythology of filling the lunar Eye, 'reversal' and the raising of the renewed sun—encapsulates this critical time of year when the sun's course changes direction, when the healed Eye is 'filled' and when life-giving seed, infused with light, flows forth from 'He who conceals the Hours'.[47]

A reversal of time at the beginning of the seventh month is ritually associated, too, with the burial of Osiris *(see page 63)*, here depicted in the top register, where four figures, each wearing the White Crown, guard the recumbent mummy of Osiris encircled by 12 disks and 12 stars *(fig.62)*.[48] A falcon's head emerging from a disk intersects this starry circle at the zenith, radiating light downwards from his Eye to illuminate the mummy beneath, gifting it with celestial light—though, in other Ramessid versions, only six stars and seven disks are included, reinforcing the connection with the seventh night hour.[49] Here, when the year reverses and everything changes direction, the earth-born Osiris is resurrected, presided over by the 'mysterious *Aker*-lions', the unified time-lords shown supporting the sun boat above, human-headed now in this celestial realm, in contrast to their animal faces in the bottom register.[50] Everything is poised between their two poles of 'Yesterday' and 'Tomorrow', and no Sethian foe must interrupt the cosmic movement along this vertical axis.[51]

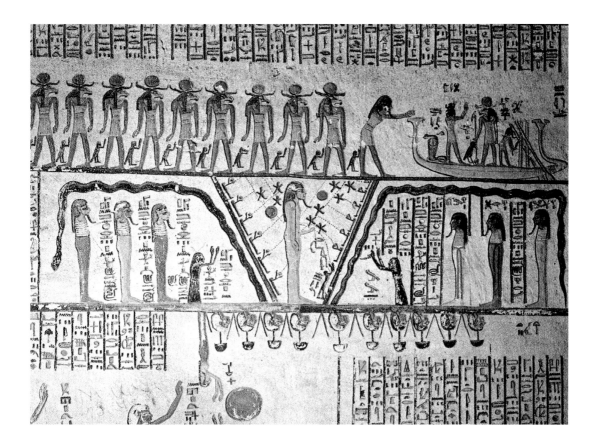

This, too, is the message of the *Benu*-bird soaring forth on the last day of the sixth month in Herakleopolis, the city dedicated to the ram god Harsaphes. The eternal bird witnesses the triumph of light over darkness, rising from the ashes as new life comes forth and time changes direction in the recurrent movement of the world.

In their own inimitable way, the Egyptians symbolically expressed what Plato would much later tell 'philosophically' in the *Statesman*—that the power to reverse direction in the peaceful 'age of Kronos' resurrects the 'earth-born' dead.[52] As noted earlier, he also says it was the ancestors who passed on memories of this 'divinely regulated' resurrection to the living at the beginning of the 'contemporary rotation' in the 'age of Zeus'— particularly precious information given the precise location of Osiris's resurrection at the top of the right wall in Ramesses VI's burial chamber. For on the ceiling directly above is Hathor enclosing the

ABOVE fig.61 Enclosed within a stylized clepsydra, a god called 'He who conceals the Hours' procreates a child from his fiery seed. Twelve hour goddesses, each holding a disk, are ranged along the clepsydra's sides, and above are 14 deities (only nine shown here), their ram heads 'turned back' towards the approaching sun boat. (Detail from the *Book of the Earth*'s middle section, Ramesses VI's sarcophagus chamber. 20th Dynasty.)

'little Re' in her womb-disk in the *Book of Day*'s first hour *(fig.14)*, the beginning of his lifetime journey from birth to old age through the 12 hours of the day, and the hour synchronized with the *Book of Night*'s ninth hour, when the Osirian ancestors come forth to receive their food *(see page 21)*.

Across on the other side of the ceiling, the *Book of Night*'s closing scene shows the male–female pair from the Hermopolitan Ogdoad, Huh and Hauhet, 'Endlessness', embodying the unceasing nature of this renewing circuit as they kneel in

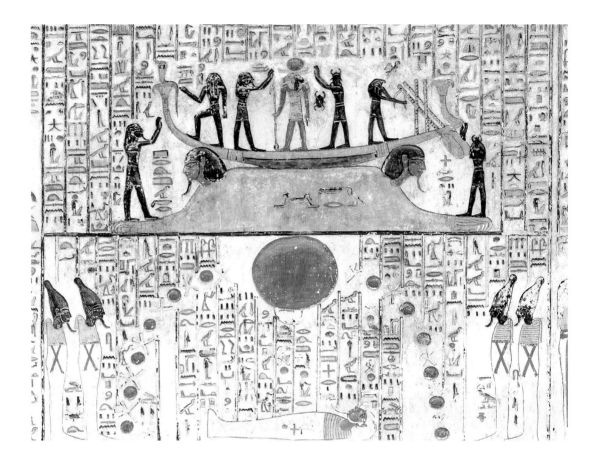

praise before a potter's wheel, a young child and two scarabs *(fig.14)*.[53]

Thus, from the sun disk held by Nun in the watery 'Place of Destruction' in the *Book of the Earth*'s bottom register to the celestial disk shining on the recumbent Osiris mummy in the top register and then onwards to the *Books of the Day and Night* on the ceiling above, there is a vertical ascent towards regeneration in the cosmic East, to that brief moment in the rising dawn when the worlds of past and present merge and meet, the beginning also of the 'contemporary rotation', as Plato puts it in the *Statesman*, ruling the development of human life from birth to death.

Plato's account seems haunted by this ancient knowledge, and, interestingly, puzzled about the source of the myth Joseph Skemp, in his translation of the *Statesman*, suspected an Egyptian origin. 'The whole question of Plato's knowledge of

Egyptian life and history', he wrote, 'is one which calls for further investigation.'[54] Undoubtedly, Plato historicized the ancient cosmic reversal, 'the most important and the most complete of all "turnings back" occurring in the celestial orbits', but at the same time he illuminated its Egyptian meaning, and, as importantly, he opened the way to understanding the pairing of the temples at Abu Simbel.[55]

GREENING EGYPT: THE ABU SIMBEL PLAN

It is as ruler of the 'divinely regulated' southern circuit that Ramesses II fronts the great temple at Abu Simbel, protecting the tiny statues of Osiris *(fig.63)* and Horus placed on the terrace at his feet and guarding the Osirian mysteries of life, death and resurrection when the annual sun is travelling in the south, 'descending' and 'ascending' along the river axis. What happens at this southern 'reversal' deeply influences the growth and development of all life, and inside the temple itself, in the first hall, are the famous scenes of the Battle of Kadesh, depicting Ramesses' defeat of the army of the Hittites, enemies who were potentially threatening this cosmic solar movement.[56]

In the second hall, beyond, the temple priests carry forth the sacred boat of Amun, moving from west to east towards the doorway, following the course of the night sun as they prepare to greet the 'great god' who redeems the world when rising forth from the secrecy of the horizon, that brief flash lighting up the temple's interior when the world is created anew. It also shines on Ramesses in the very depths of the holy mountain, enthroned together with Amun, Re and Ptah, the triad revealing the nature of divinity, and here reinforcing the king's divinized status (fig.51).

His divinization is stated, too, in the *Decree of Ptah-Tatenen* inscribed on a stela in the first hall. He is the bodily 'image of Re', the Memphite creator god confirms (see page 42), here manifesting in his form of the 'Risen Earth'.

It is revealed yet again on the temple's façade, where Ramesses commands the annual river axis with Maat, whose tiny figure he offers to the falcon-headed Re-Harakhti striding forth in the central niche above the temple's entrance (fig.64). In one hand Re-Harakhti holds a *wsr*-hieroglyph, in the other a Maat emblem, so that the whole forms a rebus of Ramesses II's throne name, 'Powerful of Maat is Re' (User-Maat-Re). Holding this cosmic name, the solar falcon invests the Abu Simbel temple mountain with the light of divine sovereignty, king and god together ensuring Maat's way prevails

ABOVE LEFT *fig.*62 Osiris's resurrection illuminated by light raying down from the healed Eye. Supporting the sun boat above are the human-headed *Aker*-lions facing in different directions. (Detail from the *Book of the Earth*'s top register, Ramesses VI's sarcophagus chamber. 20th Dynasty.)

RIGHT *fig.*63 Statue of Osiris beneath a colossal figure of Ramesses II on the north terrace of the great temple at Abu Simbel. (19th Dynasty.)

worlds meet, all tied to the mysteries of the southern 'reversal' in the solar cycle.

It is sometimes claimed that Ramesses was a secret admirer of the radical 18th-Dynasty ruler Akhenaten, and superficially he might seem to be following in the footsteps of this controversial pharaoh here at Abu Simbel. The divine fiery power and sacred light that brought everything to life in Akhenaten's city of the sun at el-Amarna seem to be aflame here also, turning everything green, causing grain to germinate and seed to flourish, generating humans and illumining their wisdom as creatures of light. The king's very public offering of Maat on the large temple's façade, his reverence for the sun and his relationship with Nefertari all might seem to be echoes of Akhenaten's reign. Indeed, Akhenaten was also keen to depict life in the palace, evident in those delightful intimate scenes on private stelae showing him with Queen Nefertiti and their daughters. Yet these seemingly 'naturalistic' glimpses into palace life served a very

and maintaining the sun's eternal movement, the enduring cycle of recurrence, along this river axis.

Yet although the scale of this solar architecture is breathtakingly monumental, divine in conception, the great temple is built alongside Nefertari's dwelling *(fig.6)*, which is oriented southeast to northwest and is connected with the royal couple's 'palace' lifetime unfolding from birth to death, attuned to the sun god's lifetime through the daytime 12 hours—the 'age of Zeus', as Plato would later say. Often described quite separately, these two temples belong together; they need each other as surely as the king needs the queen, the day needs the night, and the months need the year, for here at Abu Simbel the course of the day and the year are the time realities governing the royal couple's life as rulers of Egypt, and it is precisely the juxtaposition with Nefertari's daytime temple that anchors the great temple in a yearly cycle tied to the night hours.

In short, Abu Simbel is where the 'sun of the year' and the 'sun of the day' are brought together, the age of Kronos and the age of Zeus, the sacred space where the yearly, the daily and the human

definite purpose, representing a direct challenge, even a provocation, to all those traditional travellers in the sun boat through the hours. The transformational mysteries of Re's conception, birth and growth, maturation, death and regeneration, on which a palace lifetime had been traditionally based, were not for Akhenaten. Gone, too, in his cult of the sun was the sun god's transformational journey through the night hours.[57] Akhenaten's relationship was with the celestial sun, the Aten, and the celestial sun alone, to the exclusion of all other deities. That was the message from him to his subjects inscribed on the private stelae. His 'palace' lifetime represented a complete break with the past.

In reality Ramesses and Akhenaten were worlds apart, and had Abu Simbel been built before Akhenaten's reign, he would surely have dispensed with its twin temples, though possibly without him they would never have been built in quite the same way. For they stand as a monumental response to

his turbulent reign, anchoring a palace 'lifetime' once again to the annual 'leaving' and 'returning' southern sun, to a solar rebirth and regeneration reconcilable with the night hours, the months, and the death and regeneration of Osiris in the earth—a sun travelling to its lowest ebb at the night-time of the year, a sun wedded to the loving-destructive Eye goddess Hathor-Sekhmet.

Already Ramesses' father, Seti I, in his temple complex at Abydos, had reconnected Egyptian kingship with the *Book of Night* and the *Book of the Earth*, the ancestors and the death and rebirth passage of the sun and stars. Now, at Abu Simbel,

LEFT *fig.64* Ramesses II offers Maat to the figure of Re-Harakhti carved in the niche above the great temple's entrance at Abu Simbel. (19th Dynasty.)

BELOW *fig.65* View of Hatshepsut's temple at Deir el-Bahri, which is oriented towards the southern solstice. (Western Thebes. 18th Dynasty.)

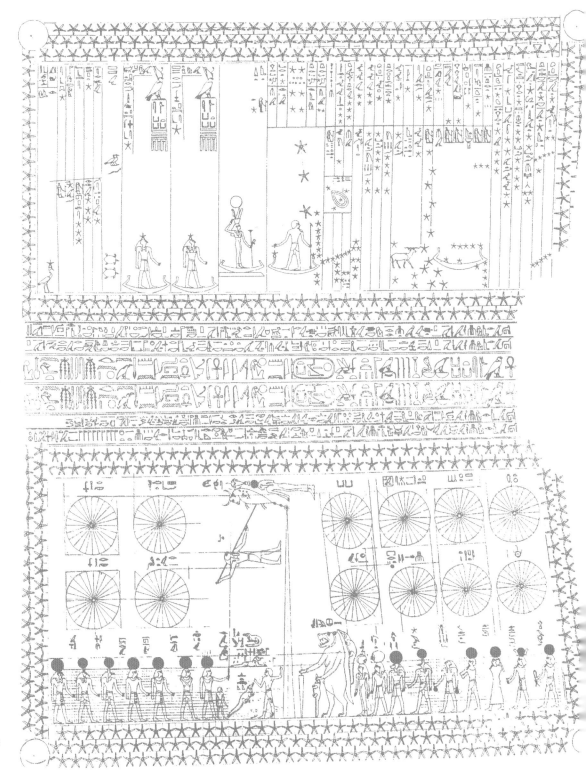

his son took this a stage further, reuniting this night-time of the year with a solar palace lifetime based on the daytime 12 hours.

In doing so, he also harked back to Hatshepsut's temple at Deir el-Bahri (fig.65), built at an ancient site long sacred to Hathor by a female pharaoh blatantly excluded from Seti I's king list at Abydos, yet one who bequeathed great wisdom to the Ramessid kings. Her inspirational temple, with its central chamber high up on the third terrace oriented towards 'southern reversal', provided a blueprint for Abu Simbel. Here, on a ceiling in one of the chambers, was inscribed the *Ritual of Hours*, juxtaposed with the night hours and chapters of the *Book of the Dead (see page 19)*. Here, too, Hatshepsut showed intimate scenes of her conception, birth and nurture in the palace, scenes replicated by Amenhotep III in the Luxor temple. Indeed, the astronomical ceiling in the unfinished tomb of Hatshepsut's close adviser, Senenmut, unequivocally sets out the time cycles ruling Deir el-Bahri. On one half of the ceiling are 12 circles representing the months, with each one divided into 24 segments, thus symbolically linking the diurnal and annual cycles, whilst figured on the ceiling's other half are the planets, together with Sothis and Orion, as well as the 'sheep' and 'boat' decans (fig.66).

Ramesses continued her work, for what he built here at Abu Simbel was an architectural template for 'living the hours' in a greening, growing cosmos of 'life'. These were twin temples not solely based on 'astronomy', but on a view of the cosmos harmonized with the royal couple's lifetime on earth. This was not the work of a megalomaniac ruler intent on personal aggrandisement, for

Ramesses knew these temples must express the relationship between the macrocosm and microcosm.[58] And it was here in Nubia that he could best reveal his relationship with the sun's annual 'ascending' and 'descending' movements and most powerfully express the mystical meaning of the southern 'reversal of time' for his cosmic sovereignty. For this was the axis around which the 'Coming Forth' season moved, lived and had its being, the axis around which the months and night-time hours also turned, bringing the victory of light over darkness and making possible the growth of all life on earth.

To be sure, these complementary rhythms of time are well hidden at Abu Simbel, but it has long been suspected that Hathor's return from Nubia in the *Goddess in the Distance* myth is a 'solstice' return.[59] Her role in 'turning back' the years and rebirth by day is also at least as old as the *Pyramid Texts*, as the king enigmatically proclaims in Utterance 405 when identifying with the Eye of Re:

> For I am that Eye of yours on the horn of Hathor,
> which brings back [*innt*] the years through me,
> I spend the night, and am conceived and born
> every day.[60]

Much later, when the veil had lifted a little on Egypt's esoteric mysteries, these time cycles were woven into the very fabric of Hathor's temple at Dendara. Hence, as the river of time flows northwards, travelling through the centuries into Graeco-Roman Egypt, it takes us to Hathor's last lovely dwelling, where her astrology-loving priests built their own temple of the cosmos, perpetuating the wisdom enshrined at Abu Simbel.

---

*LEFT fig.*66 Line-drawing of Senenmut's astronomical ceiling showing Sothis and Orion sailing in their boats in the southern half. Behind them are the planets, and further along the ceiling on the right are the 'sheep' and 'boat' decans. In the northern half are 12 circles representing the months, each divided into 24 segments and named according to its eponymous feast, with the sixth and seventh months represented on the far left. (Unfinished tomb of Senenmut at Deir el-Bahri, Western Thebes. 18th Dynasty.)

PART 2

DENDARA'S MYSTERIES: ABU SIMBEL RESONANCES

LEFT fig.67 View of some of the 18 sistrum columns in the pronaos of the temple of Hathor at Dendara. Together with the six columns on the temple's façade *(see fig.69)*, they total 24 columns in all and create musical harmony in this vast space. (First century CE.)

SIX

TEMPLE OF TIME:
FOUR FACES OF HATHOR

To travel to Dendara is to arrive at a Hathor temple utterly unlike Nefertari's womb-like cave in Abu Simbel's sandstone mountain. This is a monumental dwelling presenting from a distance an air of lonely splendour, shimmering in the heat haze on the edge of cultivated fields against a backdrop of desert hills *(fig.69)*. Located around 60 kilometres north of Luxor, it stands where the river Nile curves at Qena to flow almost due west for a while before resuming its northerly direction near Nag Hammadi, with the result that the temple's perpendicular direction to the water course is along a north to south axis.

Known to the Egyptians as *Iunet*, meaning 'pillar', the site acquired the epithet *T3-ntrt* (Tanetjeret), 'the goddess', seemingly to differentiate it from other Upper Egyptian temple cities called *Iun*, from which ultimately derived the Greek name Tentyris, which eventually mutated into modern Dendara. Generations of pharaohs built here to serve the goddess of music, love, wine, dance and fertility, including the Sixth-Dynasty ruler Pepi I, a revered ancestor king and great devotee of Hathor who was still honoured in the Ptolemaic temple founded sometime during the first century BCE, more than 2,000 years after his reign.

Precisely when it was started is difficult to determine, however, since many of the cartouches are empty, doubtless due to Ptolemaic dynastic struggles in the mid-first century, and in fact the earliest identifiable Greek king is Ptolemy XII Auletes (*c.*80–58 and 55–51 BCE), father of the renowned Queen Cleopatra VII, who is portrayed on the temple's exterior back wall with Caesarion,

her son by Julius Caesar *(fig.72)*.[1] Only after her defeat, when Egypt became a province of the Roman Empire, albeit a much prized one, was the temple finally completed under the early Roman emperors, whose names are inscribed in the pronaos at the temple's entrance.

Fronting the temple are six huge sistrum columns on the façade, each of their capitals carved with Hathor's distinctive four faces, staring watchfully to deter unwelcome visitors from entering her sacred precincts *(fig.69)*—or they would have done had not some iconoclast rendered them featureless, carefully chiselling out each eye, each nose and each mouth, and thus nullifying the *sekhem*-power radiating from the goddess's face. Whoever did this vicious act of vandalism, whilst leaving everything else untouched, certainly knew the meaning of Hathor's divine face and her power either to blast or bless.[2]

Within the pronaos itself, 18 massive sistrum columns, most similarly defaced, support the ceiling *(fig.67)*, thus, together with the six on the façade, the total is 24 columns in all, one for each hour of the day and night. Indeed, entering this vast entrance hall feels like stepping into the almanac of an ancient Egyptian star-gazer, a veritable compendium of Dendara's star lore, here blended with Hellenistic astronomy and astrology. For spanning the brilliantly painted ceiling, on both its east and west sides, is Nut's arched body, a sparkling celestial river of time, on which sail not only the Egyptian decanal stars, but also the Babylonian zodiacal signs and planetary figures in their annual circuit *(figs.68,71)*.

Over 1,000 years have elapsed since Abu Simbel, and here at Dendara there are signs of the cosmopolitan Hellenistic culture now pervading Egypt. The land has been ruled since the fourth century by a Greek dynasty of Ptolemies, who reside in their new capital at Alexandria. Knowing their political power in Upper Egypt will be considerably consolidated by the promotion of large-scale temple building, these Greek rulers have certainly been keen to depict themselves as traditional pharaohs serving the deities, and conversely, Dendara's astronomer priests have clearly been open to Hellenistic influences.

LEFT fig.68 Enclosed by the sky goddess Nut, the heavenly sun shines down on the head of Hathor placed on an edifice representing Dendara temple. On the legs of Nut is a scarab, symbolizing the zodiacal sign of Cancer, which is equated with Mesore, the last month of the Egyptian year when Re's annual birth is celebrated. This combination of 'birth' imagery mirrors the *Book of Day's* opening scene *(figs.8,10)*, though here Hathor's temple replaces the parturient goddess in the eastern horizon. In the register next to Hathor's face are Anukis and Sothis sailing in a boat, then Isis-Sothis (in the form of a recumbent cow), deities associated with the New Year and arrival of the inundation. (Detail from the rectangular zodiac on the east side of the ceiling in the pronaos of the temple of Hathor, Dendara. First century CE.)

And here in Hathor's late temple, one of the most beautiful and best preserved in all Egypt, ritual life still follows the flow of the hours, unfolding day by day, and year by year. The hours are woven into the very fabric of this sacred goddess dwelling: they are depicted on the ceiling of the pronaos, they are hidden deep underground in the southern crypt at the rear of the temple and they are enshrined in the Osiris chambers on the roof terrace above. Time unites these temple heights and depths in most extraordinary ways in a deep rhythm of life that reverberates with Abu Simbel.

CRYPT SECRETS:
THE VANISHING QUEEN

The southern subterranean crypt (Crypt South 1), or the 'secret palace' as Dendara's priests called it,[3] is one of several crypts hidden away, either within walls or beneath floors, in the innermost part of the temple. At first sight, a link with Nefertari's temple seems unlikely, since a queen is nowhere to be seen there—is in fact surprisingly absent in light of the queen's presence alongside Ptolemy XII in the two subterranean crypts on the temple's east

and west sides, and also the considerable power queens could wield in the Ptolemaic era.[4] Indeed, Cleopatra VII herself, the final Greek ruler of ancient Egypt, who is shown on the temple's exterior back wall *(fig.72)*, continued to be honoured in Egypt long after her death.[5] Yet from the Late Period onwards, there had also been a trend towards goddesses taking on queenly characteristics, so much so that Dendara's version of the famous royal birth cycle, which is based on the New Kingdom sequence in the Deir el-Bahri and Luxor temples, shows not the Egyptian queen as the royal birth mother, but Hathor herself, and the same shift applies in the crypt's 'secret palace', with its glorification of Hathor-Isis as queen *(fig.76)*.[6]

There is, though, another reason obscuring the connection with Abu Simbel, since crypts are generally considered to symbolize the *Dwat*,[7] and, whilst sometimes true, this is singularly inappropriate here, since, as Dieter Kurth has already pointed out in his much-neglected study of Pepi I at Dendara, a powerful solar current surges through this 'secret palace' from dawn to dusk, east

LEFT *fig.69* View of the temple of Hathor at Dendara. The features of Hathor's face at the top of each column on the façade have been carefully erased, though when precisely this happened is unknown. (First century CE.)

RIGHT *fig.70* Damaged face of Hathor on a gold-sign flanked by Hathor (right) and Isis (left). Both goddesses wear the Red and White Crowns, symbolizing their status as queens of Egypt. (Exterior south wall of the temple of Hathor, Dendara. Late first century BCE.)

to west.[8] Not only does it energize the most sacred part of the temple underground with solar life, it also complements the main north–south alignment of the temple above, thus creating a two-directional axis running above and below ground at the back of the temple.

In fact, there is a third important axis here, a 'vertical' axis, which was highlighted by that great scholar of Dendara, François Daumas, when he plausibly suggested the Crypt South 1 would have housed Hathor's cult image, which was taken each year to the roof terrace for her heavenly 'New Year' union with the sun, the Aten, enacted in a special kiosk located in the southwest corner.[9] According to Dendara's festival calendar, the union was also celebrated on the 26th day of the fourth month (26 Khoiak), during the Khoiak Festival inaugurating the 'Coming Forth' season.[10]

Vividly described in an inscription at the crypt's entrance, the ceremony began in the eighth hour of the day, when the king entered before Hathor, chanting invocations to 'open the face of the Eye of Re' and bring her to the roof.[11] And here in this remarkable temple, where ritual and symbolic correspondences are expressed in meticulous detail, the crypt's entrance intersects very precisely with the solar cycle's eighth daytime hour in the crypt itself, thus architecturally marking the relevant time of day *(see page 94).*

Before ascending to the roof the priests would have first taken Hathor's cult image to the *Wabet*, the 'Pure Place', for a ritual purification and adornment *(fig.74)*. Then they made their way up the western spiral staircase, surrounding Hathor with heavenly music as they took her in a columned naos to the apex of her holy dwelling. 'The sky and stars make music for you', they sang, 'the sun and moon praise you, the gods exalt you, the goddesses sing to you.'[12]

At the New Year, priests bearing precious raw minerals, masked standard bearers and musicians all rose up along this spiralling 'vertical' axis to the wide expanse of the sky above, so that Re's streaming solar rays might touch Hathor's unveiled statue in a life-giving embrace of love. All the temple images needed to be recharged, their energy periodically renewed with celestial life, and here, attracted by his daughter's glittering beauty, Re's

divine *Ba* descended towards her, infusing her cult image anew with heavenly solar light and life before it was carried back to the darkness of the crypt via the sloping eastern staircase. For a brief moment this kiosk became the *akhet*, the eastern horizon gateway between the worlds, where Re and Hathor, the heavenly and the earthly, met in a nourishing union *(fig.74)*.

Access to the 'secret palace' is via a modern flight of steps leading from an opening in the floor of the *Per-Neser* shrine, the Lower Egyptian 'House of Flame' traditionally belonging to the cobra goddess Wadjet, and one of 11 shrines located behind, and around the central barque sanctuary at the very back of the temple. An opening in the *Per-Neser*'s west wall leads into the shrine of Hathor's beloved consort, Horus of Edfu, whilst to the east is the *Per-Wer* shrine *(fig.73)*, the Upper Egyptian counterpart to the *Per-Neser*, known as the 'Great House' and originally associated with the vulture goddess Nekhbet.

Devoted as ever to ancient traditions, Dendara's priests retained these archaic Dual Shrines, which, by New Kingdom times, had also become the domain of Hathor in all her different manifestations as coronation goddess.[13] In fact, inscriptions around the *Per-Neser*'s entrance praise Dendara's irascible goddess queen and 'female sun'.

She is:

> Queen of Upper and Lower Egypt, the female sun, the noble and powerful one in Tarer [Dendara], the flame, the great one who burns enemies, who emits her burning breath against her enemies…[14]

This fiery Lower Egyptian shrine is dangerous, and whoever enters it needs to burn incense for its indwelling serpent goddess to create a peaceful green 'energy field'. Only then is it safe to enter the crypt beneath, descending ever deeper into the palace of Dendara's fourfold serpent queen.

Formed from five interconnected narrow chambers, each decorated with exquisite reliefs originally painted in brilliant colours, the crypt was normally sealed off from the rest of the temple. Deep in the earth, it was a hidden place, only accessible in ancient times to those who knew the secrets, yet it was one of the first parts of the temple to be built. Indeed, that this 'secret palace' was introduced into the building programme right at the very beginning speaks volumes about its importance to Dendara's priests, and not surprisingly, since preserved here is a radically new version of a solar lifetime, albeit a 'wisdom of the queen' harking right back to Nefertari's temple at Abu Simbel. It is time, therefore, to descend into its dark depths.

ABOVE *fig.71* The rectangular zodiac on the east side of the ceiling of the pronaos at Dendara. (First century CE.)

BELOW *fig.72* Relief of Queen Cleopatra VII, the last ruler of the Ptolemaic dynasty, depicted here with her son Caesarion by Julius Caesar. (Exterior south wall of the temple of Hathor, Dendara. Late first century BCE.)

RIGHT *fig.73* Plan of the temple of Hathor at Dendara.

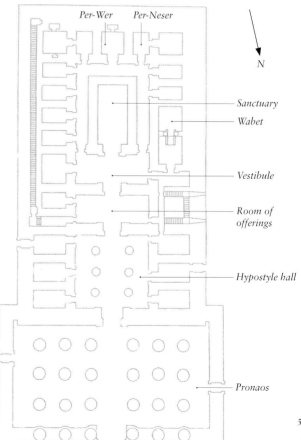

Per-Wer   Per-Neser

N

Sanctuary

Wabet

Vestibule

Room of offerings

Hypostyle hall

Pronaos

SEVEN

SERPENT IN THE STONE:
THE SECRET PALACE

One of the most extraordinary things about ancient Egypt is the longevity of its culture, which seems monolithic, even a sterile repetition of ancient knowledge. Yet though there is continuity, there is also change, marked by the priestly capacity to re-express sacred traditions, as in the Crypt South 1's highly original version of a palace 'lifetime' through the hours, here understood in terms of Hathor's fourfold nature, which is everywhere to behold at Dendara.[1]

Hathor's priests thought in 'fours'—expressed in reliefs, in texts, and most obviously in the huge sistrum columns adorned with her four faces (fig.67). They also compressed a solar 'lifetime' into the four distinct phases of 'birth', 'life', 'fruitfulness' and 'old age/burial', which were embodied, too, by the four nourishing *Kas* depicted in the *Wabet*, or 'Pure Place' where a beautiful scene of Nut adorns the ceiling (fig.74), the four male avatars of Ptah, whose animating powers infused the Memphite demiurge's body with 'life'.[2]

True to the spirit of Dendara's fourfold queen, these phases also govern her 'secret palace'. Etched into every wall, every corner, they create a new fourfold vision of the daytime journey, no longer on the scale of Nefertari's temple, or even that of the *Book of Day*, but rather allocated to small chambers ranged two by two on either side of the crypt's central 'House of the Sistrum' (fig.75). Yet beneath these surface differences, Hathor's 'green bird' wisdom still snakes through this dark, cramped crypt, and her vitality energizes this underground royal dwelling to make life fruitful. For housed in this crypt were not inanimate cult statues or objects, but bodies living in a constantly flowing, constantly changing, solar cycle of 'Becoming', moving through time in a sequence of intense moments from dawn to dusk, in a subterranean realm sustained by Dendara's fourfold queen.[3]

RAISING A TEMPLE: BIRTHING THE HEIR

Day dawns in the crypt's southeast corner, in the 'House of Harsomtus' dedicated to Horus, 'Uniter of the Two Lands', the child god worshipped at Dendara together with Hathor's musical child, Ihy; and here, exactly as in the *Book of Day*'s first hour (see page 21), there is a powerful confluence of the Osirian and solar realms, a synchronized merging of night and day, as darkness gives way to rising light.

Here Harsomtus is said to 'shine in his house during the night of the child in the nest, illuminating the land from the birth-bricks'.[4] He is betwixt and between, a son both of Isis and Hathor, the two goddesses who rule this secret palace (fig.76), a child enfolded in night and born at dawn, from a mother squatting on her birth-bricks like all Egyptian women labouring to give birth.[5]

This is the mysterious time when waking souls feel deeply united with the Osirian ancestors in the *Dwat*, the hour when past, present, and future

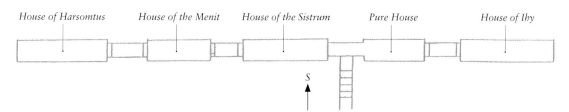

House of Harsomtus	House of the Menit	House of the Sistrum	Pure House	House of Ihy

S

LEFT fig.74 Enclosed by Nut floating in the heavenly waters, the heavenly sun shines down on Hathor's cult image in the eastern horizon. This beautiful scene relates to the ceremony of 'Uniting with the Aten' which was celebrated on the temple's roof terrace. The two trees framing the horizon, and the milk streaming down from Nut's breast, characterize this union as a nourishing source of fertile life. (Scene on the ceiling of the *Wabet* in the temple of Hathor, Dendara.)

ABOVE fig.75 Plan of the Crypt South 1, the 'secret palace', in the temple of Hathor, Dendara.

merge in the time of 'yet-to-be-born', when the worlds of the living and the so-called dead meet.

Indeed, this powerful fusion is reiterated in a text praising 'Gold, Mistress of Dendara', shining in the chamber whilst 'the ancestors are united around [*m-itrti*] her'.[6] So, too, a *Djed*-pillar is said to support Harsomtus when he appears as 'the living *Ba* in the lotus of the day boat, whose vitality [*nfrw*] the arms of the *Djed*-pillar raise as the *sšmw*-image'.[7] Just as Re's 'vitality' is raised in the *Book of Day*'s first hour, and Ramesses II touches the 'vitality' of Hathor's *menit*-necklace at Abu Simbel to initiate his ascent skywards, so this *Djed*-pillar, the ubiquitous symbol of Osiris's stability, elevates Harsomtus's vital beauty when he succeeds to the throne.[8] For Harsomtus is the renewer of the generations, praised as the primal snake surging forth from the waters of Nun, the snake god of life and reincarnation, able to reverse time, changing from old man to child, from death to birth as the 'Agathodaimon of the Ancestors', provoking growth in the whole of creation through his rising serpent energy:

> The august powerful one surging from Nun, Father of the gods creating himself, the divine Falcon who rules this land … Agathodaimon of the Ancestors, Lord of the Cult Places … the Creator who creates everything which exists, the Old Man who is renewed perennially.[9]

ABOVE *fig.*76 Hathor and Isis depicted on the east wall of the 'House of Harsomtus' in the Crypt South 1. (First century BCE.)

ABOVE RIGHT *fig.*77 Solar birth in the eastern horizon here symbolized by paired ovals, each enclosing a snake and surging forth from a lotus flower. Each oval is placed on a birth-brick, and supporting the one on the right is a *Djed*-pillar. These ovals also allude to the *Iterty*, the Dual Shrines physically located above the crypt at the back of the temple, so that, as in *fig. 68*, the dawn mother is identified with Hathor's temple, here giving birth to Harsomtus, the primordial snake god. (Relief on the south wall of the 'House of Harsomtus', Crypt South 1. First century BCE.)

Sometimes, though, a single image speaks louder than words, as here in the unusual relief dominating the crypt's south wall and wonderfully evoking Harsomtus's birth (*fig.77*). It depicts a pair of oval shapes, each enclosing a snake and surging forth from a lotus, the fragrant plant opening its petals at dawn to reveal Harsomtus in the day boat (*see above*).[10] Placed beneath one of the ovals is an animate *Djed*-pillar, symbolic of Osiris certainly, but also of Hathor, who is praised as the 'female *Djed*' here at Dendara.[11] Both support this new growth cycle, this rising child, for just as surely as Harsomtus appears as Osiris's successor, he needs Hathor's solar vitality to 'raise his beauty' at dawn.

What, though, do these strange ovals represent? They have been described as a 'gangue' in which minerals are embedded, or even 'a kind of vegetal

uterus'.[12] But strangely, the clue is hidden away in the already-quoted inscription about Hathor being surrounded by the ancestors *(see page 88)*, where the writing of the preposition 'around', *m-itrti*, includes two stylized ovals, each enclosing a snake, which together are to be read *itrti (fig.78)*.[13] Dendara's priests loved to use word-play to communicate layer upon layer of hidden meaning; they were also highly adept at uniting hieroglyphic writings and visual imagery, and here 'around' is written with hieroglyphs *(fig.78)* that read like a commentary on the visual 'oval' representations, having therefore both phonetic value and symbolic meaning.

It also resonates with the name of the ancient Dual Shrines, the *Per-Wer* and the *Per-Neser* belonging to the Upper and Lower Egyptian crown goddesses *(see page 84)*, which were collectively called the *Iterty*, sometimes written with this oval hieroglyph as a determinative, and regarded as representative of the sanctuaries of all the local deities in their respective regions of a united Egypt.[14]

Inhabiting them were also the royal ancestors, and at key moments in the pharaoh's life—notably his birth, his coronation and his regeneration during the Sed Festival—this collective body, living in the ancient cities of North and South, gathered in these shrines to welcome the reigning king who continued the royal traditions, and with whom they were united in an unbroken chain of 'life' passing

through the royal generations.[15] Here at Dendara, it is the revered King Pepi I who is twice depicted in the *Per-Wer* shrine in the main temple, not simply as a historical ruler from some dim and distant past, but rather a powerful mediating figure, tangibly present in the *Per-Wer* at the confirmation of the ruling king's power, still actively participating in the temple cult as he offers a small figure of Ihy to the goddess. Not surprisingly, the *Iterty* are also mentioned several times here in this shrine.[16]

In short, Dendara's priests utterly transformed the *itrti* hieroglyphs *(fig.78)* into these highly stylized ovals cocooning Harsomtus as the primal snake *(fig.77)*, the 'Agathodaimon of the Ancestors'. At the same time, they drew on a whole tradition of the nurturing mother goddess as 'temple' or 'house' enclosing her child 'within' her, which is encapsulated in Hathor's name, meaning 'House of Horus', the temple matrix sustaining his *Ka*-life.[17] For these Dual Shrines are built directly above the crypt in the main temple *(see page 84)*, which means Harsomtus is simultaneously being birthed by Hathor's temple, the architectural meaning being enhanced by the brick placed beneath each oval in the relief. Furthermore, bricks were not only used by Egyptian women to squat on when giving birth, they were also placed at the corners of a temple during the foundation ceremony, which was regarded, at least in Ptolemaic times, as the

'birthing of a temple'.[18] Clearly, birthing is 'building work' here in Harsomtus's chamber, for it is the Dual Shrines that bring him forth in Dendara's dawning cult world, their function reinforced by the hieroglyphic writing of 'around', *m-itrti*, in the same chamber.

'Open for me' is the plea to Hathor of Dendara in a 19th-Dynasty spell for hastening birth. 'I am the one whose offering is large, the builder who builds the pylon for Hathor, Lady of Dendara, who lifts up in order that she may give birth. It is Hathor, the Lady of Dendara, who is giving birth.'[19] This magician knows temple-building is a creative, birth-giving act, as do Hathor's priests at Dendara, to the extent that they describe every cult image, every cult object in the inscriptions, clearly stating their material composition so that a proper ritual can be performed.[20]

Perhaps on one level, too, the crypt's emphasis on 'building' also reflects the half-hearted approach and lip service the Egyptian priests generally paid to their foreign rulers in Alexandria. Skilled communicators that they were, these Dendara priests might show Ptolemy XII in the crypt's 'secret palace', might know that he was a generous benefactor of Upper Egyptian temples, but in the Ptolemaic dynasty's troubled twilight years, when it was not clear at times who even ruled in Alexandria, what mattered to them above all was 'building a temple' rather than 'building a pharaoh'. For it was the temple priests now, not the Greek pharaoh, who ultimately protected Hathor's sacred mysteries.[21]

ABOVE *fig.78* The *itrti*-hieroglyph.

RIGHT *fig.79* A rare depiction of a squatting goddess giving birth in a temple gateway (or naos) assisted by her Hathorian attendants. This relief is from Dendara and here, as elsewhere in Hathor's temple, solar birth is explicitly contextualized within an earthly setting. (Egyptian Museum, Cairo, JE40627.)

RIGHT fig.80 Line-drawing of an animate *menit*-necklace, with its counterpoise terminating in the head of Hathor, 'Eye of Re'. Perched on one of its arms, which hold out *Ankh*-signs of life, is Hathor's child. To the left Hathor holds out another *menit*-necklace towards her son Harsomtus and Ptolemy XII (not shown here), transmitting her 'attraction' in the eastern horizon. (Detail of a relief on the north wall of the 'House of the *Menit*', Crypt South 1. First century BCE.)

ENCIRCLING LIFE: HATHOR'S MAGIC

Adjacent to Harsomtus's 'birth' chamber is the 'House of the *Menit*', where the phase of 'life' is represented by a loving-destructive solar Eye goddess, seen on the north wall holding her *menit*-necklace out to the sistrum-shaking Harsomtus and Ptolemaic king. Clearly, this precious necklace, one of Hathor's nine cult objects here at Dendara, has a life of its own, since perched on a chest behind Hathor is another large *menit*-necklace, its animate counterpoise terminating in the head and arms of the goddess, carrying her child on one arm and holding out *Ankh*-signs *(fig.80)*.[22] 'Life' radiates from these green jewels, filling the room with Hathorian attraction, with that aura of magnetic love that rouses Re to renewed activity, compelling him to rise each day, empowering the charismatic sun king to draw followers to him and binding everything together in an embrace of love *(see page 38)*.[23] From the Middle Kingdom *Story of Sinuhe* to here in the Dendara temple, the story is virtually unchanged: to rule effectively, the beautiful youth in the eastern horizon needs the loving-destructive Sun Eye to empower him, and no impediment of vision must disturb the flow of life-giving energy. Thoth's healing arts have made quite sure of that.[24]

Dominating the south wall is another large *menit*-necklace, here called 'Hathor of Dendara, *Menit*, Eye of Re', its beads joined to four naos sistra, each with a uraeus serpent rearing up in the doorway of the superstructure *(fig.81)*. At the base of the two sistra on the right sails a solar boat, whilst to the left of the ensemble squats a small figure of Maat, the other great 'daughter of Re'. From one perspective these four sistra could be interpreted simply as the four supports of the sky marking the four regions of Egypt, or even the four supports of Hathor's 'New Year' shrine described in an inscription at the crypt's entrance—that is, until it is remembered these are musical instruments sacred to the unpredictable Sun Eye, rhythmically shaken in the cult both to calm the rage of irascible deities and empower initiates for heavenly travelling.

Movement means 'life' says Hermes in the *Corpus Hermeticum (see page 129)*, and Plutarch explains that the sistrum indicates 'the things which exist should be shaken and should never stop moving' *(see page 129)*.[25] Dendara's priests know this too, though visual images, not Greek words, are their preferred mode of expression. For here in this chamber dedicated to 'life', these four naos sistra,

joined to a *menit*-necklace, wonderfully express a cosmos in motion, a solar musical realm guided by Maat and governed by love, a ceaseless way of the world ruled by Dendara's powerful fourfold queen.[26]

Probably, too, these complementary scenes on the north and south walls visually incorporate the Egyptian word *pekher*, which means not only to 'travel round' the solar circuit but also to 'embrace' and 'encircle hearts'. Thus, by placing her hand through her *menit*-necklace (north wall), Hathor makes an embracing *pekher* gesture that resonates with the sun boat 'travelling round' the solar circuit guided by Maat across on the opposite wall.[27]

It also creates a powerful web of enchantment within the room, for, as Christian Jacq observed, 'The important thing in the practice of magic is to identify the thread which links everything and unites all creatures in a chain of cosmic union.'[28] And whether it be 'encircling hearts' or 'encircling the solar circuit', clearly the uniting magical thread here is Hathor's necklace, its green beads shining

on strands of light, a necklace belonging to a goddess whose attraction weaves through the 'life' world. She is the great 'encircler' in the 'gate of gold', Weret-Hekau, 'Great of Magic', in the solar circuit, who draws everything together in a living bond of love.[29]

And here once again, as in Harsomtus's chamber, all eyes are on Hathor's temple and its cult objects. Indeed, it is because Hathor sees her newly-founded house, built to accord with Ptah-Tatenen's Memphite creation, that her heart is said to be 'sweet':

> When she saw the radiance in which her house was built, raised and made new as done by Tatenen, then her heart was sweet because of her foundation.[30]

The very act of building the Dendara temple, with all its beautiful cult objects, propitiates and delights its indwelling loving-destructive Eye goddess. It is a service rendered to her in this eastern-horizon chamber of 'life'.

LEFT fig.81 Four naos sistra attached to the beads of a *menit*-necklace placed on a large collar. A solar boat is depicted at the base of the two sistra on the right, and on the left is the goddess Maat, who guides the sun boat's movements in this temple of the cosmos energized by Hathor. (Detail of a relief on the south wall of the 'House of the *Menit*', Crypt South 1. First century BCE.)

RIGHT fig.82 Three sistra depicted in the 'House of the Sistrum'. (Crypt South 1. First century BCE.)

ENCOUNTERING DEITY: SISTRUM POWER

Ascension heavenwards culminates in the crypt's central chamber, the 'House of the Sistrum', dedicated to 'entering the heaven of Re's house' and 'seeing the sun disk (Aten) in every form', a room dominated by Hathor's musical instrument, the sistrum *(fig.82)*.

In fact, adorning the chamber's walls are eight sistra, six depicted on the south wall and two on the north wall, where they are flanked by Hathor and Isis as *Ba*-birds, those heavenly birds whose destiny is to wing their way to the sky, joyfully animating their cult images below with all their *Ba*-power.

Hathor's rhythmic music energizes everything at this central southern point in the temple. Thus, in the ground-level southern crypt, the king is depicted offering to four Hathors on the south wall, paired on either side of a naos sistrum.[31] Similarly, in the central *Per-Wer* shrine there is a niche high up on the south wall, which once contained a large gold statue of Hathor and has a sistrum symbol depicted on the back wall *(fig.83)*. Directly behind this niche, carved on the temple's exterior south wall, is a large head of Hathor on a gold-sign that once served as a focus for worshippers within the temple precincts *(fig.70)*, though, as in the temple's pronaos, its features have been intentionally erased, depriving Hathor of the dynamic *sekhem*-power that would be particularly potent here in the noonday heights.[32]

It is into this '*sekhem*-sanctuary of the god' that Ptolemy XII now enters to see the divine image, to encounter the deity's gaze in an intoxicating exchange of vision. Declaring himself to be 'free from all fear', he beholds the glittering face of Hathor in the 'Places of Drunkenness'.[33] She it is who now settles on his brow as the fiery uraeus, empowering his solar sovereignty whilst uttering words spoken to Egyptian kings from time immemorial in the *Per-Wer* shrine: 'I have received your ritual with jubilation and place myself between your eyebrows.'[34] Old traditions have been retained with tenacious accuracy, and just as Ramesses offered Maat to Amun-Re in the sistrum-pillared hall at Abu Simbel, so the Greek ruler is here acclaimed as 'the son of the union of Maat'.[35] He comes, too, as Thoth's initiate, equipped with all the god's protective magic and 'perfection of the spells derived from the knowledge of the writings of Am-Tawy (Thoth)'—necessarily so, since this

LEFT *fig.83* The niche high up in the south wall of the *Per-Wer* shrine, which originally contained a large gold statue of Hathor, and has a sistrum image depicted on its back wall. Sistrum energy is particularly channelled in this innermost part of the temple, the niche being located above the southern crypt's 'House of the Sistrum'. Directly behind, on the temple's exterior back wall, is the damaged face of Hathor *(see fig.70)*.

RIGHT *fig.84* Isis seated on a throne floating in water. (Relief on the south wall of the 'Pure House', Crypt South 1. First century BCE.)

Ptolemy is also a king sworn to secrecy, who does not leave the crypt's house of truth 'revealing what he has seen'.[40] Only a few Thoth lovers of wisdom are privy to these Hathorian mysteries, and interestingly, throughout this crypt, Hathor's sistrum-shaking divine son invariably precedes the Ptolemaic king, as if to reinforce his dependence on a mediator for access to the goddess and her cult. Over and over again, and with the lightest of touches, Dendara's priests highlighted their relationship with Egypt's foreign rulers.

CALMING THE COBRA: THE FRUITUL MOTHERS

From this noonday chamber the journey flows westwards to the 'Pure House' along a connecting small passage which also intersects with the crypt's entrance leading from the *Per-Neser* above *(see page 84)*, an entrance very precisely aligned, therefore, with the eighth hour.[41] So too, the 'Pure House' itself reverberates in every way with the ninth hour 'rising for Isis', when the sun boat enters the Field of Reeds, where grain grows to incredible height and Re 'comes forth from the egg' in his glorious 'second birth' into a divinized realm of heavenly royalty *(see chapter 2)*.

And just as Nefertari's transverse chamber is a maternal place where Taweret 'gives birth to the gods', so the crypt's 'Pure House' is consecrated to Hathor as:

> Mother of the mothers, Eye of Re, the Great One, Mistress of the Sky, who comes forth at dawn every day.[42]

rite of seeing the deity requires the king's complete command of Thoth's arts.[36] It also magically reinforces what is stated about his Thoth-like performance of the same rite in the *Per-Wer* in the temple above.[37]

In fact, many of the crypt's themes reappear in the *Per-Wer* and such resonances are a deliberate ploy to unite the *Per-Wer* with the daytime solar circuit. 'The great powers' are in the *Per-Wer* 'like the horizon containing the Aten' declares a text inscribed there.[38] Manifestly, this ancient shrine is a heaven on earth, a correspondence needing also to be expressed architecturally, and something Dendara's designers achieved when they located the *Per-Wer* above the crypt's 'House of the Sistrum', connecting it with the nourishing goddess life flowing from dawn to dusk in the secret palace.[39]

For a second time in this crypt, Hathor rises at dawn, manifesting now as 'Mother of the mothers', a nourishing bearer of life shown suckling her child on the north wall. In the 'House of the *Menit*' she has risen in the East, but now, in this 'Pure House', there is a deepening experience of dawn; and the relationship between these two chambers is reinforced by their measurements, which correspond exactly.[43]

Here in this northern realm (the temple's west side being more generally associated with the north and Isis), Hathor rises in a watery uterine space, enhanced by a beautiful scene of Isis on the south wall, majestically enthroned and floating on water (indicated by the zigzag lines beneath her throne) (*fig.84*). 'Water enchantments' are now the magic surrounding these purified goddesses, a theme that is reiterated in the praise of Isis in *Isheru*, sailing on her pleasure lake, which is typical of sanctuaries dedicated to leonine Eye goddesses.[44]

This is the calm after the storm, after the heat roused for ascent, and a very different atmosphere prevails in this room. Now everything is directed towards ensuring 'fruitfulness' in every sense of the word, towards procuring 'greenness' and the peaceful *hetep* state.[45] Now, too, the moon begins to exert its moist influence, indicated by the praise of Hathor as 'Mistress of all the gods, shining with the double feathers as the Moon', gloriously illuminating the Two Lands with her beauty shimmering in lunar light.[46] So the three gods Re, Tatenen and Thoth are said to surround her protectively, a divine triad reinforcing her divinized state:

> Your father Re surrounds you with his protection,
> Tatenen adorns your body, and Thoth, the Great
> One, sanctifies your body with his spells,
> you rejoice knowing his protection.[47]

'All gods are three', and here in this divine chamber Hathor rules as 'Mistress of all the gods'. Two of this triad, Re and Tatenen, also bless her crowns during her major 'Festival of Drunkenness', celebrated on 20 Thoth in the first month of the 'Inundation' season, when she is annually consecrated as 'queen among the gods', peacefully returning from the South in harmony with the

returning inundation waters.[48] Then, too, her aromatic *menou*-drink is offered, made from grain, herbs, honey, dates and other sweet substances, a drink which 'makes the inner outer', illuminating the *Ka* and revealing the innermost secrets of the heart.[49] No wonder then that Hathor manifests here as 'Mistress of all the gods, who appears with the *Nfrt*-crown [the White Crown] in the *Nt*-crown [(Red Crown] on her head, the Mistress of Life'.[50] Adorned with the Double Crown, she reigns as heavenly queen, the 'Mistress of Life' here in this 'Pure House'.

And just as Hathor once stretched out her arm in a beautiful *Ka*-life gesture to bless the fiery uraeus serpent on Nefertari's brow at Abu Simbel (*fig.41*), so here the ancient acclamation 'Awake in peace', the incantation praising the uraeus as protector of the pharaoh, resounds in this subterranean chamber, a great paean of praise to Hathor, majestically rising in the East surrounded by Thoth, Seshat and a retinue of deities.[51] Importantly, this chamber is located beneath the

RIGHT *fig.85* The Sixth-Dynasty pharaoh Pepi I offers a tiny statue of Ihy to three Hathors depicted on the north wall of the 'House of Ihy'. A fourth Hathor (not shown here) is on the west wall, thus locating the fourfold goddess in the Northwest at the close of day. (Crypt South 1. First century BCE.)

Lower Egyptian *Per-Neser* shrine, therefore magically amplifying her serpent power, for, in both chambers, litanies are directed towards propitiating the fiery cobra. Her power, fully extended now, needs to become a gentle, not a burning, heat, if everything is to reach 'fruitfulness'. Every room in this wonderful temple has been carefully planned to bring out correspondences, and here the *Per-Neser*, the 'House of Flame' and the crypt's 'Pure House' act like echo chambers, creating a powerful serpent 'energy field' between above and below.[52]

Reaching this most royal of realms, the Greek king enters the fruitful place of second birth, born into heavenly divinity, and in return for all the offerings he brings, all his dancing and skipping before the goddess, all the jewels that adorn her, this serpent goddess who 'burns the rebels with her terror' awakens peacefully on the king's brow to bless his sovereignty.

The noonday flame of the central chamber overflows now with the maternal waters of fruitful life, shimmering with the liquid silver and gold of the mother goddesses. Everything is being brought to fruition as Hathor and Isis rise in their flowing, liquid bodies of light, guarding the mysterious flow of midheaven solar life.

PEPI'S LINEAGE: IHY'S HOUSE

This powerful life-force streams ever westwards into the 'House of Ihy', dedicated to Hathor's musical young son *(fig.85)*. Initially it seems curious he should represent the phase of 'old age' and 'burial', yet he it is who appears here in the 'Gateway of Speaking Justice' during his 'Festival of Entering the House of the Bier', and, like the sun god in the *Book of Night* who enters the West at the close of day to 'judge the inhabitants of the *Dwat* and know the condition of the Westerners',

Ihy's fate here merges with the Osirian night realm and the judgement meted out there.[53]

Significantly, for the only time in this crypt, Osiris appears, standing with the connecting passage's guardian deities as the 'Pillar in Hatmehit who destroys the companions of Seth' and offering his protection to Ihy's house, to 'his sleep and every chamber where he dwells'.[54]

As the ancient *Coffin Texts* so vividly tell, Ihy is, in fact, no stranger to night's turbulent darkness, which has to be lived through if he is to 'beget a begetting' and 'break forth from the egg'. He knows what it means to enter Nun's dissolving primordial waters, lying there as an 'inert one' in an agonizing state, his flesh blackening and decaying before renewal can begin, eventually slithering forth as a reborn 'Jackal of Light', dancing towards the lotus vessel of rebirth adorned with the jewels of his mother, Hathor.[55]

Hence framing this crypt are Dendara's child gods: in the East is Harsomtus and here in the West

is Ihy, a complementary relationship heightened once again by the exact correspondence between the measurements of their respective rooms.[56]

There is, though, another important presence here in Ihy's house: the Sixth-Dynasty ruler Pepi I, shown at the end of the north wall reverently offering a statue of Ihy to three primordial forms of Hathor *(fig.85)*—or four, since the adjacent Hathor on the west wall must surely be counted, hence the temple's fourfold goddess is located not just in the North, but symbolically in the Northwest. Once again, the precision with which this crypt has been planned is extraordinary, since, according to the *Book of Day*, it is in the Northwest that the sun god unites with his mother at the close of day for the death that brings rebirth.[57]

At the north wall's other end, the reigning pharaoh, Ptolemy XII, offers a tiny figure of Maat to Isis, 'Mother of the God', and Harsomtus.[58] From one perspective, his offering, here called the 'seed of the bull', encapsulates the plenitude of life

filling this chamber—gold, silver, lapis lazuli, turquoise, faience, incense from Punt, wood, myrrh, wine and the green stone—all signs of the wisdom goddess's inexhaustible creativity, all seeded by the bull, and with their flow into the temple's treasury safeguarded by the 'green Eye of Horus'.[59] But mention of this 'seed' also has a regenerative ring about it, bringing to mind the mystery of the 'Bull-of-his-Mother', the fertile god whose impregnation of his own mother (often identified with Isis, 'Mother of the God') ensures rebirth *(see page 32)*. Manifesting as both father and son, his power is incarnate in all the Egyptian kings, a community of rulers living in a vibrant field of 'presence', a unified 'life world' where 'then' and 'now' are merged in eternal succession.[60]

Both in the *Book of Day* and Nefertari's temple the royal ancestor king is honoured at the close of day, but whereas the Ramessid kings perpetuated a lineage reaching back to the 18th-Dynasty ruler Amenhotep I *(see page 33)*, here at Dendara the all-important chain of transmission stems from Pepi I, who is a permanent dispenser of blessings within the temple. Through the ritual they perform together, Ptolemy XII, the reigning king, and Pepi I, the ancestor king, serve in the spiritual life of the Dendara temple, the 'two in one' mystically united in the *Ka*-lifeline of Egypt's royal rulers, all of whom have shared in the mysteries of Isis and the bull god, and all of whom are united through the *Ka*-power streaming through the generations.[61]

The Egyptians also believed the heart to be the true source of the Bull-of-his-Mother's energy; hence, when Ptolemy XII offers Maat to Isis, the falcon-headed god Hekenemankh, 'the Praiser in Life', acclaims the king's radiant heart-opened life with the words:

> To you belongs your heart which is open through your radiance. Everyone praises you.[62]

'To make the inner outer' is the aim of Hathor's cult here at Dendara *(see page 95)*, and by offering this 'seed of the bull', Ptolemy makes manifest the truth of his regenerative heart.

Ending here in Ihy's 'House of the Bier', this 'lifetime' journey, constellated around the central 'House of the Sistrum', has unfolded through 'birth', 'life', 'fruitfulness' and now 'old age/burial'. It seems linear and sequential, but it is in reality a solar circular path, a feminine wisdom of life, a beauty that is born and reborn, returning the end to the beginning like the tail-biting ouroboros-snake engulfing its own origins. For across on the south wall Hathor is seen suckling her young child, accompanied by Horus the Behdedite, whilst behind them stand Dendara's child gods, the sistrum-playing Ihy, 'the green shoot of Sekhmet', and Harsomtus, 'the beautiful child of Gold'.[63]

Just as she brings forth Ramesses from the western mountain at Abu Simbel, eternally resurrecting him through her power of love, so she nurtures her child within this western chamber, where sleeping and waking, dying and living, are utterly united in this passage through the daytime hours. How apt then that it should belong to Hathor's musical child, for his music is 'life' and, like life itself, is inextinguishable. As a solar initiate would so movingly experience in an

'immortalization rite' preserved in a fourth-century Greek magical papyrus from the Theban region: 'O Lord, while being born again, I am passing away, while growing and having grown, I am dying … as you have established the mystery.'[64]

If Abu Simbel is to be believed, however, this daytime journey through the 12 hours also co-exists with the sun's annual cycle, the crucial 'reversal' of the year between the sixth and seventh months, and, interestingly, the crypt's inscriptions correlate each chamber with particular annual festivals, hence overtly tying the daytime hours to the cycle of the year.[65]

This 'day' and 'year' link is also consolidated in Hathor's union with the Aten in the 'New Year' kiosk on the rooftop terrace, but in terms of the sun's annual reversal, it is the six chambers across from this kiosk that are all-important. For preserved there is a unique version of Osiris's resurrection during the annual Khoiak Festival, and in fact the procession taking Hathor's cult image to the roof from the Crypt South 1 would have passed close to the entrance to the three western chambers at the top of the spiral staircase, thus weaving another connection between roof and crypt *(fig.114)*.

But whilst these extraordinary rooftop chambers are well known, what has gone unnoticed in contemporary Egyptology is the synchronization of Osiris's resurrection within them with the cycle of the 'hours' and the sun's annual southern 'reversal'. The Abu Simbel temples embodied these time realities in their architecture, and there must have been other temples, long since gone, similarly based on these cosmic rhythms, but Dendara's priests reinterpreted this sanctification of time in a startling new way, in text and image, even in architectural design, and also in the famous 'round zodiac' preserved in one of the chambers.

All this needs corroboration, however, for what actual evidence points to these time realities, especially in light of the Khoiak Festival's traditional celebration, not in the sixth or seventh month, but the fourth month of the 'Inundation' season in the ideal calendar?

EIGHT

ROOFTOP RITES:
THE KHOIAK FESTIVAL

These Khoiak chambers, which are situated on the roof terrace's east and west sides, are divided into two groups, each consisting of an open-air court and two adjoining rooms. Importantly, they are also placed directly above the six rooms in the hypostyle hall in the main temple beneath, a spatial and vertical location magically reinforcing their meaning, since there are distinct thematic correspondences between above and below.[1]

Covering almost three walls of the eastern open-air court are 159 columns of text preserving details of the festival (*fig.87*), which was clearly hugely important to the Dendara priesthood, who drew on a whole variety of Khoiak traditions from Osiris's different cult places, including Busiris, Abydos and Thebes, to record national and local rites going back centuries.[2] Here, in short, is a unique compendium of what was known in the Graeco-Roman era about this Osirian 'mystery which one does not see, of which one does not hear', and which, like the Hermetic treatises from Roman Egypt, involved an initiatory chain of transmission from 'a father to his son'.[3] Perhaps Dendara's priests, fearing this knowledge might be lost, decided to gather it together in such extraordinary detail because they were living in a period of great political uncertainty, which is emphasized by the empty royal cartouches throughout the Khoiak chambers.

The whole cult and character of the Khoiak Festival is steeped in anticipation of the season of 'Coming Forth', a time when Osiris manifests his resurrected life and vitality. For Osiris is the latent source of everything that sprouts forth from the earth. Everything grows from his unseen world—the nourishing crops, the fertility of the land, the renewal of the agricultural cycle. Everything that will sustain Egypt in the coming year depends on his resurrected power.

So the earth needs to be blessed, and these Khoiak rites are timed to honour Osiris's awakening life. Indeed, perhaps it was the desire of the Egyptians to honour his death and resurrection that gave their culture such exuberant beauty—their recognition that darkness, the unseen, the

renewing powers of night, all belonged to the wholeness of life.

The ritual procedures preserved in the eastern court's long inscription specify in great detail how to create two 'seed beds', one for Osiris 'Chief of the Westerners' (or Osiris *vegetans*), the other for Osiris-Sokar.

To make the seed bed for Osiris *vegetans*, the priests placed barley seeds, silt and floodwater in the two halves of a special golden mould, shaped like a human-faced mummy wearing the White Crown, and, using a gold situla, carefully watered them in a stone 'garden tank' called the 'garden of Shentayt' until 21 Khoiak. Then the two halves of Osiris *vegetans* were removed, bound together and exposed to the sun before eventually being wrapped in linen bands.[4]

In contrast, the instructions for Sokar's 'seed bed' read like an alchemical recipe, prescribing the preparation of a special 'paste' made from various ingredients, including pulverized precious stones, dates, myrrh, and aromatic substances, all carefully

PREVIOUS SPREAD fig.86 Paired *Bas* of Osiris adore the heavenly sun disk shining down on another sun disk in the horizon. (Frieze scene in the eastern court of the Osiris roof shrines, the temple of Hathor, Dendara. Late first century BCE.)

LEFT fig.87 Detail of Osiris's mummiform body beneath a section of the Khoiak Festival inscription. (Relief in the court of Osiris on the east side of the roof terrace, Dendara. Late first century BCE.)

ABOVE fig.88 Detail of Isis-Shentayt weighing the ingredients for the cult images of Osiris and Sokar in the Khoiak Festival. On the other side of the scales are the craft gods Khnum of Elephantine and Ptah of Memphis. An identical scene on the other side of the doorway shows Nephthys weighing the ingredients. (North wall of the central Osiris room, east side of the roof terrace, Dendara. Late first century BCE.)

measured, then blended together with water taken from the sacred lake and formed into an 'egg'—exactly the kind of operation Hermes Trismegistus might have had in mind in the *Asclepius* when he said that Egyptian cult images were made from 'herbs, stones and spices containing natural divine power'.[5] Here in the Khoiak rites, this precious 'egg' is placed in a silver vase covered with sycamore branches and laid on a bed in the 'chamber of the bed' for a period of time before being withdrawn from the mould and exposed to the sun.[6]

Thus, when tending these 'seed beds', Hathor's priests become like farmers preparing their fields and sowing their seeds to receive the life-giving power of water and sun, though clearly their labour operates on different levels of meaning. For whilst on one level these are 'material' procedures, each one being equated with a particular chamber, on another level it is the dismembered Osiris himself who is being ritually recreated. The process culminates in his transmutation into a divine golden body, his rebirth as the 'Glorious One' *(Akh)*. As in alchemy, material and spiritual transformation go hand in hand here at Dendara.

CALENDAR CONUNDRUMS: A SOLSTICE SYNCHRONICITY

So far, on the basis of the seed-bed instructions, all seems traditional enough—a factual account of the festival containing a considerable amount of ritual detail. But this version of the Khoiak mysteries is by no means straightforward, displaying some very unusual features, easily overlooked and hitherto unexplained, not least the mention of a ploughing rite performed on a date after the Khoiak Festival would normally have ended *(see page 115)*, and the inclusion of the famous round zodiac in one of the chambers *(fig.118)*, which is distinctly astrological in character *(see chapter 12)*.

There were two highly unusual calendar events towards the end of the first century BCE which had far-reaching implications for the Khoiak Festival and influenced Dendara's presentation of the rites. Because of the peculiarities of the Egyptian 'wandering' civil calendar, the festival, ideally celebrated over several days in the fourth month of the 'Inundation' season *(Akhet)*, in fact coincided

with the southern solstice, a unique event in the 1,461-year period of the Sothic great cycle.[7] Also, in year 25, the Alexandrian calendar was introduced in Egypt during the reign of the Roman emperor Augustus to try to standardize the year, and whilst the ancient Egyptian civil calendar still remained in use, this reform had considerable implications. For in this new Alexandrian calendar, 1 Thoth, inaugurating the 'Inundation' season, was placed over a month later, meaning the Khoiak rites now fell around the southern solstice, thus doubly reinforcing the 'wandering' civil calendar's synchronicity.[8] All of which must have created a particularly powerful experience of Osiris's resurrection at the time of the sun's annual 'reversal'.

Great astronomers that they were, Hathor's priests were not stuck in a time bubble of the past, but acutely aware of changes in the world around them, so much so that they created stunning ceilings to integrate Egyptian and Babylonian astral knowledge. Despite the empty royal cartouches making it difficult to know when exactly the Khoiak chambers were decorated, judging by the 'reversal' imagery in the western Osiris shrines *(see chapter 10)*, not only did they know about this 'solstice' synchronicity, they were keen to take the opportunity to create a particularly powerful fusion of Osirian and solar regeneration.[9] In short, in order to celebrate this Khoiak Festival-solstice synchronicity, they drew on traditional 'reversal' symbolism, blending together the Khoiak Festival rites with imagery relating to the pivotal hinge of the year between the sixth and seventh months, which had also always been tied to the night hours. *(see chapter 5)*. This fusion governs Osiris's resurrection at Dendara, and once this has been recognized, several unusual details throughout these six chambers immediately fall into place. And beyond, since whenever the priests brought Hathor's cult image from the Crypt South 1 to the roof, processing close to these western Osiris shrines at the top of the western staircase, as at Abu Simbel, they were weaving a ritual thread connecting a palace 'lifetime' by day with this great annual reversal.

Already in Ramesses VI's burial chamber, Osiris's resurrection in the *Book of the Earth* had been located beneath the diurnal cycle of the hours depicted on the ceiling *(see chapter 5)*. In the Cenotaph of Seti I at Abydos (the Osireion), the solar day and night boats had been shown above the iconic scene of Horus 'awakening' Osiris in the palace that is the introductory tableau to the *Book of Night*. This scene subsequently reappeared in Ramesses VI's tomb, again near the hour cycle. It is included in the western end chamber here on the roof terrace *(fig. 89)*, where Osiris's resurrection is synchronized with the sun's journey by day and night.[10] What is completely new here at Dendara, however, is the fusion of the Khoiak mysteries with the cosmic reversal between the sixth and seventh months, and their correlation with astrological

ideas widely circulating in the Hellenistic world *(see chapter 12)*.

In short, these chambers mark the culmination of an ancient tradition uniting Osirian regeneration, southern 'reversal' and the solar cycle of the hours, detectable already in the New Kingdom at Abu Simbel and in Ramesses VI's burial chamber, but explicitly articulated now in terms of the Khoiak Festival, and beginning in the eastern open-air court.

BELOW *fig.89* Iconic scene of Horus waking Osiris on the east wall of the inner Osiris room. (West side of the roof terrace, Dendara. Late first century BCE.)

NINE

REBUILDING OSIRIS:
FOLLOWING THE SUN

At the eastern court's entrance (fig.90), an alternating cycle between day and night is set in motion, the first five hours of the day being named on the doorway's east side, together with the protective deities guarding the king of Upper and Lower Egypt during these hours: Shu and Tefenet are in the 'entourage of Re' in the first hour; Isis and Nephthys accompany Horus in the second; Bastet and Sekhmet, together with their emissaries, are in the third hour; Horus and Thoth keep guard in the fourth; and then the 'entourage of Re', together with Horus, is in the fifth.[1] Across on the doorway's west side, nine protective deities are exhorted not to sleep through the night.[2] Thus, right at the very beginning, the scheme of these six Khoiak chambers is firmly established: the three eastern chambers belong to 'heaven' and 'day' and those on the western side to 'night' in an alternating cycle which governs the whole of the Khoiak Festival.[3] It also sets the stage for the redemption of Osiris, beginning in the open court with the plea to him to come and rebuild what has been ruined:

> Come to your place … come, raise again [s'h'] what is ruined and rebuild the cult places.[4]

This cryptic call, perhaps the closest the Egyptians could ever come to mentioning the devastation caused by Osiris's brutal murder, initiates the work of 'reconstruction'. What Seth dismembered must now be ritually gathered together again. The cycle of life must be reactivated, confusion must be returned to order, a 'building' work must be undertaken which both restores the 'cult places' and Osiris himself, and in which, paradoxically, he is both the 'builder' and the one who is 'rebuilt', both the reigning Horus son and the wounded father god. This Khoiak ceremony honours the painful, the difficult, the suffering that yet becomes the source of greening new life. After all, it is principally as Wennefer (Wnn-nfrw), 'He who is vital', that Osiris is worshipped here at Dendara, and 'vitality', in every sense of the word, is needed in this springtime of life.

Indeed, this call to Osiris to 'rebuild' is echoed in the main temple directly beneath these rooftop chambers, where, in the six columned hypostyle hall, various scenes show the Ptolemaic king performing the temple's foundation ceremony: leaving the palace, carrying bricks, hoeing the earth and purifying the edifice.[5] Just as a vertical 'energy field' is created between the underground southern crypt's 'Pure House' and the *Per-Neser* shrine above (see pages 95–6), so this 'temple-building' directly resonates with what Osiris is called to do

---

LEFT *fig.90* View of the entrance to the Osiris rooms on the east side of the roof terrace, Dendara. (Late first century BCE.)

ABOVE *fig.91* View from the east court of Osiris looking towards the entrance doorway where the first five hours of the day and night are inscribed on the side walls. (East side of the roof terrace, Dendara. Late first century BCE.)

on the roof, and, of course, with the temple-building for Hathor in the southern crypt's 'secret palace' *(see pages 90, 92)*.

Not that this allusion to 'building' Osiris is anything new. Already in the New Kingdom 'becoming a builder' belongs to rebirth in the *Book of Night*'s tenth hour, and repeatedly in the Ptolemaic *Book of Traversing Eternity* a word *k3t*, meaning 'work', 'building' or 'craft', defines the ritual processes associated with Osiris's mummification.[6]

However, 'rebuilding' Osiris needs Horus to occupy the throne of Egypt; hence, above the doorway giving access to the eastern central

---

ABOVE *fig. 92* Paired figures of Horus and Osiris enthroned at the confirmation of Horus as ruler of Egypt. On the right Re-Harakhti acknowledges Horus as four birds are released; on the left Isis presents an image of Horus wearing the Red and White Crowns to Osiris. (Relief on the north wall of the court of Osiris, east side of the roof terrace, Dendara. Late first century BCE.)

ABOVE RIGHT *fig. 93* Photographic detail of the daytime hours in the east window shown in *fig. 94*.

RIGHT *fig. 94* Nephthys shown standing between Hathor and Re-Harakhti in the sixth hour of the day. (Scene on the underside of the east window's lintel in the eastern Osiris shrines, Dendara. Late first century BCE.)

chamber, paired reliefs show four messenger birds flying before the enthroned Osiris and his son, summoned to take the message to the four cardinal points that Horus rules *(fig. 92)*.[7] Their flight promises the 'vital year's greenness', which the goddess Nehemetawy of Hermopolis also brings in the procession of deities honouring Osiris. Shown holding a tiny figure of Maat in one hand and a papyrus stem in the other, she offers a flourishing gift of greening life.[8]

Like the birth of the heir in Harsomtus's chamber in the crypt's 'secret palace' *(see chapter 7)*, here in this court Horus's accession is timed for sunrise, symbolically fused with Re's shining appearance in the East, and marked by the first five hours of the day listed at the entrance.

RESTORING EQUILIBRIUM: THE MIDDAY HOUR

Time flows on towards the central chamber, where the outstretched body of the sky goddess, Nut, divides the ceiling into two halves *(fig. 118)*: on the west side is the famous round zodiac *(see chapter 12)*; on the east side are the solar boats for the 12 transformations of Re through the hours of the day, which are associated with the *Ritual of Hours* during the Ptolemaic period.[9] Thus, juxtaposed here, as at Abu Simbel, are the day and year cycles, the rhythms of time governing this Khoiak Festival, and indeed the whole of the *Peret* season.

Time is moving, too, into the daytime sixth hour, detectable in the open east window beside the chamber's entrance, though unless someone crouches down to look at its lintel's underside, this hour is completely lost to view. For hidden away here is a scene of Nephthys in the solar boat, wearing the White Crown of Upper Egypt and flanked by Hathor and Re-Harakhti, who offers 'life' to her, with the first to the sixth hours of the day marked above them *(fig.94)*, represented by six 'hour' goddesses and six solar disks, culminating in the sixth hour called 'Vertical' (*'ḥ'yt*), or 'Standing One'. This is the midday hour, when the sun is high in the sky and Re joyfully unites with his daughter Maat. Its name resonates also with the 'building' work (*s'ḥ'*) Osiris is called to do in the eastern court *(see page 105)*. And, according to the accompanying inscription, Re-Harakhti unites with the 'daughter of the creator', his words leaving no doubt about the need to propitiate his primordial female Eye, on whom his very existence depends. 'You are in peace with me' he says, 'you are not angry against me. I am come into existence with you and your father, the Primordial Flood, when the land was yet in darkness.'[10]

Across in the corresponding west window scene, Atum, the ageing form of the sun god in the West, replaces Re-Harakhti as the giver of life to Nephthys, here wearing the Red Crown of the North and again accompanied by Hathor. Again,

the first six hours of the night are named in the disks above them.

Thus, juxtaposed in these two windows are the two extremes of midday and midnight united by Hathor herself, who is depicted within a solar disk on the underside of the central doorway's lintel. Shining as 'the luminous one who illumines the Two Lands with her rays, the female sun who fills the land with gold dust', she is the fiery uraeus who 'unites with Re' and 'glorifies Osiris with the praises of the blessed ones'.[11] Just as the exalted Egyptian queen manifests in the palace with the king as the 'Lady of All' *(see page 41)*, so it is Dendara's powerful queen who shines down on those passing through her exalted doorway into the central chamber, perfectly poised at the equilibrium of the world.

The same sense of equilibrium prevails within this chamber, where all the various 'ingredients' are combined harmoniously in order for Osiris to 'stand vertically' once again. On one level, the focus here is on the material care of the Khoiak seed beds, encapsulated by two scenes of the goddess Shentayt, the 'mistress of grain', presiding over the scales used to measure the various ingredients in the 'Temple of Gold' *(fig.88)*. If her seed life is to flourish, complete balance and right measure are required. Watching this weighing are the potter god, Khnum, and the Memphite demiurge, Ptah, the two gods who create the 'bodies' of the deities.

But this 'equilibrium' also applies to Osiris himself, whose death and dismemberment at the hands of his brother, Seth, has thrown every aspect of Egyptian life out of balance, and whether it be the assembling of the seed beds or the reintegration of his body, everything must now be reunited.

To this end, the nome deities, representing the districts of Upper and Lower Egypt *(fig.95)*, each bring a vessel to Osiris as he stands before Shentayt's garden. Each of the vessels contains the sacred relic of his body that was preserved in that particular district. The Upper Egyptian nomes are led by Hathor, the Lower Egyptian nomes by Ptah, 'Father of the Gods'. Both halves of the country are coming together now to honour Osiris; indeed, by giving him these relics, they are

symbolically gathering together his scattered body and converting it into the unified body-like state of Egypt.[12]

And, just as Hathor and Ptah are paired together in the 'hall of ascension' at Abu Simbel, where Ramesses and Nefertari travel from North to South through the whole of Egypt, simultaneously rising to the noonday zenith to bring Maat to Amun-Re *(see chapter 3)*, here Osiris's reassembled body ensures Maat's 'balance' prevails in the Two Lands. Though interestingly, unlike at Abu Simbel (and in the *Book of Day*), it is Seth's defeat, not his integration into the solar circuit, that matters here at Dendara. In an Egypt ruled by foreigners, Seth's creed of 'divide and rule' and his lust for possession and power, which threaten to extinguish the life-force and disrupt the very rhythms of life and growth, serve no useful function. This god of the

foreigners has no constructive role to play now, and is simply regarded with horror and abhorrence by Dendara's priests in their Khoiak rites *(fig.96)*.[13]

In this 'vertical' hour, everything is directed rather towards 'rebuilding' Osiris, who is implored now to 'enter his secret representations' and come as a rapacious falcon with 'shining feathers', surrounded by the '*Ba*-souls of the gods unfurling their wings to protect you' as he swoops down to enter his temple. For his 'material' body must be animated by his heavenly *Ba*-soul *(fig.86)*, infused with celestial life from above.[14]

To be sure, this descending great falcon with 'shining feathers' seems more like the triumphant Horus of Edfu, and intentionally so, given the functional identity of Osiris and Horus throughout these Khoiak chambers. Merged completely with the fate of his father, it is Horus's heavenly strength, his victory over Seth, which ultimately ensures Osiris becomes stable and vertical again, and, as confirmed in an inscription in the neighbouring end chamber, this happens in the sixth hour of the day: 'You rise ['ḥ'] in the sky in the sixth hour ['ḥ'yt] before your temple, the sky of your *Ba* on earth.'[15] Clearly, Osiris's 'vertical' revelation is attuned to noon, the hour when the fallen god stands 'high' again and his *Ba* shines *(fig.86)*.

But something else is happening in this 'zodiac' chamber, for inscribed here is also the 'Hour Watch', the traditional vigil for Osiris through the 24 hours of day and night, which perhaps took place on the night before burial at the end of the mummification rites. Importantly, it is precisely during the sixth hour of the night, the nadir of the vigil, that Osiris's bodily members are reassembled. Yet here this 'nadir' experience, said to be the time

when Geb comes with Shu to see Osiris's body being assembled, has been placed in a daytime chamber where Osiris stands vertical again, his body having been brought by the nome deities and illuminated by the noonday sun.[16]

This is part of the alternating polarity that has already been established with the counting of the hours at the entrance to the first court. Now, here in this central chamber, heights and depths, seemingly such polar opposites, have become transparent to the unified life within them, reinforced by the fact that Geb also rules the corresponding chamber directly across on the west side *(see page 118)*. Osirian night has become solar day, creating a complete unification between 'above' and 'below' on this vertical axis of the world, as Hermes Trismegistus would also later teach in alchemy's famous *Emerald Tablet*:

(see page 118)

LEFT fig.95 Detail of a nome representative holding a vessel containing a sacred relic of Osiris. (Central Osiris room, east side of the roof terrace, Dendara. Late first century BCE.)

RIGHT fig.96 Relief in the eastern court showing Seth's animal tethered to a stake between Osiris and Horus. (Late first century BCE.)

'Whatever is below is like that which is above, and whatever is above is like that which is below, to accomplish the miracles of one thing.'[17]

INCUBATING OSIRIS:
IN THE HEART'S GARDEN

Osiris has found his vertical stability, his inherent ability to stand, but he has not yet found his capacity to move and feel. He needs heart energy, for, as the *Memphite Theology* insists, it is the heart that commands the activity of the bodily members, 'speaking from the vessels of every limb'.[18] Inscriptions at the entrance to the end chamber confirm Osiris's 'heart is not tired'; rather, as the bringer of the inundation, he is 'reinvigorated', manifesting as the 'beautiful child illuminating those in the primordial waters, who is born every year'.[19] Hence, if he is truly to be 'rebuilt', if he is to grow, develop, flourish, and make manifest his god-like renewing powers, he needs to become attuned to the life-giving rhythms of his mother's heartbeat in the end chamber, for it

is ultimately from his mother that his 'not-tired' heart derives *(see page 50)*.

No specific hour is mentioned in this end chamber, but everything points to the 'ninth hour', the midheaven time when the sun boat crosses the annual river axis and Re 'comes forth from the egg' in his glorious second birth. First, it is accessed through the sixth-hour 'zodiac' chamber, indicative of a progression through the day. Second, what is celebrated here is Osiris's rebirth as a 'god', his gestation in the womb of his mother, Nut, praised here as the great sky mother 'who gives birth to the gods'.[20] Clearly, this is a divinizing chamber, corresponding, therefore, with those other ninth-hour 'chambers of the mothers', the Crypt South 1's maternal 'Pure House' *(see page 94)* and Nefertari's central chamber *(see chapter 4)*, and in fact Hathor and Isis appear here as nurturing mothers suckling their divine child.[21] Indeed, whilst the text for the ninth hour of the day on the zodiac room's ceiling is partially damaged, enough remains to indicate this is the hour of Re's 'renewal'

LEFT *fig.97* Osiris here manifests his power of movement. (Relief in the inner Osiris room, west side of the roof terrace, Dendara. Late first century BCE.)

RIGHT *fig.98* Four goddesses raise their arms aloft to support Osiris in Shentayt's garden. His gestation in Nut's womb is inscribed in the columns above him, and other scenes on the wall (not illustrated here) show him in different positions on a lion-headed bed, manifesting his power to transform as a living god. (Line-drawing of the central scene at the base of the north wall, inner Osiris room, east side of the roof terrace, Dendara. Late first century BCE.)

(*m3w.tw*) and 'rejuvenation' (*rnp.tw*), the very same words used to describe Osiris's gestation in Nut's womb.[22] In short, 'rebuilding Osiris' on this east side is completely in tune with the sun god's unfolding lifetime by day.

And what was only hinted at in Abu Simbel's 'chamber of the mothers' is now given its fullest expression in the graphic praise of Osiris incubating in his 'egg' in Nut's womb, gestating and transforming there as the 'son of the year', before being born at Thebes:

> O Osiris, who dwells in the West … the great Heliopolitan at the head of the Town of the Scarab. Your mother Nut is pregnant with you, she cares for your egg [*swḥt*] within her womb, she harmoniously fashions your bones, she rejuvenates your body, she gives life to your skin for your limbs, she dilates your vessels for your blood, your White Crown and uraeus are installed on your brow, she takes you from the mould [and places you] again on earth, like she has birthed you at Thebes, she makes your body youthful, you become young, she renews you at your time of year[23]

Nowhere else is there such a deeply embryological expression of Osiris's renewal, his experience of life in the eternal round, living in a perpetual state of 'Becoming', continually transforming from one state to another in an unbroken cycle of life. He is 'at the head of the Town of the Scarab', and hence, above this hymn on the end wall, he is shown in his various cult forms, able to stir and move, a resurrected god set in motion, whose heart is not 'tired', able to engender his son, Horus, when Isis, as bird, hovers above his inert body to conceive him from his 'divine semen'.

Life engenders life in this ever-renewing cycle guarded by the mothers, poignantly symbolized also by the eight cobra–vulture pairs, the tutelary female guardians of North and South, shown facing each other with their beaks touching whilst surrounding Osiris lying in his seed-bed Khoiak garden (*fig.98*). Here in this maternal garden, as the sun rises and sets and the moon waxes and wanes, new forms come forth in an organic, living, breathing and growing cosmos, eternally recreated, as Goethe says in his epic drama *Faust*, when his

111

trembling hero descends into the terrifying 'realm of the mothers', where everything is in perpetual 'formation, transformation, eternal Mind's eternal re-creation'.[24]

Tellingly, this embryological inscription is associated with Shentayt's seed-bed garden depicted at the base of the north wall, here being supported by four goddesses representing the four cardinal points, their feet symbolically placed in water, indicative of the watery womb world in which everything in this chamber lives *(fig.98)*. And just as Nut takes Osiris 'from the mould' when she births him as a 'god', so on 21 Khoiak the two halves of Osiris *vegetans* are correspondingly removed from the mould, glued together, bound by four papyrus bands and then placed in the sun for the duration of the day. These seeds need heat, and Osiris needs heat, which is beautifully expressed by the four depictions of him lying on a lion bed carved in the air vent in the chamber's ceiling—four figures waiting to receive the sun's life-giving rays from above *(fig.99)*. Circulating through the four cardinal directions, Re is said to shine on Osiris 'at dawn', in the 'sixth hour of the day', when he 'sets in life' and finally when he appears as a *Ba*-soul alighting as a glorious spirit to see your corpse'.[25] He is a god living in the complete solar round.

'Fire is the maturing force' says Hermes in the Hermetic text known as the *Poimandres*,[26] something Dendara's priests evidently well knew. In his 18th-Dynasty *Great Hymn to the Aten* Akhenaten had praised the sun god's power to animate the developing foetus, to send his heavenly rays deep into the womb to provide life-giving warmth.[27] Centuries later, this embryological knowledge was carved in stone here at Dendara, and later continued by the alchemical inheritors of this Egyptian tradition. Thus, in the Arabic *Epistle of the Secret*, Hermes instructs Theosebia to expose the 'egg' she is creating to the sun *(see page 240)*. The Islamic alchemist Ibn Umail calls it 'the egg of the sages', a pneumatic egg 'born in the air' which 'travels around the whole world'.[28] Other alchemists refer to this birth from the egg as the 'etesian stone' *(lithos etesios)*, the renewed child born every year, brooding in the vessel of the Philosophers' Stone.

Completely enclosed in this sealed vessel and warmed by fire, this alchemical 'egg' develops through a circulating process associated with the sun's passage through the zodiac.[29] In fact, Osiris's manifestation as 'child of the year' is made explicit here on the end chamber's ceiling, where he is shown as Orion, enclosed by the outstretched body of Nut and positioned directly beneath her nourishing, pendulous breasts. Joyfully sailing in his celestial boat, one arm powerfully raised aloft, accompanied by Isis-Sothis in her 'New Year' form as a cow, he is an 'Orion who circulates in the womb of his mother', his renewal synchronized with the annual arrival of the inundation waters associated with Sothis.[30]

The fragrant Sokarian egg, fabricated from aromatic sweet materials and enclosed in a silver vase *(see page 102)*, is also incubating during these Khoiak rites, and, according to the long inscription in the open-air court, like Osiris, it is to be extracted from its mould, anointed and then ritually exposed to the sun's rays from 19 to 23 Khoiak *(see page 102)*. The oil for this anointing is 'cooked' from 15 to 22 Khoiak, according to a recipe for anointing Sokar from the *Book of the Unguent*. To the ancient Egyptians, fragrance reveals the presence of a divine being—no wonder then that this Sokarian egg has to exude fragrant odours, since what is being born in this ninth-hour chamber is divinity itself. Nor is it a coincidence that this end chamber is situated directly above the 'laboratory' in the six-columned hypostyle hall in the main temple, where a recipe for 'making dry incense of the first quality' is inscribed at the entrance.[31] Once again, the thematic correspondences between 'above' and 'below' are fully worked out in this glorious Hathor temple, for, as scented aromas waft upwards from this 'laboratory', symbolically percolating to the rooftop chamber above, they herald the presence of a god within.[32]

This exposure of the Sokarian 'egg' to the sun from 19 to 23 Khoiak also brings colour, enacted when Sokar's cult image is painted on 23 Khoiak in the 'chamber of the bed' at the third hour of the day. First a blackish unguent is applied, then his face is painted yellow, his jaw turquoise, his eyes 'like filled eyes', his wig the colour of true

lapis lazuli, his crook and flail like 'real precious stones'.[33] Colours reveal the character of a person or a god, and shining with these verdant green-gold colours of life, Osiris-Sokar radiates solar beauty and vitality, coming to life like the seeds growing in Shentayt's garden, eternally renewed in the garden of the heart, nurtured by the same maternal forces that divinize Osiris in the womb.

All this would be later taught openly by the alchemist Cleopatra in the remarkable Greek *Dialogue of the Philosophers and Cleopatra*, in which she lyrically compared the beauty of vegetation and stones in the earth to the transformation of nature into the divine, telling the philosophers it was only when 'tested by fire' that truly beautiful colours appeared:

> When you take plants, elements, and stones from their places, they appear beautiful, but they are not beautiful because it is fire that tests all things. When they are clad in the glory that comes from fire and the shining colour, here then are some visions more beautiful, here the hidden glory and the sought-for beauty, terrestrial nature transformed into the divine when these things have been nourished in fire.[34]

Later she went on to say this 'clothing in glory' occurred when they went forth from the 'womb of fire'.[35] To be sure, the Egyptians experienced this

ABOVE *fig.*99 One of the four figures of Osiris depicted on a lion-headed bed in the air vent in the inner Osiris room. The accompanying texts describe how Re illuminates him through the four phases of the day. (East side of the roof terrace, Dendara. Late first century BCE.)

knowledge mythically and ritually, but Cleopatra's teaching here seems like a retelling of the Khoiak seedtime mysteries, making the secrets accessible now to alchemists in their own Greek 'philosophical' language. For, despite the different expression, the same wisdom shines through, the same delight in springtime beauty, eternally renewed, born from the fiery maternal womb in all its radiant colours.

Curiously, though, no progress is possible beyond the end chamber here on the east side of the roof terrace at Dendara. The daytime journey apparently abruptly stops in the ninth hour rather than continues towards sunset in the West and the 12th hour. But it is no mistake, since this highly unusual version of the festival is not attuned to the complete cycle, but to an alternating movement between the heights and depths *(see chapter 12)*, a cosmic movement now bringing the divinized Osiris-Sokar to the western chambers on the other side of the roof, descending deep into the earth and travelling with the sun to the nadir of the night and year.

PHOENIX RISING: BLACK EARTH REGENERATION

The smooth transition from the eastern chambers to those on the west side is made at the western court's entrance, in verses from the litany for Sokar which priests had already chanted in the east court, and now chant to honour the fragrant 'secret' veiled in Sokar's shrine entering the West *(fig.101)*:

> How sweet is the fragrance you love, be living, be living for eternity, your festival is forever… Come [you] who repels the enemies, Come [you] who teaches the child, you who inspires fear in the rebels, the Lord of Upper Busiris comes, he smites the rebels.[1]

All the sweet-smelling substances creating Osiris's life in the 'egg' in the eastern end chamber now exude their fragrance here in the West, heralding his joyful arrival as a divine being in the *Dwat*, the vanquisher of his enemies:

> How beautiful is your appearance, Heded, Lord of the White Crown, how sweet the emanations from your mouth, you travel in the *Dwat* perfect in your body … you do not distance yourself from the unique *Ima*-tree, your heart comes in joy.[2]

Reinforcing this transition at the court's entrance, moreover, is the repetition of an extract taken from the eastern court's long Khoiak Festival inscription describing an agricultural rite of 'fertilizing the field' performed from 12 to 19 Tybi (the fifth month of the year) in the 'field of Osiris'. It entails yoking two black cows to a wooden plough, with a ploughshare made in black bronze, driven by a ploughman dressed in green and accompanied by a green-clad child scattering seeds of wheat and barley—man and child, probably to be understood as father and son, ritually fertilizing and 'greening' the Osirian field.

Manifestly, here in this western court time is being measured by the months of the year, which are, of course, also tied to the night hours *(see chapter 5)*, though, curiously, this 'ploughing' episode is being performed in the fifth month, in other words after the Khoiak Festival would normally have ended.

It is also juxtaposed with a 'cooking' text describing the creation of special *kefen*-bread on

20 Tybi, when grain is 'moistened' and 'rises', overseen by Sokar. According to the festival inscription, all kinds of aromatic substances form this fragrant bread's ingredients, baked in a mould engraved with the parts of Osiris's body, here 16 bodily members (the number can vary in Egyptian sources), thus fusing his life with the mysterious transmutation of grain into baked bread.[3]

The long journey from ploughing and seeding the earth to the creation of nourishing food was profoundly sacred to the ancient Egyptians, an agricultural people closely in touch with the natural processes on which their very existence depended. They knew, too, that like the grain, which must fall into earth, grow and be reaped, threshed, sifted, ground, kneaded and baked in order to become nourishing bread, Osiris must suffer in order to rise and reveal his indestructible life-giving essence.[4]

Repeatedly in these western chambers he is called the 'seed' and 'heir' of his father, the earth

LEFT fig.100 The ascendant phase of the moon represented by 14 deities (only six shown here) ascending a stairway towards the Eye in a large disk placed on a papyrus stem. Praising the Eye on the right is the moon god Thoth. (West side of the ceiling in the pronaos of the temple of Hathor, Dendara. First century CE.)

ABOVE fig.101 Relief of Sokar's boat in the eastern court of Osiris. (East side of the roof terrace, Dendara. Late first century BCE.)

Plutarch's use of the word *Khemia* here, or *Chemia*, is particularly interesting as a name defining Egypt's blackness, since this word later became synonymous with the word for alchemy.[7] According to Isis, moreover, in the Hermetic treatise known as *Koré Kosmou* (or 'Pupil of the Eye of the World'), Kamephis (a form of Kamutef-Bull-of-his-Mother) gifted her with the 'perfect black' which he had learned from Hermes, and which she in turn is now giving to her son, Horus.[8] It is not Hermes who guards this Osirian mystery of the 'Black One' here at Dendara, however, but his Egyptian counterpart Thoth *(see below)*.

SIXTH-MONTH SYMBOLISM: FILLING THE EYE

What, though, have the fifth-month rites of 'fertilizing the field' and creating the '*kefen*-bread' to do with this Khoiak Festival, which, after all, took place during the fourth month and would already have ended by the fifth month?

These are certainly very curious, easily overlooked, details, though perfectly understandable if related to the great cosmic reversal of the year at the end of the sixth month and its ritual of 'Filling the Eye' *(see chapter 5)*. Far from being irrelevant, these fifth-month rites in fact mark the passage of time towards this momentous 'turning-point', which is explicitly symbolized in the scene precisely placed above the doorway leading into the central chamber on the other side of the western court.[9] Here the lunar Eye, named Osiris-Moon-Thoth, is shown within a disk supported by Shu and Thoth standing in a boat *(fig.103)*, with a procession of deities symmetrically ranged on either side of this 'periodically renewing' lunar disk, all coming to 'fill the Eye' of the raging 'bull of the new moon' with a whole host of precious stones, including chrysocolla, amethyst, turquoise and lapis lazuli. Interestingly, too, behind Thoth is a scarab, and a baboon perches on the boat behind Shu, the very same creatures housed in Ramesses II's solar shrine oriented towards the sun's southern 'reversal' on the terrace at Abu Simbel *(see page 61)*. Here at Dendara they are with the two gods who also entice the raging Sun Eye back to Egypt from the South in the *Goddess in the*

god, Geb. He is also hailed as the 'Black One' *(Kmy)* in the western court, the 'unique god among the gods',[5] bringing to mind Plutarch's description of Egypt as 'the black earth', having immediately beforehand mentioned the Heliopolitan Mnevis bull, which was required to be totally black:

> They call Egypt, since it is mostly black, *Khemia*, like the black part of the eye, and they liken it to a heart.[6]

ABOVE *fig.102* A *Benu*-bird depicted high up in a corner of the north wall in the western court of Osiris. To the right are cartouches containing the names of the planet Jupiter and Osiris. (West side of the roof terrace, Dendara. Late first century BCE.)

ABOVE RIGHT *fig.103* Scene of 'Filling the Moon Eye' at the end of the sixth month, which is precisely located above the doorway giving access to the central Osiris room. Here Thoth (left) and Shu (right) support a large disk enclosing the Eye. A scarab is behind Thoth, and a baboon perches on the end of the boat behind Shu, the same creatures originally housed in the solar shrine at Abu Simbel *(see fig.52)*. Thoth and Shu are also the deities who entice the fleeing Sun Eye back to Egypt from Nubia in the *Goddess in the Distance* myth associated with the southern solstice. (West side of the roof terrace, Dendara. Late first century BCE.)

Distance myth, with all its 'solstice' connotations *(see page 12)*.

But that is not all. Just as a person manifests the purity of the *Benu*-bird in Herakleopolis when coming for 'judgement' on the last day of the sixth month when the Eye is 'filled' *(see page 65)*, so the very same mythic combination is represented here at Dendara. For depicted high up in each of the court's four corners are *Benu*-birds *(fig.102)*, each representing an aspect of Osiris's eightfold *Ba*-soul. It is in Herakleopolis that the *Benu*-bird takes flight at the end of the sixth month, heralding the annual reversal of time in the middle of the *Peret* season *(see page 65)*, and, according to chapter 175 of the *Book of the Dead*, it is in Herakleopolis, too, that Seth finally becomes subservient to Osiris, acknowledging his right to appear on the throne of Re, the flow of blood from his wounded snout spattering the earth and resulting in a 'hacking the earth' rite *(see page 43)*. Correspondingly, the inscriptions directly beneath the 'filling the Eye' scene here at Dendara, inscribed around the entrance to the central chamber, are a direct

reminder of these events, since they describe the ritual slaughtering of the Sethian 'red bull'. Its blood, Isis says, now 'flowing in earth from the heart', fertilizing the land with its sulphuric strength, reinforcing Osiris's supremacy over his murderous brother.[10]

Placed above the 'filling the Eye' scene is a great winged disk flanked by two cartouches on each side *(fig.102)*, one proclaiming Osiris's sovereignty as the god 'who presides in the West, the Great One in Dendara', the other inscribed with the name 'Horus who illuminates the Two Lands', in other words the planet Jupiter *(see chapter 12)*.[11]

So, the mythic events characterizing the year's 'turning back' at the end of the sixth month now reappear in this western court. Hence, far from being extraneous, the fifth-month agricultural rites mentioned at the entrance in fact pave the way towards this momentous annual turning-point.

The day's heavenly waking in the eastern chambers has become Osiris's return to deep in the earth, to the nadir of the night and year, a heart-centred descent made also with the moon, since it

is the task of the moon god, Thoth, the 'heart of Re', to bring Osiris's *Ba* to his corpse, uniting him with the cosmic cycles of sun, moon and stars.[12]

According to chapter 29 of the *Book of the Dead*, which is 'a spell for a heart amulet of *Seheret*-stone', the *Benu*-bird performs the same role: 'I am the *Benu*, the *Ba*-soul of Re, the one who leads the blessed to the *Dwat*, who causes Osiris to return to earth to do what his *Ka* desires.' Protecting Osiris's return to earth here in this western court is this Egyptian forerunner of the Greek phoenix, the miraculous bird that burns itself in flames and rises eternally renewed from the ashes, and whilst Egyptian sources are silent about a dramatic rebirth from ashes, nevertheless, like its Greek counterpart, here this phoenix guards Osiris's death and regeneration in the depths of the earth.[13]

ACROBATIC GEB: OSIRIS'S REVOLUTIONS

This 'reversal' symbolism continues in the central chamber, where an extraordinary twisted-back figure of the earth god, Geb, is depicted beneath the arched body of Nut on the ceiling *(figs.104, 105)*, his somersaulting pose wonderfully evoking 'reversal' at the nadir of the night and year, as it does on the astronomical ceiling of the Roman-period temple at Deir el-Haggar *(see page 142)*.[14]

Covering the central chamber's walls are also detailed descriptions of the gates and regions of Osiris's netherworld kingdom: the seven and 15 gates of the *Dwat*, the 14 mounds of Osiris, and the secret portals of the House of Osiris in the Field of Reeds. Nothing escapes mention in this compendium of the dark earth realm he rules. Inscribed in the frieze decoration are his 74 names, each one within a cartouche, their number recalling the 74 divine manifestations of Re acclaimed in the *Litany of Re*, best known from New Kingdom royal tombs at Thebes.[15] Central to this litany is the nocturnal encounter of Re and Osiris, timed in the *Amduat* to occur during the sixth night hour, when the *Ba* of the sun god unites with Osiris's body, fusing them together as a single divine being *(fig.106)*. 'Re is the right Eye', it is stated here in the central chamber. 'Osiris is the left Eye, one god united with the other'—cosmic sun and moon unified and shining together in the depths of earth at this hinge of the night and the year.[16]

What comes out very clearly in this litany, too, is the pharaoh's desire to participate in the great celestial revolutions lived by the sun, moon and stars, moving in one continuous circuit of life, death and regeneration between heaven and deep in the earth. The same is true of Osiris here at Dendara, his festival being renewed 'like Re rising and setting', as he shines like the moon in the horizon:

> His heart is invigorated with a libation …
> his festivals renewed like Re rising and setting.
> He circulates the *Dwat*, he flies to the horizon
> as the Moon, illuminating the land like 'He of the
> Horizon in the Horizon'.[17]

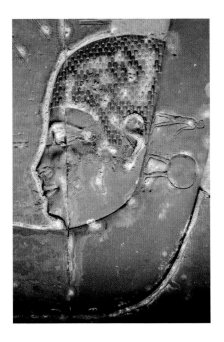

FAR LEFT fig.104 The earth god Geb, whom the Greeks identified with Kronos, somersaults to mark the reversal of the year between the sixth and seventh months on the ceiling in the central Osiris room. (West side of the roof terrace, Dendara. Late first century BCE.)

LEFT fig.105 Detail of the sky goddess Nut, who is depicted above Geb on the ceiling of the central Osiris room.

RIGHT fig.106 Unification of Osiris and Re in the sixth hour of the night. The mummiform body belongs to Osiris, the ram's head is the *Ba*-manifestation of Re. Inscribed between Isis (right) and Nephthys (left) are the words 'Osiris is resting in Re; Re is resting in Osiris'. (Wall-painting in Nefertari's tomb, Valley of the Queens, Western Thebes. 19th Dynasty.)

Indeed, his power to 'rise' and 'set' like Re is a haunting remembrance of the purified king's 'ascent' and 'descent' with the sun god already so beautifully expressed in the *Pyramid Texts*:

> You cleanse yourself in the horizon,
> and you release your cleansing in the Lakes of Shu.
> You ascend and descend, you descend with Re,
> sinking in the darkness with Nedy.
> You ascend and descend, you ascend with Re,
> and rise with the great Embracer.
> You ascend and descend, you descend
> with Nephthys,
> sinking in the darkness with the night boat.
> You ascend and descend, you ascend with Isis,
> rising with the day boat of the sun.[18]

Journeying from night to day, day to night, going up and down in the sun boat, the king lives with the sun, moon and stars in an ever-recurring cycle of ascent and descent encompassing the heavens and netherworld, the kind of unimpeded movement sought also by the astrology-loving Petosiris in his Roman-period tomb in the Dakhla Oasis when travelling with the sun down to the sixth night hour, knowing this is the moment when the year reverses *(see chapter 12)*.

Such is Osiris's life, too, here at Dendara, for it is by participating in this annual 'reversal' that he is resurrected, rising phoenix-like in the horizon as the ever-returning new moon shining in the sky: 'He wakes from sleep, he flies as the *Benu*-bird, taking his place in the firmament as the crescent moon.'[19] Or he manifests as the star constellation Orion, protected by Sothis when his *Ba* wakes.[20]

Completely united with Re in this eternal circuit, Osiris is eternally resurrected in Geb's 'earth' realm, with nothing impeding his movement at this darkest time of night and cardinal turning-point of the year.[21] And in terms of the actual Khoiak Festival itself, this central chamber corresponds with the great ritual processions that took place between 24 and 26 Khoiak, when he was triumphantly taken round the temple, following the circuit of the sun along the processional way before appearing once again in his sanctuary, 'glorified [*akh*] by the glorifications [*akhu*]', shining as the 'Glorious One' *(Akh)*, having completed his full circle.[22]

Ranged along the base of the central chamber's walls are the Upper and Lower Egyptian nome representatives, all come to honour Osiris as the god 'who wakes vital', providing Egypt with nourishment and taking possession of his kingdom in the *Dwat* whilst his son, Horus, appears on the throne of Egypt. In the noonday chamber across on the east side they were summoned to bring Osiris's scattered bodily relics, and there, too, they witnessed the unification of his *Ba* and body in the daytime sixth hour *(see page 109)*, but that unification was experienced in a very different world, a daytime world, where Osiris's falcon *Ba* hovered high in the sky in the noonday heights, swooping down to unite with his earthly body, for he had not yet become a body with a heart. Now, in the corresponding western chamber, where Re and Osiris unite in the darkest hour, the representatives of the nomes appear in the form of fecundity figures, each one pouring forth water drawn from the canals in their various districts before an offering table.[23] On the east side, they carried vessels containing Osiris's bodily relics; now they bring water—divine water to purify Osiris's resurrected body rising with Re in the solar circuit. Osiris dismembered has become an eternal Osiris, a purified *Akh*-spirit, a flowing, celestial body of light, which is revealed when midday and midnight, zenith and nadir are drawn together in a vertical shaft of solar light.[24]

Certainly, across in the eastern central chamber, Khnum characterizes the fragrant 'paste' forming the 'grain' figure of Osiris as '*akh*', evidently a material with power, not only ritually transforming 'seed', but also the very body of the resurrected god, who is now shining and fully alive here in the West.[25] And once again Dendara's priests delighted in creating spatial correspondences between 'above' and 'below', locating this starry central chamber directly above 'the chamber of the water' in the hypostyle hall in the main temple, which led out to the well supplying all the pure water needed in the temple cults. Just as fragrant odours rise up from the 'laboratory' to reveal Osiris as a god-like being in the eastern end chamber *(see page 112)*, so here a current of holy water symbolically flows between above and below, purifying Osiris's celestial body in the eternal solar circuit.

Moreover, just as the earth god, Geb, commands the Osirian king in the *Pyramid Texts* to 'raise yourself, be luminous [*akh*], and speak', so here in the central chamber Osiris becomes a 'speaking' god, which is accomplished through the *Opening of the Mouth Ritual* represented here.[26] He also receives his crown of victory, as he must if his resurrection is to accord with the 'justification' occurring when the year reverses *(see page 65)*.

Thus, on one side of the doorway lintel at the entrance to the end chamber, the pharaoh offers two Eyes to the victorious Osiris, 'heir of Geb', whose crown of justification, according to the accompanying inscription, is being given to him by Ptah, 'Father of the Gods'. In the adjacent scene, Thoth, the 'twice-great', elevates Maat, the 'seed of the bull', before Hathor, the goddess who 'detests wrong-doing, what she needs is doing the good', here complementing the role of Isis on the other side of the lintel, 'who makes young her brother with the grain in the Temple of Gold'.[27] Such a crowning is completely attuned to a creation called forth by Ptah's sacred utterances, as proclaimed also in the lintel's central inscription honouring 'the very great powers [*sekhemu*] of Dendara, the lords of commands in the sky, the earth, and the netherworld, the rulers of the gods'.[28] The silence that, according to the *Book of Nut*, reigns in the *Dwat* when the 'name of a decan is not spoken for 70 days', now resounds with the primordial power of words, for Osiris's regeneration brings a rejuvenation of the world, a new Memphite creation chanted into existence by the spirit of divine 'command'.

RAISING THE SKY: THE COSMOS ADORNED

At the beginning of the seventh month, when the festival of 'Raising the Sky' is performed for Ptah, Osiris's burial is also ritually celebrated *(see page 63)*, and there can be no mistaking this is the all-important festival in the western end chamber, as confirmed in a speech of Nephthys to Osiris inscribed at the entrance. 'The sky is raised for you on the day of your burial in Busiris', she tells him, 'and the Temple of the Phoenix stretches out its arms to receive her lord.'[29]

Within the chamber itself, the atmosphere is distinctly funereal. Anubis is shown performing the mummification rites, providing Osiris with

ABOVE *fig.107* Perched above grain watered by the cow-headed Isis-Hathor, the *Ba* of Osiris manifests his life-giving power. Behind Isis is Hapy pouring water within a rocky cavern enclosed by a snake. (Relief on Hadrian's gateway in the temple of Isis at Philae. Second century CE.)

everything he needs for burial, including a pectoral of gold for his throat and 104 amulets, all brought by Ptah to adorn his jewel-encrusted body, all carefully listed and their materials specified.

But the Osiris who manifests in this tomb-like room is no 'dead' god, for these precious stones and minerals ensure he shines as a celestial being, eternally alive as the 'power of the powers' *(sekhem sekhemu)*, who illuminates the 'one who sees him like the one who shines in gold'.[30] So, what began in the 'call to rebuild' in the eastern court now finds its completion in the germination of Osiris's golden body, wondrously shining like the sun, 'the great wonder in the cult places, the circle of gold in the cult places', radiating the sun's golden beauty, whose revelation is a transfiguring 'face to face' experience with a god who 'becomes young in his forms continually'.[31]

The ruined 'cult places' have been completely rebuilt *(see page 105)*, they are alive and living with

Osiris in the golden circle. His dismembered body has been ritually returned to eternal life. Destruction and reconstruction are at the heart of this Khoiak mystery, and, interestingly, the hieroglyphic word for 'circle' here is written with two lions, reminiscent of the *Aker*-lions guarding Osiris's resurrection in the *Book of the Earth* in Ramesses VI's burial chamber *(fig.62)*.

There is, though, another beautiful revelation in this end chamber, indicated by 44 tiny scenes in the wall frieze, each one showing Isis pouring water into the hands of Osiris's *Ba* and offering him food, as they soar together as exalted winged souls in the heavenly region. 'O *Ba* of Osiris, come, may you go forth at my voice, receive the water from my hand. I am your sister Isis who renews your body', she calls to him. Often in the previous chambers Osiris is praised as the water-bearing source, but now water is given back to him. Isis also returns life-giving grain to him, transmuted now into nourishing bread and beer through her goddess power and the devotion of her love:

> The water is for you, the bread is for you,
> the beer is for you…[32]

Like the cow-headed Isis-Sothis shown at Philae pouring water into the black earth-filled irrigation channels from which grain sprouts forth and over which hovers a *Ba*-bird *(fig.107)*, here at Dendara Isis joyfully brings rejuvenating water which both renews Osiris's exalted body and goes forth from it:

> Take the water of the inundation to revive your heart, take the primordial water which you live from every day, take the flood going forth from the sweat of your body…[33]

He is both the source of this water and the recipient of its life-giving power, the god who 'rebuilds' and is himself 'rebuilt', who gives life and for whom Isis now brings life.

In this golden room, where everything is awake, complete and unified, Osiris is praised both as the 'mummy of gold' and, in the very next breath, as the 'beautiful Ihy of gold', the sistrum-shaking incarnation of Hathor's divine child, the 'Glorious One' who is able to renew himself eternally, the golden mummy and youthful music-maker, uniting in himself old and young, death and life.[34] He and Isis live in the primal waters as the nourishing Father and Mother, the supernal well of existence from which all sustenance flows, manifesting in a cosmic dimension where everything is truly alive.

Carried aloft by the four sons of Horus, bejewelled and wrapped in four cloths coloured white, green, violet and red, his 'effluxes issuing from his body', according to inscriptions in this end room, Osiris journeys to the 'Temple of the Sistrum' before his burial in the necropolis, having bestowed the 'great office' on his 'beloved son, son of Re, Lord of Crowns'.[35] For his moment of burial is the moment when his life is reborn, eternally

'stable' and 'raised' *(fig.108)*, symbolized by the erection of the *Djed*-pillar on 30 Khoiak at the close of the festival.

It is also the moment when heaven is raised, a synchronicity with the year's reversal that has not been forgotten. For on the end chamber's ceiling are three wonderful interfolded figures of the sky goddess, Nut *(fig.109)*, each one arching her body to encompass heaven, their number encapsulating a multi-dimensional cosmos of 'three skies', or three 'realms' which Osiris has traversed in these six chambers: 1) the sky protecting the 'rebuilding' of his material body (eastern central chamber); 2) the sky of his midheaven divinization and 'rebirth' from the egg as Lord of the White Crown (eastern end chamber); and 3) the primordial night sky of the western chambers. But it is the tiny male figure enclosed by the smallest Nut goddess whose gesture is so transparent: he is lifting his arms aloft towards her arched body *(fig.109)*, ritually raising the sky anew when time reverses at the beginning of the seventh month.

Beneath this triple sky mother, Isis and Osiris sail in a boat, together with three manifestations of Harsomtus. Before the boat is another figure of Osiris, standing beside a *Iun*-pillar, the vertical cosmic axis around which these Khoiak rites constellate *(fig.109)*. Across on the ceiling's east side, a single figure of Nut encompasses the striding Osiris-Orion and the recumbent figure of Isis-Sothis as a cow sailing in their respective boats, whilst 14 divine figures ascend a staircase, at the summit of which Thoth adores the 'filled' lunar Eye, here placed on a papyrus stem immediately beneath Nut's pendulous breast *(fig.109)*. In the ceiling's central band, the falcon-headed Osiris and a *Benu*-bird sail together, followed by another *Benu*-bird perched in a boat. When the sky is raised, when cosmic 'reversal' is complete, this *Benu*-bird soars forth and the portal opens for Osiris to manifest his celestial life, earth-born and resurrected as a living *Ba*, stable and alive.

Shining in the sky as the waxing luminous moon, as the bright decanal star constellation Orion, Osiris is both the root of the netherworld and the pillar of heaven, the greening source of life for everything 'coming forth' in the *Peret* season at this pivotal

point of night and year. Correspondingly, across on the other side of the roof terrace, the eastern end chamber's astronomical ceiling shows him as Orion too, though manifesting not in the seventh month, but at the start of the New Year in the *Akhet*-season, born from Nut's womb to bring the inundation with Isis-Sothis *(see page 111)*. Thus, the two ends of the year are 'tied together' in these juxtaposed astronomical ceilings on the west and east sides in an eternal unbroken golden circle of 'life'.

ABOVE *fig.108* Osiris resurrected kneels on a serpent crowned with two *Djed*-pillars at the close of the Khoiak Festival. To the left he is again shown, here protected by the winged Isis as the king (no cartouche) offers him an *Ankh*-sign of life. (Relief on the south wall of the inner Osiris room, west side of the roof terrace, Dendara. Late first century BCE.)

BELOW *fig.109* Osiris's astral resurrection synchronized with 'raising the sky' at the beginning of the seventh month. (Line-drawing of the ceiling of the inner Osiris room, west side of the roof terrace, Dendara. Late first century BCE.)

123

Directly beneath the western end chamber, in the hypostyle hall in the main temple, is the room known as the 'Treasury', where, in the base register, fecundity figures bring precious metals and minerals from the various regions over which Hathor presides as goddess of the mining regions. The pharaoh, too, offers crowns, ornate necklaces, jewellery, and sistra to the temple deities, all the ornaments that beautify and adorn their divine bodies. Stars, gold and precious stones are all beautiful because of their luminosity. Light-filled, they shine, revealing the perfection of the cosmos, and here it is almost as though Dendara's priests are enshrining the Greek word *kosmos*. For this means both the 'harmonious order of the world' and 'ornament', 'adornment' and 'decoration', being related not only to *kosmeo*, a verb meaning 'to put in order' and 'adorn', but also to the word 'cosmetic'. To the Greeks, the well-proportioned order of the world was inherently beautiful. Beauty shone when everything was in harmony and balance; in other words, when the cosmos was 'adorned'. This same order prevailed in Egypt, encapsulated in the scenes of the pharaoh offering Maat to Hathor and Isis in the Treasury *(fig.110)*, and quite possibly contact with Greek culture during this period had influenced the conception of Dendara temple.[36]

Manifestly, Hathor's priests shared the same imaginative vision, since the end chamber's placement above the Treasury creates a vertical resonance between Osiris's starry resurrection and the adorned 'image' world below. And went beyond it, for between each scene of Isis offering to Osiris's *Ba* in the end chamber's wall frieze, they inserted three stylized *kheker*-symbols, each topped by a sun disk, a decorative motif much used in friezes in Egyptian art and architecture *(fig.111)*, but also hieroglyphically in the word for 'ornament', 'adornment'.[37] Indeed, it is for Hathor that Thoth raises a tiny figure of Maat at the entrance to this chamber *(see page 120)*, for hers is the divine loveliness adorning the world, hers the vitality glowing in all the earth substances Ptah uses to create living forms, symbolized in the ubiquitous scenes showing her offering her 'adornments' *(khekeru)* to the king *(fig.26)*, clothing him with brightness.

Allied with Hathor, Maat's cosmic way of the world, the solar circuit she guides, becomes a truth which is beauty and a beauty which is truth, and whilst at first sight it might seem strange that an

Osirian festival should be celebrated in the temple of Egypt's love goddess, who else but Hathor could beautify the verdant life sprouting forth from Osiris's unseen realm? Who else could paint Egypt with such radiant colours in the season of 'Coming Forth'? What the Greeks expressed through their etymology and philosophy, here at Dendara was ritually built in stone.

Inspired by the synchronicity between the southern solstice and the Khoiak Festival at the end of the first century, Hathor's priests gave a unique expression to Osiris's resurrection during this festival, drawing on an ancient wisdom of cosmic 'reversal' which Plato also knew (see chapter 5). Ascending and descending with the sun through the hours of day and night, his 'ruined' state has been utterly 'rebuilt' into an incorruptible celestial body, a resurrected body, for when time reverses between the sixth and seventh months, when a perpetual movement of 'descent' and 'ascent' sustains the cosmic circuit, so his 'earth-born' resurrected powers rise up from his hidden realm, causing young shoots to spring forth from Egypt's fertile black land, transforming it into a verdant sea of green. He is the 'green' Osiris, eternally renewing everything in the season of 'Coming Forth', whose ancient Khoiak Festival truly has become a solstice celebration.[38]

ABOVE LEFT *fig.110* Paired reliefs showing the pharaoh (the cartouches are empty) offering Maat to Hathor and Isis in the Treasury, which is located directly beneath the western inner Osiris room on the roof terrace. Here 'above' and 'below' are connected spatially to convey the creation of an 'adorned' cosmos when the 'sky is raised' and Osiris resurrected at the beginning of the seventh month. (Temple of Hathor, Dendara.)

RIGHT *fig.111* View of Queen Nefertari's burial chamber ornamented with a *kheker*-frieze. The deities on the columns include Hathor in the foreground and Osiris. (Tomb of Nefertari, Valley of the Queens, Western Thebes. 19th Dynasty.)

ELEVEN

HATHOR'S TURNING:
MOVING LIGHT

According to Plato in the *Timaeus*, a dialogue which includes conversations between the Greek statesman Solon and Egyptian priests, reconnecting with the harmonious circuits of the cosmos has a healing effect:

> The motions in us that are akin to the divine are the thoughts and revolutions of the universe; these, therefore, we should follow, and correcting those circuits in the head that were damaged at birth, by learning to know the harmonies and revolutions of the world, we should bring the intelligent part, according to its pristine nature, into the likeness of that which intelligence discerns, and thereby win the fulfilment of the best life set us by the gods, both for this present time and for the time to come.[1]

To Plato, health depends on living in harmony with the 'revolutions of the universe', since to do so is to remember the soul's immortality and divinity, an understanding which, he says, is lost when a soul incarnates on earth, or, as he puts it, when damage is done at birth to the 'circuits in the head'.

No ancient Egyptian would have subscribed to birth as 'damaging' in the sense that Plato meant it,

though Horus's birth was certainly fraught with difficulty, beset by the marauding presence of Seth, who was always seeking to usurp the throne and destroy Osiris's bond with Egypt. The Egyptians did, though, believe the head to be celestial in origin. In their view, the sun and moon shone as Eyes, and heads were like stars, to the extent that a person's starry rebirth could be symbolized by a head rising in the horizon between a fish-like mummy on each side *(fig.113)*. Indeed, in the western end chamber on the rooftop at Dendara, Horus tells Osiris he has come to 'attach your head to your neck and your heart is not tired'.[2] Neither birth nor the process of embodiment rupture Osiris's connection with his cosmic divine nature; this comes through his brutal death at the hands of Seth and his dismemberment, including his decapitation. Yet, as in the *Timaeus*, the remedy is the same, namely to heal his head, where true equilibrium resides, so restoring him once again to the 'revolutions of the universe'.

This cure is effected in the Khoiak Festival, and closely involves Sokar, the 'golden remedy [*pekheret*] in the temples' as he is known to the

LEFT fig.112 Grain sprouting from the celestial body of Osiris, who lies on a starry sky-sign as rays from a great disk shine down upon him. Above are the human-headed *Aker*-lions, and above them is a scene (not included here) of 'raising the sky'. (Detail from the interior of the coffin lid of Nespawershefi, Fitzwilliam Museum, Cambridge, E.1.1822. Third Intermediate Period.)

RIGHT fig.113 Bakenmut praises the astral rebirth of life in the eastern horizon. A human head emerges between a fish-like mummy on each side, illuminated by light raying down from a disk enclosing Khepri, the new sun, and the ram-headed Atum, the ageing sun. Behind Bakenmut are stars and heads. (Detail from the papyrus of Bakenmut, Egyptian Museum, Cairo, S.R.VII.10231. Late 21st Dynasty.)

LEFT *fig.114* View looking towards the New Year kiosk from the entrance to the western Osiris shrines on the roof terrace at Dendara.

(see page 119). For the remedy he needs to heal his suffering is his reintegration into the solar circuit, into the cosmic 'revolution' at the beginning of the seventh month.

It is healing, too, which the dead lying 'bound and afflicted in Hades' need in the *Dialogue of the Philosophers and Cleopatra*. Earlier in the dialogue Cleopatra has told the philosophers how plants and stones need to be tested by fire in order to reveal their beautiful springtime colours *(see page 113)*. Now she describes how the soul calls to the light-filled body, 'Awake from Hades! Rise up from the tomb … the *pharmakon* of life has entered into you', just as a scene in the western end chamber shows Horus calling Osiris to 'awake' when he brings life to him *(fig.89)*. Like Osiris in the Khoiak rites, the dead in Hades revive through the healing water of salvation flowing in the dark realm. They rise forth clothed in 'various glorious colours like the springtime flowers, and the spring itself rejoices and is glad at the beauty which surrounds them.'[4]

Cleopatra knows the dead in Hades lie 'fettered and afflicted', and, indeed, imprisonment in the earth is an age-old danger familiar also to the ancient Egyptians, an ever-present threat for those seeking heavenly transformation and rebirth. For example, Utterance 254 of the *Pyramid Texts* tells how the Osirian king sinks deep into the earth and sees Re imprisoned there, then released from his fetters by means of an amulet enclosed in red cloth. 'You see Re in his fetters', he is told, 'you praise Re when he is released from fetters by means of the Great Amulet in its red cloth, and the Lord of Peace will give you his hand.' Immediately, the text goes on to declare that the king 'has attached his head to his neck', which culminates in his virile epiphany as 'Bull of the sky', too powerful to be restrained by his enemies.[5] By attaching his head, he restores himself to the solar circuit and attains his heavenly throne 'in the presence of Seth', healed by the magical power of 'the Great Amulet in its red cloth', here a true '*pharmakon* of life'.

Egyptians.[3] Interestingly, the word *pekheret*, which may be the origin of the Greek word *pharmakon*, meaning 'remedy' or 'charm', derives from the triliteral *phr* root, from which also come words meaning 'to embrace' and 'travel round the solar circuit' *(see page 92)*, as well as 'circumambulate' the walls of a sacred precinct, something Osiris certainly undertakes in the Khoiak Festival

OSIRIS IN MOTION: SISTRUM MUSIC

But it is not just male deities who bring healing in the Khoiak Festival. Hathor-Sekhmet is praised as 'Lady of the red cloth' in ancient Egypt, and here at Dendara, Osiris, the 'renewed bull', is said to travel in the solar day and night boats with the 'Eye of Re, protecting Osiris in this place, the female sun shining in the horizon every day'.[6] Her rage and radiance, her 'sympathy' and 'antipathy', flow through the whole of these Khoiak 'greening' rites, turning the light around throughout Osiris's journey.

Interestingly, on the very day when Osiris completes his ritual circumambulation of the temple, 26 Khoiak, and is fully integrated into the solar circuit (see page 119), according to Dendara's festival calendar, Hathor is taken in procession to the rooftop terrace for her union with the heavenly Aten, and hence passes close to the western Khoiak chambers at the top of the western staircase (fig.114).[7]

Reinforcing her presence in this 'rebuilding' of Osiris, moreover, are the six sistrum columns in the hypostyle hall in the main temple, which symbolically support the six Khoiak rooftop chambers. For these columns are architectural renderings of Hathor's sacred musical instrument, played by daughter goddesses to create cosmic harmony and propitiate the anger of irascible solar deities, including Hathor herself, turning them from 'red' to 'green', from destruction to life. Writing in the first century CE, at a time when the Dendara temple was still functioning, Plutarch tells how the sistrum is used to repel Typhon and restore fettered nature to movement:

> The sistrum also indicates that the things which exist should be shaken and should never stop moving... For they say that with the sistrum they repel and ward off Typhon, meaning that when decay confines and restricts nature, the power of creation sets her free and restores her by means of movement.[8]

In the Crypt South 1, four sistra wonderfully evoke a cosmos in motion (fig.81). No wonder then that Osiris is praised as Ihy, Hathor's sistrum-shaking child, in the western end chamber on the roof terrace (see page 122), or that he is taken to the 'Temple of the Sistrum' immediately before burial at the close of the Khoiak Festival (see page 122), honouring the instrument that Plutarch says restores everything to movement. Though standstill might threaten at the perilous time when the year reverses, this is no static world in which he lives. Rather it is a solar circuit moved by the rhythmic music of a radiant goddess who is 'mistress in the boat of millions, with the beautiful face and sweet lips'[9] and whose sistrum-shaking power weaves a circling web of love around these six rooftop chambers, magically supported by the six sistrum columns in the hall beneath.

According to Hermes in Book 12 of the *Corpus Hermeticum*, movement is the 'energy of life', and this Greek incarnation of the Egyptian god Thoth also tells his pupil Tat that in the divinized cosmos there is nothing 'that does not live', since how can 'any part of the incorruptible be corrupted or anything of god be destroyed?' Then, drawing on a craft language rich in allusions to the metallurgical art of smelting and dissolving metals, Hermes reiterates that everything 'divine' is constantly transforming, dissolving and reforming in order to become new:

> Dissolution is not death but the dissolution of an alloy. They are dissolved, not to be destroyed, but to become new. And what is the energy of life? Is it not motion? In the cosmos, then, what is motionless? Nothing, my child.[10]

Nothing is static, nothing motionless, in this divinized cosmos that is a 'plenitude of life'. Nothing dies in the 'recurrence of eternity', since everything is alive through recurrent movement. So often it is doubted the Hermetic Greek treatises contain genuine Egyptian knowledge, yet this very same wisdom is enshrined right here in the Dendara temple. The expression may be mythical, not 'philosophical', rich in imagery and symbolism, but this Hermetic teaching is perfectly in tune with the Khoiak mysteries, in which Thoth, the Egyptian Hermes, plays a prominent role. For Osiris lives as a god in an eternal cosmos, a cosmos in motion, a divinized world where there is nothing 'that does not live', where forms are forever dissolving and

reforming to reappear anew in the greening season of 'Coming Forth'.

And even the smallest object can convey this Hathorian musical power, as seen, for example, in the decoration of a tiny blue-glazed rectangular vessel dating from the New Kingdom and now in the Louvre Museum.[11] On its longer sides, each measuring no more than seven centimetres, are scenes of desert animals and a man and calf sailing in a skiff, with another man shown behind them holding a rope attached to a net within which five birds are trapped, symbolic of the destruction of cosmic enemies (here embodied by the marsh fowl).[12]

Sekhmet's death-dealing presence pervades this landscape, complementing the Hathorian scenes that are incised, again with consummate skill, on each end of the vessel. Thus, depicted on one end are the seven Hathors *(fig.115)*, who are well known for their role in predicting a newborn baby's fate and length of life and are here processing towards a woman who is rhythmically shaking her sistrum before them, their leader holding a papyrus sceptre to show that she comes 'propitiated' and 'green'.

Across on the other end, juxtaposed with these seven Hathors, are two fecundity figures positioned beneath a lotus-flower necklace and enclosing a large fish, a *Tilapia Nilotica*, which is closely associated with Hathor as goddess of rebirth *(fig.116)*.[13] Stars, too, begin life as fish, seven of them at any one time gestating in the sky goddess's placental waters, swimming in her great lake of renewal *(see chapter 12)*, and here this imagery, intimately linking fish, birth and Hathor, transforms this beautiful object into a uterine vessel where life is continually renewed, safely protected from enemies by the loving-destructive Eye goddess. No wonder the woman shakes her sistrum before the seven Hathors.

And the same music is still playing in the Dendara temple, still sustaining the regeneration of life, ceaselessly energizing the movements of the stars, moon and sun, the music of the beautiful goddess with her four faces, goddess of eternal delight and queen of the solar circuit in all her raging, greening glory.

ABOVE LEFT fig.115 Scene showing seven Hathors processing towards a female figure shaking a sistrum. The figures are incised on one side of a rectangular calcite box. (Louvre Museum, Paris, E25298. 18th Dynasty.)

ABOVE RIGHT fig.116 Scene on another side of the box showing a fish between two fecundity figures.

TWELVE

ASTROLOGER PRIESTS: OSIRIS'S ZODIAC

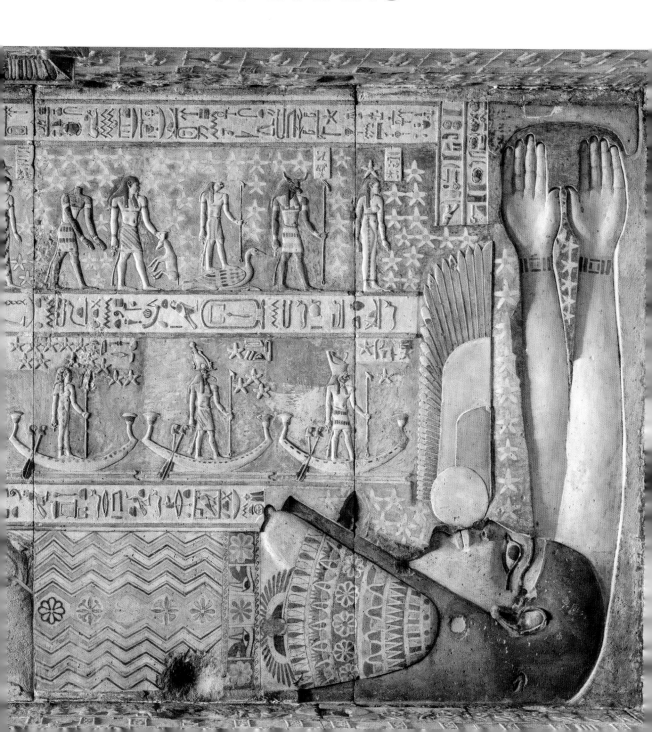

No journey through the Khoiak chambers would be complete without returning to the famous round zodiac carved on the ceiling of the central eastern chamber *(fig.118).* (The one there now is a replica; the original is in the Louvre Museum.) Indeed, the question arises as to why this spectacular sky map, which combines the 12 zodiacal signs with the decanal stars and other Egyptian imagery, was located in this particular room, and why juxtaposed with the solar boats marking the 12 daytime hours across on the other side of the ceiling, thus bringing together the cycles of day and year?

What is striking is that this zodiac has never been related to the Khoiak Festival, its interest for

scholars deriving rather from the astronomical, or astrological, data that might be gleaned from its complex imagery. Yet it provides a perfect counterpart to the ritual process in the chambers beneath—that is, if it is viewed 'astrologically', not as an 'astronomical' map marking the actual constellations in the sky in the middle of the first century BCE. The very fact that it is called the 'sky of gold' in the surrounding inscriptions hints at a symbolic vision of the 'immortal' or 'divine' heaven, not observational astronomy.[1]

Due to the establishment of the Ptolemaic dynasty in Alexandria, which saw an incredibly rich exchange of religious, scientific and philosophical knowledge with the Hellenistic world, including

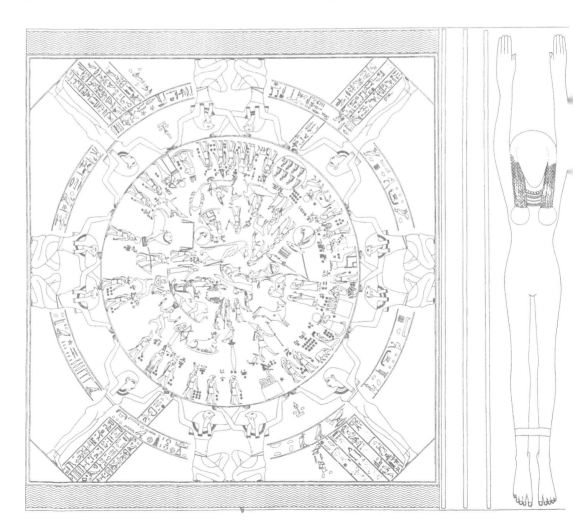

Babylonian and Persian astronomy and astrology, Egypt became renowned as the home of astrology.[2] According to Diodorus of Sicily, writing in the first century BCE, the Egyptians observed the influence of the planets on all living things. He even claimed the 'Chaldeans of Babylon' owed their fame as astrologers to the knowledge they learned from the Egyptian priests.[3] Be that as it may, the spread of astrology in Egypt is confirmed by a group of early first-century CE Demotic horoscopes from Thebes. One of these provides the earliest known complete list of the astrological 'places' (called 'houses' in modern astrology).[4] Unknown in Babylonian astrology, these 12 'places', based on the 24-hour solar cycle, divide a person's lifetime into spheres

of activity, enabling astrologers to evaluate the influence of the sun, the moon and the planets on these different areas. Wherever this system of the 'places' first developed (and Egypt cannot be ruled out), its introduction into Theban astrology provides an interesting backdrop to Osiris's lifetime in the Khoiak Festival, lived beneath a zodiac and harmonized with the 'hours'.

EXALTATION AND FALL: THRONES AND PRISONS

In their authoritative corpus of Egyptian zodiacs, Otto Neugebauer and Richard Parker recognized the astrological significance of Dendara's round zodiac.[5] They pointed out that the five planets known to the Egyptians (Mars, Jupiter, Saturn, Mercury and Venus) were personified as gods and placed close to the zodiacal signs that defined their 'exaltations', or *hypsomata* in Greek, which is associated with the Greek word *hypsoō*, meaning to 'raise something up' or 'exalt', and closely related to *hypsos*, meaning 'height'. Accordingly, in both ancient and modern astrology, when planets are in their particular zodiacal sign of 'exaltation', they are considered to be especially powerful, their strength augmented. They also have corresponding zodiacal locations denoting their 'depression' (*tapeinoma*), or 'fall', as it is known in modern astrology, a kind of descent into Hades, where they

PREVIOUS SPREAD *fig.117* The sky goddess Nut prepares to swallow the sun at sunset. Included in the group moving towards her arms is a 'headless' deity in the upper register. The boat beneath him belongs to the 'sheep' decans.(Detail from the rectangular zodiac on the east side of the ceiling in the pronaos of the temple of Hathor, Dendara. First century CE.)

LEFT *fig.118* Line-drawing of Dendara's round zodiac, here juxtaposed with the solar boats for the 12 transformations of Re during the daytime hours, the first two boats being allocated to Hathor and Khepri as the deities of the first hour. In the centre, dividing the day and year, is the outstretched body of the sky goddess Nut. This boat iconography emerged in the Ptolemaic period as a version of the *Ritual* of *Hours* and also appears in the temple of Edfu. (Ceiling of the central Osiris chamber, east side of the roof terrace, temple of Hathor, Dendara. Late first century BCE.)

are in their weakest, or lowest, position, before periodically returning again in the opposite direction towards 'exaltation'. These two extremes of 'exaltation' and 'fall' enable astrologers to evaluate a planet's influence in a person's birth chart depending on its location. They also define the characteristic pattern of planetary movement, for, as Hermes says in the *Asclepius*, 'In heaven time runs by the return of the coursing stars to the same places in chronological cycles … order and time cause the renewal of everything in the world through alternation.'[6] What led to the development of this important astrological doctrine is unclear, however. It is not even known when and where it first happened, though it possibly derives from Babylonian planetary lore.[7]

Be that as it may, Dendara's priests evidently knew the doctrine, and, intriguingly, in an astrological Greek papyrus from Egypt, probably from the second century CE, these two alternating extremes are called 'thrones' and 'prisons':

> Their thrones are the signs upon which they are exalted and have royal power, and prisons wherein they are depressed and oppose their own powers.[8]

The text then continues by listing planetary 'exaltations' and 'depressions' in accordance with Hellenistic astrology.

In the *Pyramid Texts*, 'imprisonment' in the depths of the earth is a very real threat faced both by the sun god and the Egyptian king seeking regeneration *(see page 128)*, and judging by this unique 'thrones' and 'prisons' terminology, it migrated into Egyptian astrological lore. Certainly, such terminology is applicable to Osiris's regeneration in the Khoiak Festival, for it is precisely in the 'zodiac' chamber that he is extolled in the noonday 'vertical' hour, the time when the sun is at its greatest height, and when his body is reassembled to create a unified Egypt *(see chapter 9)*. Not only does the Egyptian word for this noonday hour (*'ḥ'yt*) resonate with the rebuilding (*s'ḥ*) work Osiris is called to undertake in the eastern court *(see page 107)*, it also resonates with the word *s'ḥ*, meaning 'to be installed on a throne', reinforcing the room's sense of 'enthronement' and Osiris's status within it.[9]

Conversely, across in the western chambers Osiris experiences a potentially 'imprisoning' descent into Geb's earth realm, travelling to the nadir of the night and year, when darkness seems to triumph before miraculously turning once again towards renewed strength and light. And it is surely no coincidence that Jupiter is the particular planet named in the western court above the scene of 'Filling the Eye' at the end of the sixth month *(fig.102)*, since this Egyptian sixth month corresponds also with Capricorn, the most southerly zodiacal sign, which is not only traditionally associated with the southern solstice, but is also the sign in which Jupiter's astrological 'depression' or 'fall' occurs. Clearly, the intent is to synchronize Jupiter's planetary 'fall' with Osiris's descent into the *Dwat*, especially with that critical 'turning-point' of the Egyptian year when everything reverses direction. In fact, of all the five planets depicted in the round zodiac, it is Jupiter's alternating movements between Cancer and Capricorn—the zodiacal signs of his 'exaltation' and 'fall', and also the signs of the northern and southern solstices—which most closely correspond with the pattern of Osiris's death and regeneration in these Khoiak rites.

But this is not simply about astrological 'fall'. For in the western central chamber, Osiris, and all the starry beings resurrected with him in the *Dwat*, now turn in the reverse direction, towards heavenly ascent with the sun, a recurrence of time perfectly symbolized by Geb's somersaulting pose on the chamber's ceiling *(fig.104)*.

Explicitly counting up to the sixth daytime 'vertical' hour in the zodiac chamber across on the east side *(see page 107)*, and symbolically counting down to the last day of the sixth month across on the west side, Dendara's priests marked the two points of extreme 'height' and 'depth' in the diurnal and annual cycles, showing that what concerned them was, in the words of Hermes in the *Asclepius*, 'the renewal of the world through alternation', or, put in astrological terms, 'exaltation' and 'fall'. Furthermore, by integrating Osiris's death and resurrection with this doctrine, they created a contemporary astrological framework for the Khoiak rites.

ABOVE *fig.119* Detail from *fig.118* showing the imagery near Capricorn and Aquarius, the time of year associated with astral 'dissolution' and rebirth in the round zodiac.

This all-important alternating pattern also explains why the eastern chambers stop in the daytime ninth hour rather than continuing through to sunset and the 12th hour. True Egyptians that they were, Dendara's priests knew that if Osiris was to be truly 'exalted', he needed his beating heart—in other words, his life in the 'heights' had to encompass the sixth to the ninth hours, for it is the ninth hour that truly marks Egyptian 'exaltation' or 'thrones'. It is in the central *Ka*-chamber at Abu Simbel that Nefertari is crowned and 'exalted' after attaining the zenith *(fig.41)*; likewise, Hathor is 'exalted' in the ninth-hour 'Pure House' in the Crypt South 1 *(see page 95)*. So, too, *Book of the Dead* vignettes specifically show the thrones of Osiris and 'Food' in the Field of Reeds, the field that the sun god enters in the ninth hour for his second birth *(see page 27)*. This is the hour, too, when Osiris is divinized as 'Lord of the White Crown' in the eastern end chamber, nourished and birthed by his mother Nut *(see page 111)*. In short, Egyptian 'exaltation' encompasses the sixth to the ninth hours.[10]

CAPRICORN AND AQUARIUS: REGENERATING STARS

It is beyond the scope of this book to study the round zodiac's complex symbolism in every detail, but its different areas are clearly allocated to different spatial regions, West and East being specifically marked by hieroglyphs on the outer rim. Significantly, in the zodiac's Southeast area, Sothis, the star of Isis, is represented as a cow in a boat, located close to Leo, the zodiacal sign associated with her heliacal 'New Year' rising in mid-July after a disappearance of 70 days *(fig.118)*. Also close by is the planet Jupiter in its sign of 'exaltation' in Cancer, whilst striding triumphantly near Taurus the bull is Osiris-Orion, separated from Sothis by a papyrus stem supporting a falcon crowned with the Red and White Crowns, a perfect image of Horus 'upon his greenness'.

What is particularly striking, too, is the cluster of imagery in the Northwest associated with the signs of Capricorn and Aquarius—the equivalents of the ancient Egyptian sixth and seventh months *(fig.119)*. Here, in the decanal band around the zodiac's periphery, eight kneeling figures, with their arms tightly bound behind their backs, are shown enclosed in a circle. In other words, they are 'imprisoned', a state defining Egyptian astrological 'fall' *(see page 134)*. Above them is the planet Mars, called 'Horus the Red', standing on the goat of Capricorn and thus appearing in his astrological sign of 'exaltation', a combination instantly bringing to mind that critical reversal at the end of the sixth month when the Eye is filled and the Sethian enemies overcome *(see chapter 5)*.

Symbolically reinforcing 'fall' here in the Northwest, moreover, is the strange headless animal between Capricorn and Aquarius, which probably represents the star group called the 'Headless One' in the constellation of Capricorn listed in Classical astronomical sources.[11] In fact, a headless man is shown among the sunset group of astral deities processing towards Nut's mouth in the eastern rectangular zodiac on the ceiling of the pronaos *(fig.117)*. Reminiscent of the 'Headless One' shown centuries before at the close of the *Book of Day (fig.11)*, he is placed further along the register from Aquarius *(fig.71)* and next to a deity

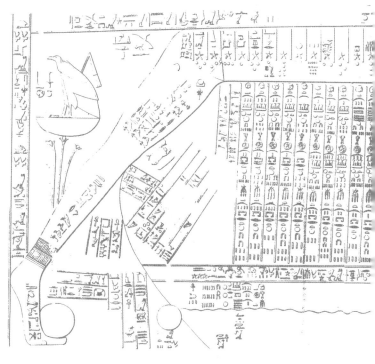

RIGHT fig.120 Shu raises the sky. Inserted above him, between the names of the sheep decans Seret (right) and Sawy-Seret (left) in the decan list inscribed across Nut's body, is a brief text about the rising of Sothis in the first month of *Akhet*. Based on a pattern of dissolution and renewal, of rising and setting, in a year divided after 180 days, the prototype for this cycle is the alternating relationship of Sothis and the sheep decans. Here the air god symbolically raises aloft their complete cycle. This cycle is also tied to the night hours (on the ceiling's east side is the *Book of Night*); and to Re's lifetime, whose rising in the Southeast is described in texts by Nut's legs. (West side of the ceiling in the transverse hall of the Cenotaph of Seti I at Abydos. 19th Dynasty.)

seemingly about to slaughter the animal in his hand. However, Hathor's playful priests, with their extraordinary capacity to think visually, would also have known this headless animal in the round zodiac is the hieroglyphic sign used in the Egyptian word *ḫnw*, meaning 'within'.[12] Hence on one level it conveys both the 'interiority' and 'headless' state endured by all star bodies undergoing regeneration in the *Dwat*, including Osiris *(see page 127)*, but, knowing this headless animal represents the 'skin of a goat', they would doubtless have delighted in including it as a particularly apt Capricornian image, adding to the layers of meaning.

Immediately to the left of the bound captives is a 'duck' *(s3)*, or possibly a goose, Geb's totem animal, which sometimes replaces a duck in words containing *s3*. Here, when combined hieroglyphically with the adjacent ewe *(srt)*, it conveys the name of the decan Sawy-Seret, one of the 'sheep' decans, and the neighbouring decan of Seret 'ewe' in the decanal star lists. Above the circle enclosing the captives in the round zodiac is the traditional figure of Aquarius, holding two vases from which water gushes down onto a fish. Such a water-pouring gesture, so minutely portrayed here in the zodiac,

not only echoes that of the Nile inundation god, Hapy, the Egyptian god traditionally associated with the sign of Aquarius and the seventh month *(fig.107)*.[13] It also parallels the water-pouring performed by the fecundity figures when Osiris himself is purified in Geb's earth realm in the western central chamber, so he is renewed and ready to appear on the horizon as the moon and Orion, fully united with Re and integrated within the solar circuit *(see pages 118–19)*. It also, of course, evokes the annual rebirth of the decanal stars, swimming in a great lake of renewal as fish when they are reborn in the *Dwat (see below)*, which is beautifully encapsulated here by Aquarius pouring water onto a fish.

Interestingly, the *Book of Nut* specifically associates the 'New Year' rising of Sothis with the 'sheep' decans, Seret and Sawy-Seret. This cosmological 'book' is preserved in the Cenotaph of Seti I at Abydos, where it is juxtaposed with the *Book of Night* on the other side of the ceiling, and subsequently reappears in Ramesses IV's tomb at Thebes, again with the *Book of Night*.[14] Its texts were evidently still important enough to be commented on extensively in Roman-period papyri

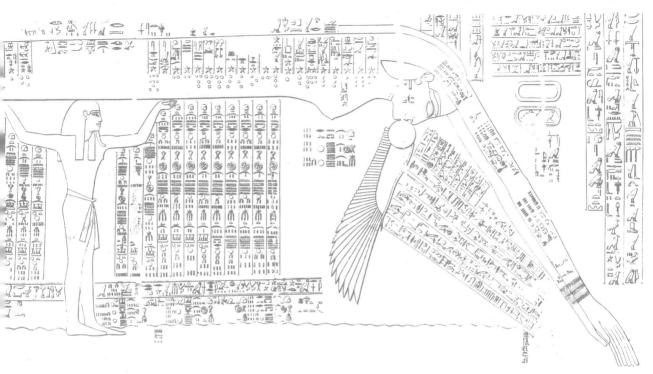

(see below). It shows Shu raising aloft the sky goddess, Nut *(fig.120)*, her arched body covered with a list of the decanal stars and surrounded by texts dealing with their annual motions—living and dying, rising and setting, ascending and descending with the sun—in this heavenly circuit she rules.[15]

'It happens that one star dies and another star lives every decade of days' is said of these 36 star groups that rise successively on the eastern horizon just before the sun, appearing at ten-day intervals and dividing the year into 36 periods of ten days (with the addition of five epagomenal days), inaugurated by Sothis rising again after her 70-day period of invisibility. Like all living beings, these decans have their own lifecycle, 'rising' and 'setting' like the sun, being born, living, dying and then disappearing into the *Dwat* for regeneration, entering Nut's body in the Northwest and coming forth again reborn in the Southeast at dawn. The texts also state that seven decans are in the *Dwat* at any one time, undergoing purification and letting all their impurities fall to earth, then entering a great lake of renewal created by falling tears and swimming around as fish in the life-giving waters before rising again with the sun as living stars born of Nut.[16]

In terms of Dendara's round zodiac, however, it is the terse inscription in the space above Shu's head *(fig.120)* that holds especially priceless information:

What is done in the first month of *Akhet* in accordance with the coming forth of Sothis.

Something is happening here that corresponds to the rising of Isis-Sothis at the beginning of the 'Inundation' season, though the fact that this text is inserted very precisely between Sawy-Seret and Seret suggests these 'sheep' decans are also involved. Not only does the text create a break in the list of decans *(fig.120)*, it also implies there is a division of time between these two particular decans

Nothing more is said, but when this line reappears in a later Demotic commentary on the *Book of Nut*, it is explained that 'all these sheep stars are in the sky in the first month of *Akhet* at the rising of Sothis; that means it is so that the rising of Sothis occurs in the first month of *Akhet*.' Then, after leaving an intentional gap, the commentator goes on to confirm that 'no decrease occurs at the beginning of the year with Re.' Furthermore, Re is 'on the way of the stars which the book [called] "Dissolution" has named, stating:

Sothis—18 stars are behind her and 18 stars in front of her.'[17]

Discursive language was never the style of the ancient Egyptians, steeped as they were in a mythic experience of the cosmos, but here this commentary highlights the symmetrical relationship existing between Sothis and the 'sheep' decans. They are in the sky during Sothis's rising. Then, after confirming no star travelling with Re is lost, this unknown scribe turns to the division of the year after 180 days, marked by 18 stars behind Sothis and 18 in front of her (each of the 36 decans representing a ten-day period).[18] Here this division is also associated with a book called 'Dissolution'. It is 'dissolution' that Aquarius also conveys in the round zodiac at Dendara when pouring water onto a star fish, close to the area inhabited by the 'sheep' decans (fig.119).

Thus, though, on one level, Shu's 'raising of the sky' at Abydos (fig.120) might seem related to Sothis's 'New Year' rising in the first month of *Akhet*, the text's precise location between Sawy-Seret and Seret, together with the Demotic commentary, suggests he is raising a complete year tied to Sothis, the sheep decans and the hours, something Seti I would have wanted to reinstate in the aftermath of Akhenaten's reign (see page 75). In fact, Senenmut specifically highlighted the 'sheep' decans along from Sothis and Orion on his astronomical ceiling (fig.66), juxtaposing them also with the 'boat' decan as if to reinforce the sense of continual, and uninterrupted, movement in the solar circuit. Clearly, they held particular significance for him, presumably because they were the distinctive stars related to the year's division after 180 days, just as Sothis marked the 'new year' at the other end of this alternating cycle. In fact, at one end of the rectangular zodiac on the east side of the ceiling in the pronaos at Dendara, close to Nut's legs, are Sothis and Orion (figs.68,71), whilst at the other end, among the deities approaching Nut's mouth, are Aquarius and a headless figure, with the boat for the 'sheep' decans shown directly beneath him (figs.71,117). It stands to reason there would have been particular decans synchronized with annual 'reversal' after 180 days, and, in light of the prominent 'ram' symbolism during this

period (see chapter 5), the 'sheep' decans would have been particularly in tune with this crucial time associated with astral regeneration.

All of which explains why a 'sheep' decan is placed in the round zodiac's decanal band in the Northwest section (fig.119), diagonally across from Isis-Sothis in the Southeast, close to Capricorn and Aquarius marking 'dissolution' and regeneration, precisely the time when Osiris is in Geb's realm in the Khoiak western chambers. He is not shown, though, in the zodiac's 'fall' region, nor would he be, given the zodiac's location in the 'noonday' chamber, where he is in the 'heights' and consequently depicted as Osiris-Orion close to Isis-Sothis at the time of her midsummer rising. It means, though, that when they are 'exalted', other decans are in 'fall'. 'One star dies and another star lives', says the *Book of Nut*, and here evidently in 'fall', or more precisely being renewed by Aquarian water, preparing to rise in the eastern horizon at dawn, is the complementary star decan.

Here then is an astrological chart for Osiris's regeneration in the Khoiak rites, utterly embedded in the Egyptian experience of the decanal stars and based on the age-old alternating pattern of eternal recurrence. In short, by traversing these celestial heights and depths with the circling starry bodies, he enacts a purifying ritual of death and regeneration, an annual journey of manifestation, invisibility and reintegration, in which sun, moon and decanal stars are all united in one continuous cycle, participating in the life eternal. And somehow, even though they had to compress this rich astral symbolism into the tiny space allocated to each zodiacal sign, Hathor's priests managed to integrate this starry way with astrological 'exaltation' and 'fall'. Tellingly, too, this 'sky of gold' is located in the noonday east chamber, where the sixth hour of the day is inscribed in the east window and the sixth hour of the night in the corresponding west window (pages 107–108), reinforcing the sense of an 'alternating' pattern.

Importantly, too, juxtaposed with the zodiac across on the other side of the ceiling are the solar boats for the 12 hours of the day (fig.118). Thus, just as the great temple at Abu Simbel is dedicated to the sun's annual course tied to the southern

ABOVE *fig.121* Detail of the six hour goddesses towing the solar boat during the sixth hour of the night (only five goddesses are shown here). In the register beneath, an Eye goddess, holding a naos sistrum, heads a procession of deities facing towards a cow on a shrine and moving northwards in the opposite direction to the sun boat above. Here this two-directional movement reinforces Petosiris's power to 'ascend' and 'descend' with the sun in the annual solar cycle. (Wall-painting on the west wall in the first room of Petosiris's tomb, Qaret el-Muzzawaqa. Late first/early second century CE.)

turning-point, and Nefertari's temple to the daytime rhythm of the 12 hours *(see chapter 5)*, here, inscribed on the zodiac chamber's ceiling, are these same complementary rhythms of time associated with the springtime season of 'Coming Forth'. In short, what was implicit at Abu Simbel is now explicitly set out here at Dendara and integrated with astrology.

'Do you not know, Asclepius' asks Hermes, 'that Egypt is an image of heaven or, to be more precise, that everything governed and moved in heaven came down to Egypt and was transferred there? If truth were told, our land is the temple of the whole cosmos.'[19]

In their own inimitable way Hathor's priests say the same. For the alternating movements of the cosmos—the 'exaltation', 'fall' and renewal of the astral bodies—are perfectly mirrored in Osiris's 'rebuilding' in the Khoiak rites. Their temple is indeed a temple of the whole cosmos, an eternal 'heaven on earth'.

HEIGHTS AND DEPTHS: PETOSIRIS'S ZODIACS

During the Roman period this astrological knowledge evidently spread far beyond the Dendara temple, for it recognizably governed the decoration in Petosiris's extraordinary two-chambered tomb in the necropolis at Qaret el-Muzzawaqa in the Dakhla Oasis, dating to the late first or early second century.[20]

Here, on the west wall of the first room, six hour goddesses, led by Maat, are shown towing the night boat of the sun, their number encapsulating the decisive sixth hour of the night, whilst among the companions behind the boat is Thoth standing on a lioness *(figs.121, 122)*. Across on the east wall, to the right of the doorway, Osiris presides over the weighing of Petosiris's heart at the judgement in the afterlife, whilst directly above this scene is the figure of a deity 'raising the sky', now damaged, but enough remains to show his ritual gesture evokes the festival celebrated at the beginning of the seventh month *(fig.124)*.[21] Perched on a *Wadj*-plant behind Osiris, moreover, is a Horus falcon; and the adjacent scene on the south wall shows Thoth and Horus riding on a sphinx and protecting the Moon Eye *(fig.124)*. All this imagery creates a magically-charged setting for Petosiris's regeneration with the sun god at the beginning of the seventh month.

On the other side of the doorway, Petosiris, fashionably dressed in contemporary Roman costume, powerfully turns his face towards the viewer, and towards the sun's nocturnal descent across on the west wall *(fig.123)*, or would have

done had its features not been deliberately excised. 'May you take wing as an ibis, may you alight as the alighting falcon, without your *Ba* encountering any obstacle in the *Dwat*' is the concluding wish inscribed beside him.[22] There are enemies about, they are feared, and what the 'justified' Petosiris desires, above all, is unimpeded power to move freely between the heights and depths, 'ascending' and 'descending' without hindrance when the sun boat reverses between the sixth and seventh months.

And lest there be any doubt as to the relevant time of year, painted on the ceiling above is a round zodiac, encircled by a crocodile and snake *(fig.125)*, and showing a falcon placed in the starry area beyond the zodiacal circle, close to Capricorn and Aquarius, the very same bird in which the shapeshifting Petosiris manifests when he 'descends' into the *Dwat*.[23] Wonderfully expressed here, in a private tomb far away in the Western Desert, is another Egyptian 'descent' and 'renewal' marked by Capricorn and Aquarius, an astrological 'fall' and regeneration.

In fact, the Roman-period astronomical ceiling, which originally adorned the ceiling of the

LEFT *fig.122* Detail of Thoth standing on a lioness in the procession behind the sun boat.

RIGHT *fig.123* An unusual full-face portrayal of Petosiris, dressed in Roman-style costume and holding a scroll. The accompanying text refers to his 'ascent' as an ibis and 'descent' as a falcon. (Wall-painting on the east wall [north section] in the first room of Petosiris's tomb, Qaret el-Muzzawaqa. Late first/early second century CE.)

BELOW *fig.124* Petosiris's 'justification' before Osiris, which is located beneath a damaged scene of a figure 'raising the sky', and placed near a scene of Horus and Thoth protecting the Moon Eye on the south wall. (East wall [south section] in the first room of Petosiris's tomb, Qaret el-Muzzawaqa. Late first/early second century CE.)

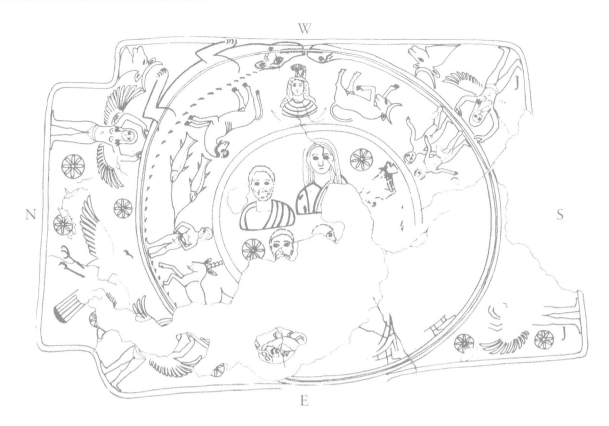

ABOVE LEFT fig.125 Line-drawing of the zodiac representing astrological 'fall' on the ceiling of the first room in Petosiris's tomb. It is encircled by a crocodile and a snake, and a falcon is depicted in the outer rim, close to the signs of Capricorn and Aquarius, the equivalents of the Egyptian sixth and seventh months. As the bird associated with 'descent' *(see fig.123)*, the falcon here marks the time of year associated with Petosiris's regeneration. (Qaret el-Muzzawaqa. Late first/early second century CE.)

ABOVE RIGHT fig.126 Line-drawing of the zodiac expressing astrological 'exaltation' on the ceiling of the second room in Petosiris's tomb. Wall scenes in this room relate to the ninth day hour. (Late first/early second century CE.)

sanctuary in the nearby Deir el-Haggar temple, similarly symbolizes this critical moment, not only including the six goddesses towing the sun boat, but also, as at Dendara, a somersaulting figure of Geb.[24] Importantly, too, even though this astral map is enclosed by the sky goddess, Nut, its registers, despite variants in the imagery, are based on the

same template for annual 'reversal' as the *Book of the Earth* on the right wall of Ramesses VI's sarcophagus chamber *(see chapter 5)*.[25]

Painted on the ceiling of Petosiris's inner chamber, by contrast, is a complementary 'daytime' zodiac *(fig.126)* and whilst all its details cannot be elaborated here, enough needs to be mentioned to indicate its 'exaltation' themes. On its west side is a solar boat with Horus at the helm, adored by four baboons. Across on the east side is a winged *Wedjat*-Eye, a scarab and an anthropomorphic deity with four ram heads, perhaps symbolizing the four *Bas* of Re and his universal procreative power, as celebrated in the *Book of Day*'s first hour *(see page 151)*.

Importantly, the zodiac signs run from Aries to Virgo on the zodiac's south side, and Libra to Pisces on the north side, an arrangement corresponding more generally with their zodiacal division into 'northern' and 'southern' signs as measured from the equinoctial signs of Aries and Libra respectively. Thus, according to the early

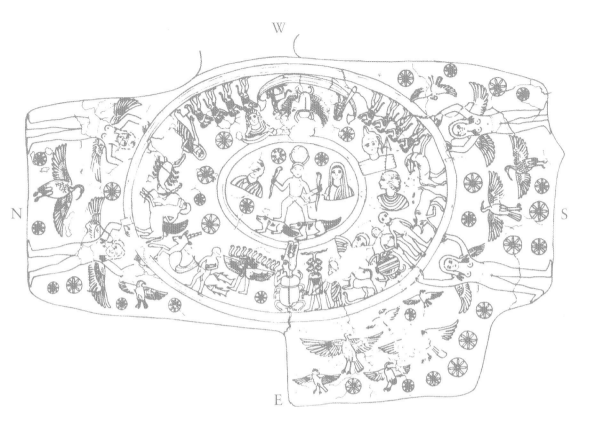

first-century poet Manilius in his Latin *Astronomica*, the six stations of the day begin with the Ram (Aries), the six of the night with the Balance (Libra).[26] Evidently the same principle underlies Petosiris's zodiac, for the arrangement of the signs divides the zodiac into its 'day' and 'night' sides, as indicated also by the stars and moon in the Libra half.

In the zodiac's central circle are a male and a female bust, possibly planetary figures representing Saturn (Kronos) and Venus (Aphrodite), from whom, according to Plutarch, 'all things take their birth' in an alternating cycle of 'sleeping and waking'.[27] They flank a figure of Horus on two crocodiles, his head adorned with a sun and moon as he brandishes a snake in each hand. Like 'Kairos', the young child of Kronos, whom the Greeks identified with the Horus child and called the 'right moment' or 'fullness of time', Horus appears here as the victorious saviour of the world, making all things new, the youthful incarnation of the 'old' god and accomplisher of the full circle of the year in the unceasing cycle of periodic

regeneration. [28] Like the crowned falcon perched on a papyrus stem in Dendara's round zodiac *(see page 135)*, he manifests at the heart of this exalted turning circle of life, uniting sunrise and sunset, North and South, night and day, all held together in perfect balance and equilibrium, as they are in Dendara's 'vertical' sixth-hour chamber, where the goddess Shentayt holds the scales in balance and the Lower and Upper Egyptian nome deities, the northern and southern representatives of the country, bring the relics of Osiris's material body to create Egypt's body-like state anew, protected and unified by Horus. Here, too, the hours are counted up to the sixth hour of the day and the sixth hour of the night, time itself being held in complete equilibrium *(pages 107–108)*.

Nevertheless, true Egyptian 'exaltation' also includes that midheaven ninth-hour 'rising for Isis' *(see page 135)*, and here the wall decoration in Petosiris's inner chamber perfectly evokes the *Book of Day*'s ninth hour, thematically complementing the zodiac on the ceiling. Here, Isis, is shown

143

pouring water for Petosiris in the form of a *Ba*-bird, bringing the flowing eternal waters of life *(fig.127)*. Behind her is a solar scarab placed on a gold sign, with a moon disk in front and a sun disk behind. Here, too, is Sokar's boat, and a rearing cobra wearing the Red Crown, behind whom is a hippopotamus goddess, probably Ipet, who, according to late texts in her temple at Karnak, is the light-bearing mother of Osiris and the decans, the 'Great One who gives birth to the gods', identified with Nut.[29] Three different forms of Osiris, paralleling his three different forms in the *Book of Day*'s ninth hour *(fig.20)*, graphically encapsulate his divinity.[30] Earth's fruitful abundance is beautifully represented by a fecundity figure and a field goddess bearing gifts close to fruit-bearing trees and grain *(fig.127)*. Moreover, a scene of the 'weighing of the heart' adjudicated by Thoth on the south wall, and another showing Petosiris's entry before Osiris on the east wall, confirm Petosiris's status as a 'justified one', which is essential if he is to sail in the sun boat with Re during the ninth hour *(see page 29)*. In his tomb's first chamber, he has been regenerated when the year reverses between the sixth and seventh months, and now he experiences 'exaltation' in the midheaven daytime heights, as Isis-Sothis rises again to inaugurate a 'New Year' and Horus holds the re-created world in perfect balance in the zodiac above.

In short, the Egyptians in the Graeco-Roman era were very capable of integrating astrological 'exaltation' and 'fall' into their Osirian resurrection mysteries, and indeed, judging by Dendara's round zodiac and Petosiris's tomb decoration, were very keen to do so. The roots of this development, however, go back a long way, for the astrological 'thrones' and 'prisons' terminology is detectable already in the Old Kingdom *Pyramid Texts*, in the juxtaposition of the king's 'enthronement' in the 'Field of Reeds' with his 'purification' in the *Dwat*:

> O king, you are the son of a Great One. You are
> purified in the lake of the *Dwat*, you take your
> throne in the Field of Reeds.[31]

Moreover, though during the New Kingdom there was no attempt to make calculable connections between planetary positions and human affairs, as in later astrology, there were, nevertheless, stirrings of that 'astrological' outlook that would come to such full-blown fruition centuries later in the cosmopolitan melting-pot of Hellenistic Egypt.[32]

Central to this astrological reinterpretation are the *Books of the Day and Night*, with their 'wisdom of the hours', which Hatshepsut inscribed in her glorious temple at Deir el-Bahri, with its central sanctuary oriented towards the annual southern 'reversal'. Her devotion to Hathor is everywhere to behold, and whilst she undoubtedly drew on Middle Kingdom traditions, it was she, together with her adviser, the great star-gazer Senenmut, who established new sacred architecture at Thebes, firmly rooted in a geocentric solar theology and attuned to a lifetime following the sun through the course of the day and year.[33] Little could they know that, centuries later, due to the challenge of Hellenism and encounter with Babylonian astrology, this knowledge would reappear, re-expressed in radical new ways.

It could be said that astrology helped to bridge the gulf between ancient Egypt and the wider Hellenized world, for the Egyptian priests were clearly able to meet the changing conditions of their age, both absorbing contemporary ideas and adapting them to ancient cult knowledge. Moreover, this knowledge evidently became available to private individuals for their own personal salvation, and in Petosiris, proudly dressed in his Roman-style costume and holding his papyrus roll like a scribe of Thoth, we see exactly the kind of sophisticated Egyptian who might have transmitted Egyptian wisdom to the wider world, at ease in the multicultural milieu of the Roman Empire yet steeped in ancient tradition—another Petosiris to add to that famous astrological lineage associated with King Nechepso and the priest Petosiris, to whom Greek astrological writers, particularly those inspired by Hermes, attributed the invention of their revelatory art.[34]

Astrology has always gone hand in hand with alchemy, yet though it is well known that the Egyptian decans continued in astrological literature, reappearing in Hermetic treatises and subsequently reaching medieval Europe, as well

as India and East Asia, it has gone unnoticed how Hathor's transformational wisdom similarly travelled far and wide in alchemy, or the 'Hermetic art', as it is known.[35]

Yet one of the chief guardians of seed life in the Khoiak rites is none other than Sokar, the metalworking god of Memphis. And even Hathor herself is earth-born, so it is said at Dendara, being created in the interior of the lotus at the dawn of time from the effluences flowing from the sun god's eye. These fall on the earth to become a beautiful woman called 'She who shines like Gold, the Eye of Re'—the kind of birth Sylvie Cauville had no hesitation in describing as 'alchemical'.[36] Hathor presides over the mining regions, and, indeed, her Treasury is placed directly beneath the Khoiak end chamber on the west side of the roof terrace, filled with the countless gifts she bestows from the depths of the mountains (see page 124).

The solar circuit Hathor rules is steeped in metallurgical lore, and long after her Dendara

ABOVE fig.127 Wall-painting of Isis pouring water into the hands of Petosiris's *Ba*, with two 'field' deities bearing offerings shown beneath them. This imagery relates to Petosiris's second birth in the nourishing Field of Reeds during the daytime ninth hour. (West wall in the second room of Petosiris's tomb, Qaret el-Muzzawaqa. Late first/early second century CE.)

temple closed, her wisdom of metals continued to inspire Hermetic alchemists. Like astrologers in reverse, these alchemists turned their gaze towards the cosmic life of metals growing like seeds, glittering like stars, in the earth. They were also keenly interested in transmutation, how 'material' bodies became 'glorious' spirits. Importantly, they honoured the Egyptian temples as the source of their wisdom. How this came to be means delving into Hathor's ancient metallurgical mysteries, which are closely tied to the years and days, to the rhythms of time sustaining all life on earth.

PART 3

COPPER IN THE CRUCIBLE: METALLURGICAL MYSTERIES

LEFT fig.128 Metalworking scenes showing workers employing foot-bellows to increase the temperature of the furnace and removing the crucible from the furnace using a two-person carrying 'shank'. On the right, they pour the molten metal, probably copper or bronze, into a many-spouted mould, perhaps to make temple doors. (Tomb of Rekhmire, Western Thebes. 18th Dynasty.)

THIRTEEN

HATHOR'S TREASURE:
HEAVEN OF COPPER

To think of ancient Egypt is to think of gold, for it was a land where, according to a letter from the ruler of Mitanni to King Amenhotep III, 'gold was as plentiful as dust'.[1] It is as 'Gold' that Hathor glitters in Amenhotep's first Sed Festival at Thebes, shining in the metal symbolic of divinity, immortality and eternal starry life.[2] However, not only is she praised as 'Gold of the gods' and 'Silver of the goddesses' in her Dendara temple, she is also identified with another metal, called 'true bi3':

> Eye of Re, Uraeus, who radiates light, Gold of the gods, Silver of the goddesses, true *bi3* of the Ennead…[3]

Coming after gold and silver, it might be expected that *bi3* here means 'copper', which was already being worked by the ancient Egyptians in the First Dynasty. Yet it is not so simple, since, whilst it is sometimes translated as 'copper', it is also claimed that *bi3* primarily means 'iron', or 'ferrous metal', and that the word *ḥmty* designates 'copper'.[4] Described as 'one of the most involved problems of Egyptian lexicography',[5] this terminology has become mired in controversy, and if *bi3* were not closely connected with Hathor's role in the solar circuit, it might be best to leave this copper conundrum well alone—though to do so would mean losing the copper thread linking ancient Egypt with alchemy.

COPPER ENIGMAS: LINGUISTIC KNOTS

There is nothing unusual in something being known by two names in ancient Egypt. There are, for example, two distinct words for the heart (*ib, ḥ3ty*), and for the sistrum (*zššt, sḫm*). Malachite, too, has two names, *sšmt* and *w3ḏ*, the latter also designating other green stones, due to its root meaning 'green'.[6] Similarly, in Coptic, copper is called *barōt* and *homent* (or *homt*), a hint in itself that a similar dual naming probably existed in earlier times.[7]

Indeed, that *bi3* can mean 'copper' is evident at the copper-working site of Ayn Soukhna on the west side of the Gulf of Suez, where a Middle Kingdom inscription refers to a large mining expedition which brought back 'turquoise, *bi3* and all the beautiful products of the mountain'. Here in this locality, where actual furnaces for copper production have been discovered, *bi3* must surely mean 'copper', as is rightly recognized in the publication.[8] In contrast, when a similar inscription occurs at Magharah in the southern region of Sinai, the word after 'turquoise' is written with the copper ideogram usually read *ḥmty*.[9] Clearly, the Egyptians could designate copper in different ways.

In fact, Émile Chassinat had no hesitation in associating *bi3* with 'copper', pointing out that during the Old Kingdom, when there was no iron technology, *bi3* must mean 'copper', not 'iron' or 'ferrous metal'. To be sure, Chassinat recognized that the word later acquired 'iron' connotations, possibly, he thought, because both metals were obtained through a smelting process, but he emphasized that throughout pharaonic Egypt, *bi3* meant 'copper' and its alloys.[10] Certainly, in traditional African metallurgy, often no distinction is made between copper and iron, copper being referred to as 'red metal', or even 'red iron';[11] and given that iron ores occur together with copper in Sinai, and iron is also a by-product of copper-smelting, the Egyptians probably also saw a natural affinity between these two metals, hence naming them both *bi3*. Unlike the Near East, however, Egypt was notoriously slow to adopt the use of iron and ferrous metallurgy, iron objects being relatively rare before the middle of the first millennium, though during the New Kingdom some came as gifts from foreign rulers.[12] Copper, on the other hand, was worked from very early times.

LEFT fig.129 Votive stela at Serabit el-Khadim dedicated to Hathor, 'Lady of Turquoise', her typical epithet in this mining region.

ABOVE fig.130 The *bi3*-hieroglyph.

Thus, to explain *bi3* primarily in terms of iron seems singularly incongruous, especially when it follows 'gold' and 'silver' in Hathor's epithets at Dendara. It is difficult to imagine how iron, a metal not noted for either solar or luminous qualities, which, according to Plutarch, the Egyptians called the 'bone of Typhon', and which alchemists associated with Mars, could possibly capture the spirit of such a glittering sun goddess.[13] Copper, in contrast, with its rich colouration and metallic lustre, is full of her dazzling qualities; and, not surprisingly, if *bi3* is interpreted as 'iron' or 'ferrous metal', scholars are confronted with the implausibility of their translations in the context of Hathor.[14] There is much at stake here, though, for not only is *bi3* a Hathor metal, it is also intrinsic to the *Book of Day*, and, of course, copper is at the core of Graeco-Egyptian alchemy *(see Part 4)*. Hence, recognizing *bi3* as an Egyptian name for copper is fundamental to unearthing Hathor's treasure, the all-important metal putting alchemy-seekers on the scent.

TRANSFORMING MALACHITE: THE RED AND GREEN

It was as the green 'Lady of Turquoise' *(Mfk3t)* that Hathor gave her precious copper treasure to Egyptian mining expeditions when they journeyed to the inhospitable desert regions of Sinai, an area

also called *Bi3*, though most mining inscriptions left by these expeditions mention turquoise only. Timing was everything here, since this 'unstable' stone's beautiful colour 'waned' in unfavourable conditions, changeable like Hathor herself, the 'lady of the beautiful colour'.[15] This copper-working area was charged with her sacred power, and dedicated to her at Timna in the Arabah region of Israel was a small Ramessid temple, strategically located in the very centre of the ancient copper-mining area, built beneath the overhanging rock of one of 'King Solomon's Pillars'.[16] So too, in the southwest of Sinai, an area Beno Rothenberg evocatively called a landscape of 'mining and prayer',[17] a Middle Kingdom scene at Magharah shows King Amenemhet III standing before Thoth and Hathor in gratitude 'for the bringing of turquoise and copper [*ḥmty*]' mentioned in the accompanying inscription.[18]

Like countless metalworkers before and after them, the Egyptians must have marvelled at the metallurgical process producing copper, seeing how malachite *(fig.131)* had to be crushed, pulverized and broken down before being heated in a small furnace to produce beautiful flowing red metal completely separated out from its original matrix, a transmutation of matter perfectly encapsulated in the two root meanings of *bi3*: 'separation' and 'treasure', 'miracle', 'precious thing'.[19] Obtaining this precious metal by means of fire must have seemed like a revelation of Hathor herself, since this wondrous malachite substance displayed the very same red and green colours she manifested as the loving-destructive Eye goddess.[20] Not only did malachite eventually produce red copper, it also, when crushed, provided the Egyptians with their striking green eye-paint. Copper oxide ores, such as azurite *(fig.131)*, were also used to create the beautiful blue-green colours of pottery glazes, and the Egyptians must have seen how red copper itself, when exposed to air, eventually developed a thin layer of green rust or patina on the surface, giving the metal a particularly attractive appearance. In every way this metal and its ore resonated with Hathor's oscillating green-red nature.

Hence, whenever Hathor holds out her sacred *menit*-necklace to the Egyptian king, threaded with

ABOVE *fig.131* Malachite, the source of copper for the ancient Egyptians, here with traces of azurite.

its green-coloured beads, and with a counterpoise which actual later objects often show to have been made of bronze, she holds it also as guardian of these sought-after treasures, and, indeed, all the 'marvels' from the mining regions. 'I give you the mountains charged with what is in them, the two mountain chains birth for you their products', she tells the pharaoh at Dendara, and here the word for the mighty 'mountain chains' (mnty) recalls an Egyptian word for 'thighs' (mnty), suggestively linking together 'mountain' and 'goddess' birthing.[21]

Mountains are like great caverns or wombs. They are said to be 'pregnant' with the minerals gestating within them, and, like a pregnant woman, they bring forth their hidden treasure. Late Period mining terminology particularly plays on this sense of a mountain as a great womb, a hidden place of creation and emergence, encapsulated by a word, ḫ3t, designating the network of passages within a mountain, which resonates with the word ḫt, meaning 'womb' or 'matrix'.[22] Both the mine and the womb give birth, both are linked to the human lifecycle within a gestating world, and anyone venturing into the mines must therefore first propitiate their indwelling deities through offerings.[23]

RISING IN COPPER: METALLURGICAL MYSTERIES

Wherever metallurgy is practised in traditional cultures, it is shrouded in secrecy, and ancient Egypt is no exception, though a cryptographic text in Ramesses VI's tomb at Thebes, oriented towards the *Book of Day*'s iconic scene of the 'little Re' gestating in Hathor's womb *(figs.8, 10)*, lifts the veil a little on Egypt's metallurgical lore, giving out secrets rarely disclosed about the sun god's birth.[24]

Ramesses VI seems to have been a king particularly steeped in Hathor's copper mysteries *(see chapter 15)*; indeed, when this text was inscribed earlier in Ramesses III's temple at Medinet Habu (albeit not written cryptographically), the accompanying scene shows not Hathor giving birth to the sun, but rather the king and baboons worshipping the sun child perched on the prow of the solar boat, thus completely masking Hathor's crucial presence.[25]

What makes this cryptic text so special, however, is its allusion to the sun god's birth in a 'copper firmament' *(bi3)*:

> Singing by these gods in the following of Re
> when he shines in the eastern horizon of the sky.
> This is the Lord of the Palace.
> They are the ones who raise Maat for Re.
> Red *Ba*, Green *Ba*,
> *Ba* of Shu, *Ba* of Khepri,
> The four faces on one neck…
> They come into being daily
> when he comes forth in the copper firmament
> [m bi3]
> whose name his mother does not know…[26]

Amazingly, here is the Egyptian sun god rising in a 'copper firmament' *(m bi3)*, a red-green sun god with 'four faces', manifesting as 'Red *Ba*' and 'Green *Ba*', the very colours copper also displays, a god born from the womb like all humans, whose very existence depends on his vitalizing mother, with whom he grows and transforms in the eternal cycle of 'Becoming', graphically symbolized by the winged scarab soaring above mother and child.

He is born in that mysterious dawn time between night and day, the first hour called 'She who raises the Beauty of her Lord', *(see page 19)*, an hour too mysterious for either his name or that of his mother to be spoken. This mystical birth zone is shrouded in secrecy, so requires a text correspondingly veiled. Yet what is at stake here has been completely obscured by the identification of the mother as the sky goddess, Nut, rather than the copper goddess, Hathor *(see page 20)*.

It might be tempting to dismiss Re's 'rising in copper' here simply as a colourful metaphor, without genuine metallurgical meaning, but that would be to miss copper's link with power to grow and transform *(see chapter 14)*, a key theme in this 'rising dawn' first hour.[27]

Should any doubt still linger, however, about the copper connotations of the *Book of Day*'s first hour, it is only necessary to look at a Second Intermediate Period stela now in the British Museum. The stela's owner, a man called Sobekhotep, is shown with his wife before an

offering table and above them is inscribed a unique text which not only explicitly names Hathor and Thoth, but enigmatically refers to 'concealing copper' when Re rises in the eastern horizon:

> Praising Re in the eastern horizon of the sky when copper [bi3] is
> veiled [ḥ3p], and the gods are in the happiness of Hathor and the
> vitality of Thoth.[28]

Nothing more is said, nor is this brief text, with all its echoes of the *Book of Day*'s cryptographic praise of Re, known from anywhere else. The choice of the verb *ḥ3p*, though, is certainly very revealing, since this word, which is written with the 'cloth' determinative, not only means 'to veil', 'to hide' and, as an adjective, 'secret', 'mysterious', but its 'cloth' determinative also suggests some kind of 'clothing' or 'adornment' is happening in this secret horizon world. This realm is 'adorned' because it is light-filled, beautiful and harmonious, a cosmos in which, according to the *Ritual of Hours*, the truth-loving goddess Maat is also 'raised up' in the first hour.[29] But it is 'clothed', too, because Re is stirring into life, set in motion by an Eye goddess who is the instrument of his solar energy and power in the world, a copper goddess who vitalizes him in this mysterious *Akhet*-region, radiating life-filled light as she turns night into day and empowers her gestating sun child to transform and grow. 'Clothed with copper', this cosmos becomes a cosmos in motion, a cosmos in birth *(see below)*, a cosmos needing Thoth, too, on Sobekhotep's stela, whose wisdom ensures the copper Eye goddess radiates peacefully in the rising dawn *(see pages 20–21)*.

Hence it is as an 'adorned' initiate that the copper-loving King Pepi I steps into the solar boat to row Re across the sky to the West in the *Pyramid Texts*. Clothed in his 'leopard-skin', and with his 'sceptre in his hand', he travels through 'the doors of the copper firmament of the starry sky' as the sun god's companion.[30]

Judging by these cryptic texts, Egypt's copper secrets were highly protected knowledge, though they also reflect an age-old metallurgical tradition beautifully captured in the following spell from the

Coffin Texts:

> You are the shining of every day, which clothes
> the copper [wnḫ bi3].
> The perception [si3] of Re who is vital [nfr]
> with the uraei.[31]

'Clothing copper' creates the luminous sun god's 'perception'. He is 'vital' and alive, rejuvenated daily with his cobras, a god who is in his 'egg', as the *Coffin Texts* say, shining in his disk in his 'copper firmament':

> O Re who is in his egg [swḫt.f], who shines
> in his disk …
> who radiates in his horizon, who swims [nbb]
> in his copper firmament.[32]

Here the verb 'swim', describing Re in his 'copper firmament', reverberates with a word meaning 'to melt metal' (*nbi*) or 'cast objects in metals',[33] thus instantly transforming the sun god's 'shining in his disk' into a metallurgical activity—an 'eye of rising sun gazing on the world' as African copper glowing in the crucible has been so evocatively described.[34] No wonder then that a New Kingdom smith is depicted in the tomb of Puimre at Thebes working molten metal in a crucible shaped like a horizon-sign *(fig.133)*, for when it shimmers in this red-hot vessel, it truly appears like a fiery sun, a metallurgical miracle bringing light to the world.

'There is a shining Ptah has forged in his copper firmament', says chapter 64 of the *Book of the Dead*, 'whilst Re laughs.'[35] Wherever metalworking is practised, it is a deeply emotional activity, an experience of the heart arousing wonder, joy and happiness, and here causing even the Egyptian sun god to 'laugh'.

Elsewhere Ptah is said to forge a 'metal' king who is the very image of the sun god on earth. 'Re is in his body, who came forth from Re, whom Ptah-Tatenen created', declares the *Decree of Ptah-Tatenen* for Ramesses II, which is inscribed on a stela in the great temple at Abu Simbel. 'I fashion your body with electrum, your bones with copper [ḥmty], your flesh with meteoric iron [bi3 n pt]', Ptah tells the king.[36]

All the energy of the Memphite divine smith has gone into the 'birthing' of this 'golden-copper-iron' sun king, an image made in the likeness of Re himself, whom Ptah joyfully takes into his arms in a life-giving 'embrace of gold', a king created, too, alongside a palace temple where Hathor holds out her vitalizing *menit*-necklace to him *(fig.26)*, 'raising his beauty' at dawn directly across from Ptah on the opposite wall *(see pages 42–3)*.

LEFT fig.132 A fallen capital from a column in Hathor's temple at Serabit el-Khadim in Sinai where she was worshipped by Egyptian mining expeditions.

BELOW fig.133 Detail of a foundry-worker working with metal in a crucible resembling the sun rising in the horizon. (Tomb of Puimre, Western Thebes. 18th Dynasty.)

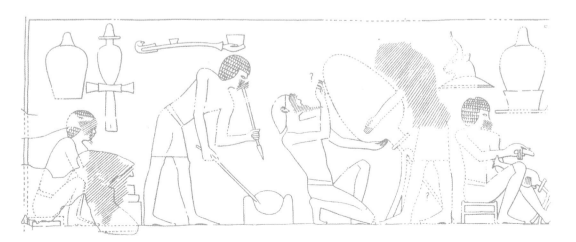

COPPER IN MOTION:
A COSMOS OF LIFE

This copper wisdom lives on in Hathor's temple at Dendara, not least in the Crypt South 1, which enshrines a solar lifetime from dawn to dusk (*see chapter 7*). Venturing into this 'secret palace' (*'ḥ št3*) hidden away in the earth seems more like entering a dark mining shaft than a royal dwelling, and probably intentionally so, since the word *št3* resonates with Late Period words denoting 'mine' (*št3*) and 'womb matrix' (*št3t*).[37]

Particularly instructive are the two dominant reliefs in the crypt's 'House of the *Menit*', the one on the north wall showing a humanized large *menit*-necklace named Hathor, with life-bearing arms extending from its counterpoise (*fig.80*), and, importantly, its material clearly specified as 'black copper' (*ḥmty km*) inlaid with gold'.[38] Across on the south wall, the corresponding scene shows another *menit*-necklace, its beads here joined to four naos sistra, with a solar boat sailing at the base of the two sistra on the right (*fig.81*). Again this group is named Hathor, but here 'copper [*ḥmty*] inlaid with gold' is their material. What this scene tells about Hathor's role in energizing the cosmos has already been discussed (*see pages 91–2*), but these metallurgical details say even more. For once again this is clearly a 'copper and gold' solar circuit, a metal cosmos in motion, filled with desire and vitality, guided by Maat and governed by love.

The specific allocation of 'black copper and gold' to the North (which in terms of Egyptian spatial symbolism is equated with the West) and 'copper and gold' to the South (equated with the East), truly turns this Hathorian room into an alchemical 'life' chamber. For the colour of the 'black copper' *menit*-necklace associates it with 'darkness' and 'night', with Hathor's power to change night into day, death into life, when she enlivens her child in the eastern horizon. Indeed, perched on the *menit*-necklace is Hathor's mercurial son, Ihy, who, according to the *Coffin Texts*, lies motionless as 'an inert one' in Nun during the night. There in the watery abyss he has 'rotted and smelled', his flesh blackening and decaying before he eventually rises out of this stench and foulness to shine forth as fragrant

as Hathor herself, with his mother's *menit*-power residing deep within his body:

> My intestines are the beads of her *menit*,
> which my mother Hathor
> places round her neck…[39]

Like copper itself, which has to undergo 'destruction' and 'death' in the crucible in order to be transformed, Hathor's child must undergo dissolution in the watery abyss before coming forth alive, adorned and renewed, ready to travel through the vast expanse of a copper firmament symbolized across on the south wall by the four copper and gold sistra. For without Hathor's love, without her copper and her adornments, nothing moves, transforms or grows—as Ramesses VI well knew, a king who, in his sarcophagus chamber, showed the loving-destructive goddess giving birth to Re in her copper firmament and vitalizing his journey of life (*figs.8,10*).

Nowhere do the ancient Egyptians explicitly set out their metalworking beliefs—everything has to be gleaned from veiled hints in highly coded imagery, myth and text—but this connection between copper and the sun god's lifetime is perfectly consistent with copper and gold beliefs elsewhere. According to Ana María Falchetti in her study of copper and gold among the indigenous metalworking cultures of Central and South America, these metals belong in 'mythic cosmological schemes' which explain the primordial organization of the world as a 'gestation in the universe' linked to the 'human life-cycle', and she emphasizes how the properties of metals should be seen in the context of life continuity, transformation and regeneration.[40] Her words hold true for the ancient Egyptian *Book of Day*, at the heart of which is a sun god and copper Eye goddess living in a metallurgical 'gestating cosmos', a unitary world of change and development, in which seeds, metals, humans and all earthly life participate, transforming through their various lifecycles.

This world of copper was already known though to the Sixth-Dynasty pharaohs, who laid the foundations for Dendara's copper wisdom, and to whom the next chapter now turns.

FOURTEEN

HORUS'S WINGS:
COPPER'S REMEDY

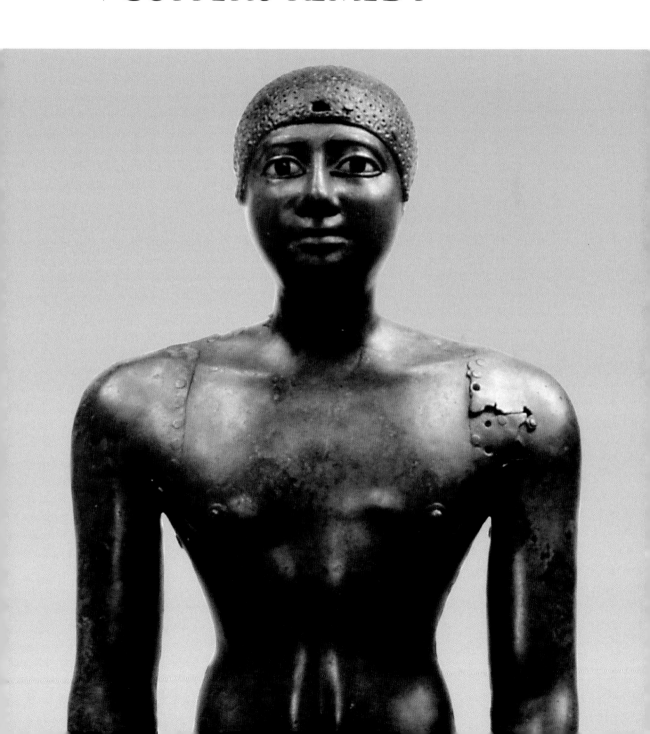

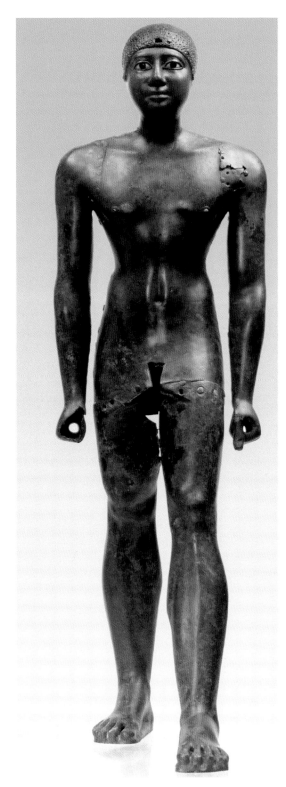

One of Dendara's most revered ancestral kings was the Sixth-Dynasty ruler Pepi I *(see page 81)*, and just how technically skilled Egyptian coppersmiths already were during his reign is perfectly illustrated by two copper statues from the temple complex at Hierakonpolis, the earliest royal metal statues to have survived from ancient Egypt.

Made from several sheets of almost pure copper, hammered into shape and riveted together, both were extremely corroded and lying in pieces when first unearthed, though the larger statue's identity has been confirmed by an inscribed copper plaque originally fastened to its wooden base, which referred to Pepi's Sed Festival. Curiously, the smaller statue, when discovered, was enclosed in the larger statue's torso and displayed a series of holes pierced in its neck *(figs.134,135)*, perhaps for attaching a now-lost Horus falcon, its wings outstretched to protect the king's head, as in other Old Kingdom statuary.[1]

Judging by the *Decree of Coptos*'s reference to a copper statue of the long-lived King Pepi II, said to be made of 'Asiatic copper coloured like gold', copper must have been highly prized by the Sixth-Dynasty rulers.[2] Though it is only due to meticulous restoration work that these copper kings shine anew, attracting and dazzling onlookers by their sheer luminosity, their eyes inlaid with obsidian and the fragments still remaining of the gold leaf that once covered their feet and fingernails.

WINGLESS BIRD: HEAVENLY FLIGHT

What copper meant to these kings can best be gleaned, however, from the remarkable Utterance 669 inscribed in the pyramids of Pepi I and Pepi II at Saqqara, though given the poor state of the walls in Pepi I's pyramid, it is a miracle this strange text survived at all there.

It begins by mentioning a festival held in the five epagomenal days at the year's end, including rebirth in the 'nest of Thoth', but very quickly switches to a conversation between the gods, Isis and Nun, the primordial god of the watery abyss, about Horus's problematic birth. For though Isis has given birth to him, he is apparently limbless.

Consternation reigns, since he remains utterly inert, unable to move freely or become active. Confronted with the mystery of how life holds together, Isis says to Nun:

> I have birthed him for you. I have taken him from the mould and fully ejected him for you, but he has no legs and no arms. So with what can he be tied together?[3]

Just as copper creates dynamic movement in alchemy (see chapter 18), and in the African Dogon myth of creation (see page 200), here in ancient Egypt, the remedy is copper:

> Bring that copper [bi3] for him which is on the prow of the Henu-boat, by means of which he will be tied together…[4]

Copper is the metal charged with 'tying' Horus's body together, the 'glue' empowering him to 'develop' and transform. Evidently, to the Egyptians 'tying', or 'knotting', secures

development (kheper) and growth, expressed, indeed, by Queen Ankhesenamun much later when she ties the cord attached to Tutankhamun's winged scarab pectoral in a scene on the small Golden Shrine (fig.136). And here in Utterance 669, after the arrival of copper, all seems well and good:

> Look he is born, look he is tied together, look he develops [ḫpr].[5]

PREVIOUS SPREAD fig.134. Detail of of Pepi I's statue from fig.135.

LEFT fig.135 Copper statue of Pepi I from Hierakonpolis. When discovered it was enclosed in the larger statue of the king (fig.139) and holes in the neck were perhaps for attaching a protective falcon behind the king's head. (Egyptian Museum, Cairo, JE33035. Sixth Dynasty.)

ABOVE fig.136 'Tying' symbolizes 'Becoming', here expressed by Queen Ankhesenamun's gesture as she knots the cords attached to Tutankhamun's winged scarab pectoral. (Relief on Tutankhamun's small Golden Shrine, Egyptian Museum, Cairo, JE 61481. 18th Dynasty.)

Yet immediately another question is posed, namely how the 'egg' *(swḥt)* is to be broken, a task clearly needing the metallurgical art of 'Sokar of Peju', the archetypal divine smith at Memphis. He it is who, having 'smelted his harpoons and carved his prongs', now comes with his pointed tools 'to break the egg and divide the copper':

> Look the king develops, look the king is tied
> together, look the egg has been broken for him …
> the king flies up and alights on the plumes of his
> father, Geb.[6]

Here, quite clearly, it is copper and Sokar's skills that ensure the Horus king's flight and elevation on Geb's ancestral plumes—his appearance as Geb's heir.

This reliance on copper is etched into Utterance 667A of the *Pyramid Texts*, with its cryptic allusion both to 'piercing' copper and the king's heavenly enthronement:

> O King, you have not died the death, you have come
> to life amongst them, the Imperishable Spirits …
> he has appeared on the lake, on his throne, after
> he has pierced the copper *[bi3]* together with his
> Akh-spirit…[7]

Renowned as the Egyptians were for their adherence to tradition, it is nonetheless astonishing to find Horus's 'copper' transformations beautifully depicted centuries later on the north and south walls of Harsomtus's chamber in the Crypt South 1 at Dendara. On the north wall (the side symbolically associated with night and the West), he appears in the form of an immobile and wingless bird *(fig.137)*, 'the falcon with concealed wings' as he is described in the accompanying text.[8] On the south wall, he is winged *(fig.138)*, his legs and feet are visible, and he is called the 'multi-coloured of feathers on the *serekh*', in other words the falcon ubiquitously shown surmounting the rectangular frame symbolizing the royal palace-façade, the *serekh*, enclosing the Horus name of the ruler in the palace.[9]

But that is not all. Just as the copper materials of the *menit*-necklace and four sistra are meticulously stated in the adjacent House of the *Menit (see page 154)*, so they are here, too,

in Harsomtus's chamber dedicated to the primal snake god *(see page 88)*. For before each falcon is an erect snake rising forth from a lotus flower in a boat. On the north side the snake is said to be made of 'copper' *(bi3)*, the boat and the lotus of 'gold', whilst across on the south side the lotus and snake are 'gold' and the boat is 'copper'.[10] In Utterance 669 Horus needs copper *(bi3)* to ensure his flight, and so does Harsomtus here at Dendara.

Nor do the parallels stop there, since, according to the room's southern frieze inscription, Harsomtus is 'shining in his house on the night of the child in the nest', which harks directly back to 'Thoth's nest' mentioned in Utterance 669 all those centuries before *(see page 156)*.[11] If he is to soar heavenwards, he needs wings, he needs Hathorian copper to vitalize his life, create movement and 'tie' his limbs together. Wherever union is needed, Hathor's copper love flows, and here at Dendara it transforms the immobile child of Isis into a

'living *Ba*, the Lord of Life', praised as 'Khepri on the lotus', a bird able to change, to grow and move freely.[12]

In fact, in the 19th-Dynasty temple of Seti I at Abydos, Isis explicitly confirms Hathor's instrumentality in her child's formation, telling her Horus son (here Ramesses II) that whilst he has come forth from her as an established king, it is Khnum who has moulded him with his hands, Ptah who has welded *(nbi)* his body and the four Hathors who have nurtured his vitality. Their nurturing role is graphically depicted in the accompanying scene, which shows them suckling four figures of Ramesses alongside Isis tenderly carrying him on her arm.[13] Fully alive and empowered by Hathor, he is surely to be understood here as a shining metal statue, quite possibly even a copper statue.

Clearly there was a long tradition of copper-working in Egypt, and even though more than 2,000 years separated Utterance 669 from Harsomtus's chamber, Dendara's priests were still drawing on the same mythic knowledge when they portrayed the ancient copper wisdom in the form of the two beautiful falcons. No wonder then that Pepi I appears in the crypt's westernmost chamber *(fig.85)*, not just as a great temple benefactor, but also an inspirational king steeped in copper wisdom—knowledge of a winged and wingless bird which, centuries later, startlingly reappears in Islamic alchemy and is explicitly connected with

ABOVE LEFT *fig.137* A wingless Horus falcon depicted on the north wall of the 'House of Harsomtus' in the Crypt South 1 in the temple of Hathor at Dendara. (First century BCE.)

ABOVE *fig.138* The complementary winged falcon across on the south wall. (First century BCE.)

the ancient Egyptian temples *(see chapter 16)*.[14] In short, Utterance 669 must rank as a key metallurgical text, the earliest to give voice to this ancient Egyptian copper lore.

COPPER AND EGG: THE GOLDEN FALCON

The whole atmosphere of Utterance 669's 'tying together' episode seems steeped in the Egyptian ritual world of metalworking and its beliefs. It is even perhaps mirrored in the statues from Hierakonpolis made from copper sheets riveted together with nails, in other word 'tied together' *(figs.135,139)*. Later metal statues, too, could be assembled from separate components. Bronze statues of Thutmose III, for example, had separate arms which were attached to square dowels projecting from the shoulders, with the joins then concealed—a method of assembly again involving 'tying together'.[15]

Utterance 669's cryptic reference to Sokar coming with his sharp tools to 'break the egg' for the Horus falcon has all the ring of a metallurgical 'birth' process, particularly since traditional smiths in cultures worldwide liken pouring metal into moulds to conception, the mould replacing the maternal womb, especially in the lost-wax casting process.

It is unclear whether Old Kingdom smiths, like their counterparts in Mesopotamia, used this method for creating larger statues, but it was certainly in use by the New Kingdom, and involved first making a perfect model in wax, then completely encasing it in a fine clay mixture, leaving it to dry and heating it so the wax ran out, leaving a mould into which the metal could then be poured.[16] Like an embryo in the womb or a bird incubating in the egg, the divine image lay hidden within its mould, which then had to be broken open in order to 'birth' the beautiful form hidden within, ready to be polished and finished, cleansed like a newborn babe. Just as in Hindu lost-wax casting, this 'birthing' must have been a highly ritualized process in ancient Egypt, a sacred embryological art surrounded by great secrecy and ritual prescription, which is perhaps glimpsed in this cryptic reference to Sokar 'breaking the egg'.[17]

Be that as it may, in Utterance 669 two key words, copper' *(bi3)* and 'egg' *(swḥt)*, govern Horus's transformation into a winged falcon. And whilst not overtly solarized here, by the time this royal bird reappears at Dendara he has clearly been integrated into a solar lifetime, a shift already detectable in the Middle Kingdom *Coffin Texts*, where non-royal individuals also experience his copper transformations. For not only do the spells there directly mention the 'egg' and 'bringing copper' from the *Henu*-boat to 'tie' Horus's limbs, but the goal of ascension is said to be a celestial throne of 'malachite'.

The blessings of abundant nourishment also await those individuals able to reach the Field of Offerings. 'The bread of the Field of Offerings is eaten, being the food of the Turquoise deity, and the young girls are happy', says Spell 682. The mood is Hathorian, distinctly solar and 'green' in spirit, and, just like the sun god's annual birth 'from the egg' in the *Ritual of Hours*, the joyfully enthroned sky walker travels the heavenly ways of Re and his two Maat goddesses, attuned to 'eternity circulating on this day of achieving the year'.[18]

Indeed, in the *Ritual of Hours* and *Book of Day* this dynamic between 'copper' and 'egg' helps to define the sun god's all-important 'human' and 'divine' aspects, which, from the Middle Kingdom onwards, or even earlier, characterize his twofold nature, as well as the kind of solar circuit he rules. According to the *Myth of the Destruction of Humanity*, Re, ruler of humans and deities alike, is a sun god subject to all the vicissitudes of earthly life, including ageing, and is utterly dependent on his loving-destructive Eye, Hathor-Sekhmet, when humans rise up in rebellion against him;[19] while a myth preserved in the later *Bremner-Rhind Papyrus* intertwines the peopling of the earth and the birth of deities, linking the creation of the firstborn Heliopolitan divine pair, Shu and Tefenet, with the creation of humans from the tears of the Eye.[20] Through his Eye, the creator is directly related to life on earth, and this same human–divine relationship, mediated by the solar Eye, weaves through the *Book of Day*'s solar circuit, highlighted by the juxtaposition of 'copper' (*bi3*) and 'egg' (*swḥt*).

Thus, during the first hour Re rises with Hathor in her copper firmament (*bi3*), gestating in the womb like all humans and participating in a mortal existence closely related to earthly life (*fig.8*). And just as copper empowers Horus's flight in Utterance 669, so in the second hour, according to the *Ritual of Hours*, Re comes forth 'with his wings' (*see page 22*), blessed with the power to move and transform, which is highlighted by the winged scarab flying above his parturient mother. Not only does his copper birth vitalize human life on earth, but he is also born 'from the egg' (*swḥt*) in the divinizing ninth hour (*see page 28*) and then, in the tenth hour, praised as the powerful 'Gold of the Stars' (*see page 30*).

In this goddess-ruled metallurgical circuit, as dawn copper turns to midheaven gold, Re manifests his twofold aspects, the earthly and heavenly, the mortal and divine—a copper–gold combination beautifully symbolized by a falcon statue

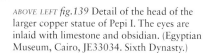

ABOVE LEFT *fig.139* Detail of the head of the larger copper statue of Pepi I. The eyes are inlaid with limestone and obsidian. (Egyptian Museum, Cairo, JE33034. Sixth Dynasty.)

ABOVE *fig.140* Restored copper and gold falcon from the temple of Horus at Hierakonpolis. When first discovered the falcon's copper body was still intact, but exposure to the air caused the copper to disintegrate. It probably dates from the Sixth Dynasty, though the gold crown may be a later addition. (Egyptian Museum, Cairo, CG 52701 and 14717.)

discovered at Hierakonpolis, its head and crown made of gleaming gold and its body all copper (*fig.140*).[21] Certainly, later Graeco-Egyptian alchemists, who repeatedly stated their knowledge came from the Egyptian temples, ascribed 'human' qualities to copper (*see chapter 18*), and there is every reason to suspect the ancient Egyptians did likewise, as reflected here in the sun god's daytime transformations. He is born first in copper, a metal both luminous and corruptible, like all earthly life, a metal also facilitating his heavenly ascent, and then again in gold, the incorruptible metal forming the flesh of deities.[22]

But it is not only Horus-Re who needs Hathor's copper—evidently Osiris requires the same remedy, as witnessed in the *Book of the Earth* depicted on the left wall of Ramesses VI's sarcophagus chamber (*fig.141*). The extraordinary rendering of Osiris's 'earth-born' resurrection on the right wall, when the Moon Eye is 'filled' during the annual 'reversal' between the sixth and seventh months, has already been explored (*see chapter 5*). Yet, according to chapter 140 of the *Book of the Dead*, two Eyes shine during this period of cosmic change, meaning the Sun Eye, Hathor-Sekhmet, is instrumental, too, in this great reversal.[23] After all, Utterance 405 of the *Pyramid Texts* tells how Hathor as the solar Eye 'brings back' the years (*see page 77*)—a role beautifully conveyed much later in Petosiris's tomb at Qaret el-Muzzawaqa.

Thematically all the scenes in Petosiris's tomb's first chamber, which include the sun boat being towed by the six night-hour goddesses, encapsulate solar 'reversal' and regeneration at the end of the sixth month (*see chapter 12*). Directly beneath the sun boat, however, and not mentioned before, there is a procession led by a goddess with the head of a disk enclosing a *Wedjat*-eye, all moving (except a jackal-headed god) in the opposite direction to the sun boat above (*fig.121*). Tellingly, this Eye-goddess

is also placed beneath Thoth riding on a lioness (*fig.122*), evocative of his role to entice the Sun Eye back to Egypt in the *Goddess in the Distance* myth, with all its 'solstice' connotations. She also holds a sistrum towards a cow standing atop a shrine within which a falcon perches on a horizontal mummy.[24] 'May you take wing as an ibis, may you alight as the alighting falcon' is inscribed beside Petosiris across on the east wall (*see page 140*), here portrayed full-face so that his gaze is directed towards the alternating movements opposite. And as the sun boat in the upper register moves southwards, led by Maat towards the exit of the tomb, so the procession beneath, led by the Eye goddess, moves northwards towards the alighting falcon, uninterruptedly moving along this crucial axis to ensure Petosiris's regeneration with the sun.

Clearly, annual 'reversal' needs the Sun Eye, as does Osiris in Ramesses VI's burial chamber. And, just as her copper flows for Horus-Re in the *Book of Day*'s first hour on the chamber's ceiling, it flows, too, when Osiris is regenerated in the *Book of the Earth* on the left wall beneath. This Eye goddess, praised at Edfu as the 'Lady of Life of the Two Lands who causes death', is both the destroyer and restorer of life in this secret earth realm. Hers is the copper life pulsating when the year reverses, the heartbeat of time in these chthonic depths.[25]

Hence these rare scenes require a closer look, not just for their ancient Egyptian copper wisdom, but because here is a foretaste of the copper alchemy that will emerge in Egypt centuries later, a rare expression of the hidden tradition that will subsequently become the Hermetic art. From time to time this Egyptian knowledge does surface briefly, but it seems that Ramesses VI wanted to set out a complete form in his burial chamber, thus making his the most overtly 'alchemical' of all the chambers in the Valley of the Kings and a crucial link in alchemy's chain of transmission through the ages.

FIFTEEN

SOLVE ET COAGULA:
EGYPT'S WAY

What initially attracts a copper-seeker to the left wall in Ramesses VI's sarcophagus chamber is the womb disk surmounted by a female face towards the far right of the central section. She is unnamed, but, as in the *Book of Day*'s iconic dawn scene on the ceiling above *(fig.8)*, peering above this fiery disk is Hathor's unmistakeable full face, here flanked by two uraeus serpents called 'She who devours' and 'Flame' *(fig.142)*. To the right, a ram-headed god holding a sceptre watches intently as Atum and a deity called 'Seizer' grasp snake-like arms extending from the disk, whilst behind him, on the extreme right of the register, are the arms of Naunet enclosing a red sun disk—feminine arms, since this watery womb world requires Nun's female partner, complementing his role across on the lunar right wall *(see page 67)*.[1]

Beneath this group a black scarab is seen bursting forth from a red disk *(fig.142)*. Further along the register, to the left of Hathor, another black scarab does the same, here with his wings fully extended, having evidently been transformed into a winged being, as Re confirms in the accompanying inscription:

> The *Ba* of Re speaks above his disk, he gives orders to those who are with him, concerning the mysteries of the one who is in the sky, when he emerges after having come into being as the 'one belonging to his light', his two arms having become the two offspring of Khepri who is pleased with his two wings … the Great God makes his transformations … then darkness veils them.[2]

Being born here are Khepri's limbs, his 'two arms', and for a brief moment, before 'darkness veils' everything again, the sun god illuminates the scarab's mysterious transformations, which are, paradoxically, also his own. Clearly, like Horus in Utterance 669 *(see chapter 14)*, Khepri needs to 'grow wings' and become luminous, for then he is alive, able to transform, move and fly heavenwards, and, not surprisingly, he is said to be 'pleased with his two wings'.

KHEPRI'S ARMS: HATHOR'S GLUE

Where, though, is the copper 'tying' everything together in this animating process? The answer surely is revealed by Atum and the 'Seizer' grasping the snake arms rising forth from Hathor's red womb disk and holding this volatile goddess fast. For a similar womb disk, with Khepri flying above it, is shown in the *Book of Day*'s first hour on the ceiling above, enclosing Re as a tiny foetal child when he 'comes forth in copper' *(fig.8)*, and in the second hour the sun god rises 'with his wings' *(see page 22)*.

Indeed, the gesture of Atum and the 'Seizer' recalls a metallurgical scene in the 18th-Dynasty tomb of Rekhmire at Thebes, where two metalworkers, using a two-person carrying shank, are shown carrying a crucible filled with molten metal away from the flames *(fig.128)*. Similarly, in an Old Kingdom metallurgical scene, when the metal is ready for pouring, an overseer leaning on his staff instructs the metalworkers to 'seize' the crucible in order to pour the metal into moulds, in all likelihood copper at this early date. His words are quite literally replicated in the name of the 'Seizer' here in Ramesses VI's burial chamber.[3]

In alchemy, and in many traditional metalworking communities, the vessel containing molten metal is regarded as a womb *(see page 160)*, and, as Eugenia Herbert noted about copper smelting in Africa, 'it is very close to parturition'.[4] Judging by this scene of Hathor in Ramesses VI's tomb, the same held true in ancient Egypt, for if

Khepri is to transform, if his 'two arms' are to become wings and he is to be 'tied together', he needs these two gods to 'seize' Hathor's luminous crucible of life and her numinous copper to flow from her womb vessel.[5]

Usually such knowledge was too secret to be given out, and certainly no other version of the *Book of the Earth* includes this graphic revelation of Hathor's copper matrix, though, in their way, the various readings associated with the hieroglyphic sign used to write the word for copper, *bi3*, point to this elusive 'metallurgical' tradition. It is thought to depict a 'water-filled vessel' *(fig.130)*, though it needs but a glance at Gardiner's hieroglyphic sign-list to appreciate the uncertainties surrounding the various readings.[6] It is read *bi3* in the word for 'copper' *(see chapter 13)*, but it also substitutes for the female sexual organ in a word for 'woman', 'wife' (*ḥmt*), albeit with the different phonetic reading *ḥm*, though why this sign has these different phonetic readings of *bi3* and *ḥm* remains unexplained—as does its substitution for the female sexual organ in words for 'womb', 'vulva' and 'cow', seemingly to be read *idt*. In short, clustered around this complex hieroglyphic sign of a 'water-filled vessel' are words, apparently etymologically unrelated, but associated with 'copper', 'womb', 'woman', 'water' and 'cow'.[7]

Interestingly, African metalworking communities similarly connect copper and water. 'Water and copper are of the same essence', writes Herbert, citing a comparison between shimmering copper, 'fish-filled water' and 'the pupils of the eye' in Kotoko mythology.[8] So too, in a Kuba copper myth, the primordial sister, Mweel, discovers copper ornaments in the river Kasai and then disappears into it. Her praise-name links together 'water', 'woman' and 'copper', as in ancient Egypt, with the addition, in her case, of 'tattoos'.[9] A Burundi smith designates molten copper with a word usually meaning 'the coagulated blood of a cow' in order to convey fertility.[10] In light of these African associations, quite possibly the *bi3* hieroglyph's varied meanings reflect the highly coded metallurgical language surrounding ancient Egyptian copper-working, which was never simply a 'technical' operation.

PREVIOUS SPREAD *fig.141* View of the *Book of the Earth* on the left wall of Ramesses VI's sarcophagus chamber. The two figures of Nut arched across the ceiling enclose the *Book of Night* (right) and the *Book of Day* (left). (Valley of the Kings, Western Thebes. 20th Dynasty.)

ABOVE LEFT *fig.142* The unmistakeable face of Hathor, the copper goddess, appears above her womb-vessel, which is flanked by two uraeus serpents. Atum and the 'Seizer' are holding fast her serpent-like arms watched by a ram-headed elder. On the far right are the arms of Naunet (the female counterpart of Nun), and the scene beneath shows a winged black scarab bursting forth from a disk, manifesting power to transform in this secret earth realm. (Details from the *Book of the Earth*'s central section on the left wall of Ramesses VI's sarcophagus chamber. 20th Dynasty.)

FIRE AND WATER: HEART OF FLESH

Certainly, Khepri's animation in the *Book of the Earth* is shrouded in mystery, its copper meaning having to be gleaned from tiny scraps of evidence. It also belongs within a complete Osirian way of transformation, starting in the very centre of the bottom register. Here, in the deepest stratum of the earth, an inert female figure named the 'Corpse of She Who Destroys' lies within a semi-ovoid sarcophagus *(fig.143)*, her name resonating with the corresponding masculine 'Place of Destruction' on the opposite wall, the realm of Nun and the divided *Aker*-lions *(fig.57)*, her body all black save for her gold-coloured breast, a hint of the divine life latent in her heart region.[11] Even though she is the largest single figure in the whole composition, no other version of the *Book of the Earth* shows her apart from this one in Ramesses VI's tomb. Above her are six praising figures in their individual mounds, all coloured black and gold, and all with their legs invisible, sunk in the earth, as if immobilized in this secret realm. On the far right, female figures guard a heart whilst knife-wielding deities attend to two human heads and pieces of flesh, cooking in cauldrons heated by a fire-spitting head *(fig.144)*.[12] Here in the inaccessible recesses of the earth, in the punishing darkness, fire is dissolving mortal bodies back into their separate fleshly parts.

Across in the bottom left-hand corner stands the mummiform Osiris as 'chief of the Westerners', wearing the White Crown and protected by two ram-headed gods who grasp the defeated Apophis snake encircling his shrine *(fig.145)*. Flanking Osiris are also two chthonic deities called 'Corpse of Tatenen' and 'Corpse of Geb', sunk in an earth mound with their feet invisible, clearly also

---

ABOVE fig.143 The goddess called 'She Who Destroys' enclosed in a coffin-shaped receptacle. (Detail from the *Book of the Earth*'s bottom register on the left wall of Ramesses VI's sarcophagus chamber. 20th Dynasty.)

ABOVE RIGHT fig.144 Heads and flesh being boiled in cauldrons heated by a fire-spitting head. Between the cauldrons two goddesses protect a heart, probably Osiris's. (Detail from the *Book of the Earth*'s bottom register on the left wall of Ramesses VI's sarcophagus chamber. 20th Dynasty.)

RIGHT fig.145 Encircled by the fettered Apophis snake, Osiris stands between the 'Corpse of Geb' (right) and the 'Corpse of Tatenen' (left), all three deities being shown immobilized with their feet in the earth. Above them goddesses bind enemies. (Detail from the *Book of the Earth*'s bottom register on the left wall of Ramesses VI's sarcophagus chamber. 20th Dynasty.)

deprived of movement. Indeed, as the sun god says in the text inscribed above, arms need to bend now, shoulders need to function:

> …O you Image, Mysterious of Transformations beneath the feet of the Mysterious One. Bend your arms, raise your shoulders. See I travel above the corpses of the mysterious place, my *Ba* crossing it … as I give birth to myself.[13]

Above the shrine, four goddesses bind the arms of four kneeling enemies, each with a fiery flame burning on his head, while a fifth prisoner is similarly held captive by goddesses. Like so much else in the *Book of the Earth*, their bound state, though, is highly ambivalent. Whilst on one level it conveys their impotence to disrupt this transformational process, it also points to what is needed—limbs to be 'tied together', 'arms to bend', bringing life and movement to this stuck, motionless realm.

And now it becomes clear why Hathor's copper matrix and Naunet's watery arms should appear above 'She Who Destroys' in her sarcophagus, for, if bodies are to stir into motion, if everything is to be reunited, Hathor's liquefying copper is surely needed.

What is also needed is for Khepri to become a moving, winged being, not only to regenerate Osiris, but also the sun god himself, who is deeply involved in this regenerative 'earth' process. Re, too, possesses a mortal body of flesh, and if he is to be renewed, says the *Amduat*, he must go deep into the interior of the earth. 'You live, O flesh in earth, which is sacred for you', he is told in the *Amduat*'s second night hour. His 'flesh' is 'in the earth', the flesh that 'belongs to heaven', hence he must illuminate the deep earth as Khepri:

> Come, Re, you live in your name 'Living One', Khepri, foremost of the *Dwat* … illuminate the darkness so that your flesh lives and renews itself.[14]

Here in Ramesses VI's tomb, where, paradoxically, Re is both the light-bringer and the god whose transformations are being illuminated, the renewal of his 'flesh' is graphically enacted. He is indeed 'giving birth' to himself and simultaneously ensuring Osiris's own regeneration.

Intriguingly, the early fourth-century Egyptian alchemist Zosimus of Panopolis experiences a remarkably similar renewal of 'flesh' in his famous 'visions'. He vividly describes how he falls asleep and sees a phial-shaped altar presided over by a

priest called Ion, who tells him that he has endured 'intolerable violence' at the hands of a sword-wielding man, who drew off the skin of his head and 'mingled the bones with pieces of flesh', burning them in a fire, so that his body would be transformed and become a spirit. Zosimus then sees Ion spewing out all his flesh and his eyes turning to blood.

Alarmed, he wakes, but then falls asleep again and sees this same phial-shaped altar, filled now with bubbling water, in which innumerable people are being burned, but remain alive nevertheless. He is told this terrifying vision is 'the entering, the going forth and the transformation', the 'maceration' that people who want to obtain 'excellence' must undergo in order to become spirits fleeing bodies.

ABOVE LEFT *fig.146* Judgement scene from a Roman-period tomb at Akhmim/Panopolis showing Horus and Anubis supporting the balance; to the right is the female 'Devourer' before a large cauldron containing a tiny skeleton-like figure. In the early fourth century, the alchemist Zosimus of Panopolis vividly records his visions of punishment and regeneration in cauldrons presided over by the priest Ion, who changes from copper to silver to gold.

ABOVE *fig.147* Cocooned in an egg-like oval, and illuminated by the sun, Horus emerges from the body of Osiris, making manifest his father's regenerative *Ba*-power now that he has been set in motion by Hathor's copper. (Detail from the *Book of the Earth* on the left wall of Ramesses VI's sarcophagus chamber. 20th Dynasty.)

He then sees a 'Man of Copper' holding a tablet of lead (the alchemical metal of Osiris). This is the man who had vomited flesh, but he now presides over this bubbling water as 'the one who sacrifices, and the one who is sacrificed'. He too, like Re in ancient Egypt, is the agent of his own rebirth, and later Zosimus is told this 'Man of Copper' will become a 'Man of Silver' and eventually a 'Man of Gold'.[15]

Dreaming before the phial-shaped altar, this influential alchemist from Panopolis experiences how all must first be dissolved before being reunited in the Man of Copper's bubbling water—transformations mirrored centuries before in Ramesses VI's tomb, and hinted at in a Judgement scene in a Roman-period tomb at Panopolis showing a tiny figure in a cauldron beside the scales *(fig.146)*.[16]

COSMIC REVOLUTIONS: THE MYSTERIOUS LADY

What is desired by the people 'burnt but living' in the Man of Copper's vessel is to become 'spirits fleeing the body', and here in the *Book of the Earth* Osiris desires this too, for above the scene of Hathor's copper matrix his black body is shown cocooned in an oval egg-like vessel, illuminated by a sun disk and protected by Isis and Nephthys *(fig.147)*.[17] Everything has been 'tied together' and enlivened in Hathor's copper realm, and now Horus the Behdedite is seen rising forth from Osiris's body, his birth a revelation of his father as a fully-functioning *Ba*, here surely to be understood as alive within his 'egg' and set in motion.[18]

ABOVE fig.148 Osiris stands erect in the 'concealed chamber', on either side of which a deity holds aloft a vessel to catch the blood pouring down from decapitated enemies, who are held upside down by a male figure above. Beneath are Anubis and another god guarding a mysterious box, probably the canopic chest for Osiris's embalmed viscera. (Detail from the *Book of the Earth*'s top register on the left wall of Ramesses VI's sarcophagus chamber. 20th Dynasty.)

In the adjacent scene to the left, two gods raise a tiny human-headed *Ba*-bird wearing the White Crown called 'the *Ba* of Osiris'. He is held aloft on a plinth like a cult statue, a spirit bird now able to flee his body, possessing complete freedom of movement, able, like Horus in Utterance 669, to rise up and alight with Geb.

And it is the 'Corpse of Geb' who, together with Osiris's *Ba*, praises the resurrected god within a snake-encircled shrine shown in the right-hand corner of the top register *(fig.148)*, with the jackal-headed Anubis and a human-headed figure beneath them, protecting a mysterious box, probably the canopic chest containing Osiris's embalmed viscera.

Above the shrine, a head and arms are seen descending from the roof, each hand holding a 'flame' hieroglyph to burn the feet of the inverted headless enemy on either side, whose blood pours down into a cauldron beneath.[19]

'Who is this?' asks Zosimus of Panopolis near the end of his terrifying alchemical 'visions' when he sees a man dressed in a long white robe celebrating the 'dreadful mysteries' at the phial-shaped altar, and he is told: 'The priest of the inaccessible places, he wants to put blood into bodies, give eyes to the one who has not, and resurrect the dead.'[20]

His ancient Egyptian predecessors were far more reticent, carefully veiling their secret mysteries, yet here blood also flows, to resurrect Osiris. Throughout the *Book of the Earth* there is an ambivalent dynamic between destruction and life, standstill and movement, punishment and prosperity, and once again, just as the negative 'binding' of prisoners in the bottom register signals, if reversed, the positive 'tying together' process needed to bring transformation and movement *(see page 168)*, so this blood streaming from enemies, if transposed to Osiris, alludes to his regeneration through the heart. That heart, so carefully guarded in the female realm of 'She Who Destroys' in the bottom register *(fig.144)*, here pulsates with new life.[21]

Osiris has become like the divinized king in Utterance 723 of the *Pyramid Texts*, whose 'flesh has been born to life' amidst the stars now that he possesses the 'copper bones' and 'golden limbs' of a divine body that cannot decay:

> Raise yourself on your copper bones and golden limbs. This body of yours belongs to a god. It cannot putrefy, it cannot be destroyed, it cannot decay… Your flesh has been born to life and you shall live more than the stars in their season of life.[22]

His divine body, celestial in origin, belongs to the stars, as the *Book of the Earth*'s adjacent scene to the left also makes patently clear. Flanked by a fire-spitting cobra on each side, a tiny figure is shown standing upon a great sun disk, his head placed at the apex of a semicircle of 12 sun disks and 12 stars emanating from the hands of two goddesses, Amaunet, the 'Hidden Lady', on the left and Atenet on the right, the female manifestation of the heavenly solar disk, the Aten *(fig.149)*. Above him shines a smaller sun disk, and, according to the accompanying text, he is standing 'atop his horizon so that he might guard the hours following his *Ba*'.[23] Just as the corresponding scene across on the right wall portrays Osiris encircled by 12 disks and stars *(fig.62)*, here this resurrected celestial god travels in the starry ring of the ever-circling hours, in the goddess mysteries known only to those able to take flight from their bodies and become spirit birds. Positioned directly above 'She Who Destroys' at the bottom of the wall, this scene completes the central vertical axis, the mystical ascent to the stars, moving from dissolution to regeneration, destruction to life, and requiring the soul to 'grow wings' in order to live in this immortal, divine realm.[24]

A similar scene once adorned Ramesses III's sarcophagus chamber. Now lost, it was copied by an artist working for Jean-François Champollion, fortunately, since its two disks contain crucial hieroglyphs missing from Ramesses VI's version *(fig.151)*. Within the upper disk is inscribed 'Aten', the name of the 'heavenly' sun, and in the larger lower disk, which is encircled by the double ouroboros-snake, is the name of Ramesses III, 'Ramesses, ruler of Heliopolis', with the 'Ramesses' part of his name written in such a way that it can also be read: 'It is Re who continuously gives birth to him.'[25]

Just as the macrocosmic sun, the Aten, illuminates the heavens, so Ramesses III shines as the regenerated solar king ruling in the ancient sun city of Heliopolis, where the two suns are united at the confluence of heaven and earth. Clearly, the aim is to connect the heavenly sun and the earthly sun, and the importance of this cannot be emphasized enough. For later Graeco-Egyptian alchemy is also founded on a 'two suns' wisdom, which was given

its classic formulation by Zosimus of Panopolis in his *Book of Sophe* and included the 'earthly' sun's association with purified copper *(see chapter 19)*. Had the brilliant decipherer of the hieroglyphs not copied these two suns in Ramesses III's tomb, however, not only would their ancient Egyptian meaning have been lost forever, but also their continuity in later alchemy.[26]

To the left of this beautiful scene stands a goddess called 'Mysterious Lady', with a human-headed serpent rearing up on either side of her, praising the sun disk and *Ba*-bird she holds *(fig.150)*.[27] Between these serpents and her lower limbs are a snake and a crocodile, creatures of 'fire' and 'water' respectively, which much later encircle the zodiac depicted in Petosiris's first chamber in the Dakhla Oasis *(fig.125)*, protecting the solar wheel in a room dedicated to regeneration when the year 'reverses'. Here is a 'time' tradition harking back centuries, re-expressed to be sure in zodiacal terms, but already present in Ramesses VI's tomb.

Furthermore, according to the text inscribed above this 'Mysterious Lady':

> The head of the Mysterious Lady is in the upper *Dwat*, her legs are in the lower *Dwat*, whilst the two *Bas* traverse her body.[28]

Just as the mummiform figures across on the right wall successfully span these two realms during the 'reversal' between the sixth and seventh months associated with 'filling' the Moon Eye *(see page 69)*, this female guardian does the same at this critical juncture of time in the Sun Eye's goddess realm.

And who the 'two *Bas*' traversing her body might be can be gleaned from the frontispiece of the *Litany of Re*, where a snake and a crocodile are depicted above and below a sun disk enclosing Khepri and the ram-headed sun god—the renewed and ageing forms of Re in the complete solar circuit.[29] Across on the right wall, these two *Bas* are with the Moon Eye in an earth realm dominated by male deities *(see page 69)*; here on the left wall, the loving-destructive Eye goddesses are all-powerful.[30] 'The greatness of her flame and heat are in her', says the text beside the Mysterious Lady. Hers is the incandescent serpent energy moving these two *Bas* between the above and below when the year

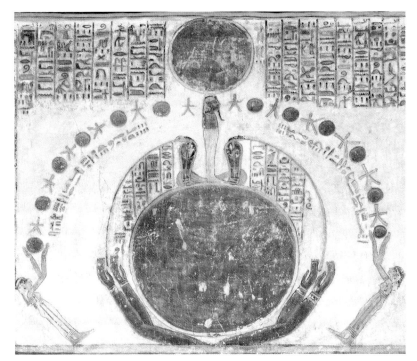

LEFT fig.149 12 disks and 12 stars leave and return to the hands of two goddesses, here guarding the hours and standing on either side of a huge sun disk with another disk shown above them. (Detail from the *Book of the Earth*'s top register on the left wall of Ramesses VI's sarcophagus chamber. 20th Dynasty.)

RIGHT fig.150 Here the 'Mysterious Lady' guards the cycle of eternal cosmic movement and regeneration. Near her legs are a snake and a crocodile, the two creatures also encircling a zodiac in Petosiris's later tomb at Qaret el-Muzzawaqa *(see fig.125)*, which is associated with the sun's annual reversal between the sixth and seventh months. (Detail from the *Book of the Earth*'s top register on the left wall of Ramesses VI's sarcophagus chamber. 19th Dynasty.)

RIGHT fig.151 Two suns depicted in Ramesses III's tomb, as copied by Champollion, but no longer preserved. Inscribed in the lower sun disk is the name of the earthly ruler, 'Ramesses, Ruler of Heliopolis'; in the upper disk is the name of the heavenly sun, 'Aten'. In the early fourth century CE the alchemy taught by Zosimus of Panopolis was similarly founded on 'two suns' and he explicitly honoured the Egyptian temples as the source of alchemical knowledge. (Tomb of Ramesses III, Valley of the Kings, Western Thebes. 20th Dynasty.)

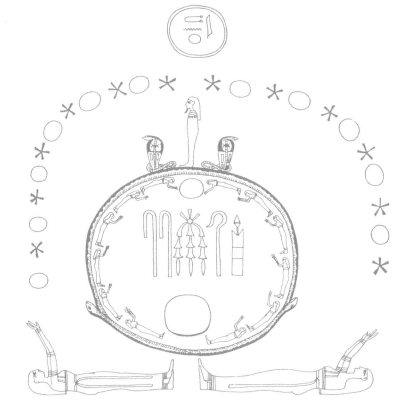

changes direction, fully roused now to subsume
their opposites into the mystery of eternal
'Becoming'. Continually turning the light around
her cosmic orbit, she eternally regenerates the sun
god, moving him through her recurrent cycles of
destruction and renewal, ruled by fire and water,
the source of eternal life in this ceaseless cycle.[31]

In the *Amduat*'s second-hour text, quoted above,
Re is told the 'renewal' of his 'flesh' means 'you live
in your name'—the name that, to the Egyptians,
defines a person's eternal essence and 'identity'
(see page 54).[32] Correspondingly, the *Book of the
Earth*'s closing vignette, to the left of the
Mysterious Lady, mythically converts these words
into a sublime visual symbol *(fig.152)*. Here two
figures of the king are depicted bearing a winged
scarab aloft, its wingtips supported by Isis and
Nephthys, whilst inscribed between the royal
figures are five golden cartouches, those bearers of
the royal mystical 'names', proclaiming the essence
of his divine nature, his power to participate in the
eternal 'Becoming' of the sun. Above the group, as
on the opposite wall, is the solar boat supported by
the *Aker*-lions, its occupants here the ram-headed
sun god praised by Thoth and Khepri, celebrating
a regenerated cosmos forever new, re-formed and
brought forth through the power of the 'Word'.[33]
The sun god's renewal of his 'flesh', his presence
deep in the earth, has brought Osiris's regeneration
too, and by virtue of his solar descent, mortal life
has been recalled to the light of its divine origins in
a goddess-ruled cosmos permeated with gold.

Transformation in this great work moves through
a mortal realm, where everything is first dissolved
before 'growing wings' and soaring to the divine, to
the numinous moment when time touches eternity
and eternity touches time in a solar cosmos forever
in motion, forever made new—a world in which
Hathor's copper love is the 'glue' binding everything
together in the heart of the earth, 'tying together'
this cosmos through the heartbeat of her love.

Nothing is destroyed in this eternal cosmos,
everything is alive in one continuous world, teaches
Hermes in Book 12 of the *Corpus Hermeticum*,
drawing on the metallurgical metaphor of the
'dissolution' and 'renewal' of an alloy to illustrate
his teaching *(see page 129)*. For 'there is nothing in

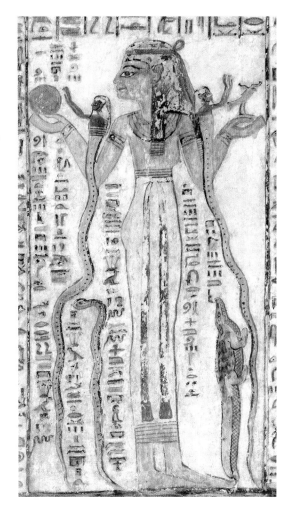

the cosmos that does not live', and, by participating
in the 'whole recurrence of eternity', nothing dies but
rather is continually dissolved, renewed and restored.
To Hermes, the essence of eternity is 'identity',
and as long as this cosmos 'holds together', it is
eternally alive. For everything is dissolved, he says
'not to be destroyed but to become new'.[34]

Some have doubted that this Greek incarnation
of Thoth transmits genuine ancient Egyptian
wisdom—his teaching seems too 'philosophical',
too Greek inspired, for it ever to have come from
ancient Egypt. Yet the correspondences are right
here in the sarcophagus chamber of Ramesses VI, a
king, who, of all Egyptian rulers, showed Hathor's
'uniting' copper wisdom to be at the very core of

ABOVE *fig.152* Final scene of the *Book of the Earth* on the left wall of Ramesses VI's sarcophagus chamber, showing the re-creation of the cosmos in the eternal cycle of 'Becoming'. (20th Dynasty.)

his regeneration, a king who showed Thoth, too, praising the sun god with Khepri in the mystical realm of 'identity'. For Ramesses knew Hathor's copper mysteries of the heart, mystically experienced in cosmic space and the revolutions of time, forever ensured his resurrection in the eternal cycle of 'Becoming', an earth-born resurrection Plato would later immortalize in his solstitial 'age of Kronos' *(see chapter 5)*.

MELTING BY FIRE: GLUING BY WATER

'*Solve et coagula*, dissolve and coagulate', say alchemists in their famous maxim summarizing the alchemical work as a process of 'dissolution' and 'undoing', followed by 'joining' and 'combining' to create a unitary whole. The Athenian fifth-century philosopher and theurgist Proclus says the same in his commentary on Plato's *Timaeus*, when declaring that everything is created through fire and water:

> For melting and welding are necessary for the production of things whose parts are like each other, the latter being provided by moisture and the former by heat, for everything is melted down by fire and is glued together by water.[35]

What Proclus has in mind, of course, is Plato's brief 'metallurgical' account in the *Timaeus*, in which he describes the creation of a mortal 'dissolvable' body from the four elements, formed from multiple parts that need to be welded together into a unity 'with many rivets too small to be seen'. When they are unified, a body is created within which 'the revolutions of the immortal soul' are enclosed, 'inflowing and out-flowing continually'.[36] As Proclus says, once this body has been 'coagulated' by the uniting of the different parts, it becomes 'endowed with life, and once endowed with life, assimilated to the immortal soul', which, unlike the appetitive soul born with the body, continually flows in and flows forth and is able to exist independently from the body.[37] True to the spirit of Plato's 'age of Zeus' in the *Statesman* *(see page 64)*, Proclus then goes on to describe the growth and development of human life from infancy onwards, including the tribulations of the human, passionate soul.

Steeped though he was in the metallurgical terminology of 'melting' by fire and 'gluing together' by water, however, the alchemical maxim *solve et coagula*, 'dissolve and coagulate', is unlikely to have originated with him.[38] More likely, in order to amplify Plato's terse metallurgical allusion, he borrowed his terminology from the alchemists, especially as the same principle underlies Zosimus's 'visions', in which the processes of 'moistening' and 'drying', 'mixing'

and 'unmixing', 'division' and 'union' lead to the transmutation of nature.[39]

Why stop at the alchemists, though, since the operations of 'fire' and 'water' are clear to see in Ramesses VI's sarcophagus chamber, its walls covered with unique scenes painted in black, red and gold on a white background, the four colours of alchemical transformation? The same dynamic principles are at work here, for fire is the 'dissolving' force in the realm of 'She Who Destroys', and copper and water the 'glue' binding everything together in Hathor's fluidic world above. 'Dissolving through fire' and 'uniting' through water' are the way of all flesh in the solar Eye's deep earth territory. This regeneration also involves a copper goddess who is instrumental in the great wheel of the sun god's life, death and rebirth through the diurnal hours shown on the ceiling above, the heavenly turning that matures everything living on earth. And as Re circulates through his lifetime from birth to death through the day, his journey is completely grounded in Osiris's hidden earth realm beneath, attuned to the year's reversal at the end of the sixth month—as Dendara's priests would later wonderfully express when they integrated 'rebuilding' Osiris into the 'hours' in their Khoiak seed-time rites *(see Part 2)*, knowing, too, that without Hathor's animating copper love, without her loving-destructive power, nothing transforms or moves or grows.

'What destroyed you and separated your spirit from your body?' the alchemist Theosebia asks Zosimus in the Arabic *Book of Pictures*, in a dialogue accompanying a scene of a dead man lying on the ground. Replying in words reminiscent of his Greek 'visions', Zosimus tells his 'mystical sister' she is responsible for his death, reducing him to 'ashes' and turning him from gold to silver to black:

> You took away my splendour and you turned me into silver, after I had been gold before. You dressed me in black, the lowest of all the colours, and then you turned me into ashes.[40]

But he then goes on to tell her:

> But then I will be resurrected, and I am bound to be resurrected, to come back to life and to become better than I was before.[41]

Called 'Queen of Egypt' in this beautifully illustrated manuscript, copied in 13th-century Egypt, Theosebia's role here harks back to the loving-destructive solar Eye's wisdom of 'dissolve and unite'. Metallurgical and Memphite in practice, transformational and healing in spirit, it flowed through alchemy long after the Egyptian temples had closed and was still touching hearts in medieval Egypt—though what has long lain hidden is the huge spiritual debt owed to ancient Egypt. For this wisdom is inscribed in Ramesses VI's burial chamber, to be sure not yet explicitly formulated in terms of alchemical principles, but nevertheless sustaining the king's regeneration. Lying in his sarcophagus, Ramesses would have been completely enclosed in his own alchemical vessel, protected by Nut guarding the diurnal circuit of 'waking' and 'sleeping' on the ceiling, illuminated by the Eyes shining on the right and left walls when the year 'reversed' after 180 days, and surrounded by an archetypal process of 'dissolving' and 'uniting' through fire and water which would later be recast in alchemy's 'great work'.

It is high time, therefore, to explore what alchemists have to say about ancient Egypt, life, death and regeneration, copper and transformation, since for too long they have been typecast as false seekers after gold, severed from the ancient wisdom that so inspired them, when all along they were genuine lovers and transmitters of Hathor's copper mysteries.

175

PART 4

ALCHEMY'S MIRROR: EGYPTIAN REFLECTIONS

LEFT fig.153 Ibn Umail's visit with friends to an Egyptian temple at Busir in the early tenth century. They are shown eagerly pointing towards a seated sage holding a stone tablet, watched by a woman pictured at a high window. Above them fly nine eagles, (though only six birds are included here). (Topkapi Sarayi Müzesi, Istanbul, Ms. A2075. 14th century.)

SIXTEEN

ALCHEMICAL TEMPLES:
EGYPT'S LEGACY

On his stela from Panopolis/Akhmim, an Egyptian priest, living in the time of the Roman emperor Hadrian in the early second century, makes a poignant plea to the devotees of Thoth and the 'team belonging to the Ibis' who see his monument: 'Set your heart on what is therein, do not forget the text collection, make copies of it.'[1] Perhaps he foresaw the fate of the wisdom he so treasured, consigned to oblivion, so the story goes, once Egypt's decline gathered pace in the Roman Empire. The Isis temple at Philae, known as the last stronghold of goddess worship, was closed by the emperor Justinian sometime between 535 and 537 CE, and, superseded by Christianity, Egypt's glorious culture, spanning more than 3,000 years, finally came to an end, its cults irrelevant, disowned even, in the changing world of late antiquity.[2]

Yet things are never quite so simple, for wisdom rarely vanishes, but rather leaves long-lived traces, especially when ritually imprinted on the memory day after day, year after year, as happened in ancient Egypt. Complicating such a neat ending, moreover, are the strange alchemical treatises thought to have emerged in Roman Egypt sometime around the first century CE. Indeed, from Zosimus of Panopolis in the early fourth century to the Islamic alchemist Ibn Umail in the tenth and the physician, musician and Rosicrucian alchemist Count Michael Maier in 16th-century Europe, the message has been the same: their wisdom comes from the Egyptian temples. Perhaps, after all, the Panopolitan priest's anxious plea did not go unheard.

It is, though, Ibn Umail's account of his two visits to an Egyptian temple (*birbā*), included in the introduction to his most famous work, the *Silvery Water and the Starry Earth* (*Al-Māʾal-waraqī wa-l-arḍ an-naǧmīya*), that most brilliantly encapsulates this legacy.[3] Known to European alchemists as Senior Zadith, the 'Righteous Elder', or simply 'Senior', this influential alchemist vividly describes how he saw a statue of a wise sage in the temple, seated and holding a stone tablet covered with symbols. In fact, from the details he gives, it has been conjectured he must have gone to an Asclepius temple somewhere in the Saqqara region

and seen a statue of Imhotep, the renowned architect of the Step Pyramid at Saqqara and later patron of alchemists, who was identified with Asclepius.[4] His visit was subsequently beautifully illustrated in an early 14th-century Islamic manuscript now in Istanbul (*fig.153*).

Islamic visitors to medieval Egypt were wont to marvel at the ancient wonders they saw there, but Ibn Umail's vision of the sage's stone was no mere traveller's tale, since 'finding the stone' is the goal of the alchemical quest, just as it was in the ancient Egyptian 'New Year' ritual confirming the reigning pharaoh's power.[5] And what he saw inscribed on this stone tablet were the mysteries of alchemy, symbolically divided into their two operations: the 'Whitening' phase on the right half and the 'Reddening' phase on the left.[6]

Extraordinarily, though, not only are a gold circle and a silver crescent moon represented in the right half, but also two conjoined birds with the full moon near them.[7] One bird is winged, the other wingless, and, according to the brief inscription beneath them, the winged female bird is here giving the male wingless bird the power of flight: 'The female is the spirit extracted from the male, carrying it, flying away with it.'

In the tablet's left half are 'two suns' shining onto a circle beneath. One sun has two rays, symbolizing the 'two in one', the other a single descending ray, and, according to Ibn Umail, they form three lights, representing 'water, air and fire' unified as the 'three in one'.[8]

Whilst it is easy to get lost in Ibn Umail's elaborate reflections, with all the veils of mystification he weaves around his temple visit, this imagery is remarkably in tune with ancient Egypt—disconcertingly so, given the 'two suns' depicted in Ramessid royal tombs (*figs.149,151*) and the winged and wingless Horus birds in the

LEFT *fig.154* Ibn Umail's visit reinterpreted in *Aurora consurgens*. Here the Egyptian temple has become a Christian church, and the three visitors point to a golden vessel placed on a column. Unlike in *fig.153* the lady at her window is not included. (Zentralbibliothek, Zurich, Ms. Rhenoviensis 172, f.3r. Early 15th century CE.)

southern crypt at Dendara, empowered for flight by Hathor's copper *(figs.137, 138)*. Elsewhere Ibn Umail associates the conjunction of sun and moon with 'lead-copper', thus incorporating the two metals alchemically associated with Osiris and Hathor-Aphrodite (or later Venus) respectively.[9]

TECHNICAL ISSUES:
THE GNOSTIC THEORY

To later alchemists, Ibn Umail's visit perfectly expresses their tradition's roots in ancient Egypt, so much so that an illustrator of the Latin medieval treatise *Aurora consurgens* turns the Egyptian temple he visited into a church, thus Christianizing this alchemical heritage *(fig.154)*. 'The only ones who know the benefit you seek', says Ibn Umail, 'are the people of the Egyptian temples',[10] reinforcing what Zosimus of Panopolis and Olympiodorus had already stated centuries before about alchemical knowledge being guarded by Egypt's priests *(see below)*.

Yet, apart from a few dissenting voices, notably some Egyptologists, and depth psychologists influenced by the 20th-century Swiss psychoanalyst Carl Jung, for whom alchemy mirrored the individuation process, the critical consensus nowadays is they were simply deluded. Metallurgical and dyeing recipes, it is repeatedly stated, were the catalyst for alchemy's development, not temple knowledge;[11] and such 'technical' recipes are preserved in two Greek papyri known as the Leiden and Stockholm papyri, which are the earliest alchemical treatises so far discovered, dating to the late third or early fourth century, and from Egypt.[12] They give a good idea of the crafts practised during that period, as do the recipes in the *Four Books* ascribed to Democritus the alchemist, though these are only known in a fragmentary form from much-later Byzantine treatises.[13]

Such recipes, combined with the theories of the Greek natural philosophers and spiced with a further religious dimension provided by the Gnostic-inspired Zosimus of Panopolis in the early fourth century, created alchemy. In other words, firmly planted within the matrix of Greek philosophy, alchemy was 'in the first place a

LEFT fig.155 Vulture ceiling in the temple of Ramesses III at Medinet Habu, Western Thebes. The nine birds shown in *fig.154* are adaptations of the nine vultures traditionally shown flying across Egyptian temple ceilings. (20th Dynasty.)

RIGHT fig.156 Detail from *fig.207* of a winged and a wingless bird in al-'Irāqī's *Book of the Seven Climes (Kitāb al-aqālīm al-ṣab'ah).* (British Library, London, Add. Ms. 25724, f.50v. 18th century.)

technique, then it became a philosophy without ceasing to be a technique',[14] a view ultimately paving the way for its relegation to little more than a 'pseudo-science', a forerunner of what would become modern chemistry—and, inevitably, a scenario denuded of the ancient Egyptian temples.[15] To be sure, the influential Dominican scholar and priest Father A.J. Festugière acknowledged that the ancient Egyptians preserved technical craft secrets in their temple archives, transmitting them from father to son, but he pointedly maintained they were incapable of pursuing a 'reasoned' method, since they were unable to think philosophically. Hence only the Greeks could have provided alchemy's theoretical framework.[16]

So dominant is this Greek-Gnostic lens that alchemy's Egyptian vision has been completely eclipsed, yet this is not the message of the ancient alchemists themselves—and, as any anthropologist well knows, what native practitioners have to tell is always the preferable story. In fact, right at the beginning of his treatise known as the *Final Count*, the Egyptian-born Zosimus provides a much

broader alchemical canvas when revealing the Egyptian 'arts' to his 'sister' Theosebia.[17] Interestingly, he is absolutely clear that Democritus's craft recipes, the 'noble arts' as he calls them, belong to alchemy's exoteric knowledge, and that there is another tradition Democritus was forbidden to reveal: the ancient Egyptian 'arts':

> All the kingdom of Egypt, O lady, depends on these two arts: that of the propitious ores and that of the natural ones … In effect, the so-called divine art, in other words the art that for the most part is founded on doctrinal and teaching principles, has been entrusted to the guardians [of the temples?] for their sustenance, and not only that art, but, at the same time, also the four noble arts and the manual procedures.[18]

In short, what Zosimus is saying here is that the so-called 'technical' treatises, or 'four noble arts' as he calls them, which nowadays are almost universally regarded as the origins of alchemy, belong to a much more extensive body of Egyptian priestly and metallurgical knowledge—the 'divine art'. He also says this 'divine art' provides the principles on which alchemy is based, evidently knowing an alchemical tradition that is ancient Egyptian to its core.

He also tells Theosebia that these 'arts' are a tightly controlled privilege of kingship:

> Some people reproach Democritus and the ancients for not having mentioned these two arts, but only those arts called noble. This reproach is futile because they could not do otherwise, being friends of the kings of Egypt and glorified through holding the first ranks among the prophets. How could they have openly set out knowledge against the kings?[19]

Only the Jews, Zosimus says, clandestinely wrote about these matters, and then only the exoteric aspects, notably knowledge of the 'gold mines' transmitted by Theophilus, son of Theogenes, and descriptions of the 'furnaces' in Maria's treatises.

Curiously, an inscription in the 'House of Gold' at Dendara makes a similar distinction between 'technical' and 'sacred' knowledge, stating that, whereas uninitiated craftsmen fabricate the statues for the temple cults, it is the initiated priests involved 'in the very secret work' in the House of Gold who guard the knowledge—a strange declaration in an Egyptian temple, but a hint perhaps that Dendara's priests were concerned their knowledge might be leaking out beyond the temple walls.[20]

Artisanal procedures, wherever they are practised, are usually closely-guarded secrets, and Egypt was clearly no exception, all mining and metallurgical operations being under the jurisdiction of the king.[21] But that Jews might have transmitted some knowledge is plausible enough in light of the Babylonian exile in the sixth century BCE, when some Jews went to Egypt, eventually settling in towns along the Nile from Elephantine in the south to Alexandria on the shores of the Mediterranean. They were certainly interested in metalworking; and, judging by the priestly titles of a man called Chaiapis, a Phoenician belonging to the Jewish quarter in Memphis in the third century BCE, he was able to serve in some capacity in the Memphite temples (*fig.157*).[22] His might be a rare example,

but it shows access, however restricted, was possible for non-Egyptians, despite the vociferous prohibitions against it in temple inscriptions. In the mixed milieu of Hellenistic Egypt there must have been ample opportunities for cultural interaction, and whilst the transmission from the temples remains opaque, the names of early alchemists, including native Egyptians, show they came from a cross-section of the population.[23]

Even so, at the very beginning of the *Final Count* Zosimus is adamant Egypt's 'divine art' was never publicly transmitted. He also knows it provided the 'theoretical framework' for the practical recipes, the 'technical' kind of alchemy, or 'noble arts', preserved in Democritus's recipes and elsewhere. It is all the more extraordinary, therefore, that his statement should have been so

LEFT fig.157 Stela (now lost) of the Phoenician Chaiapis from Saqqara. (Staatliche Museen zu Berlin, ÄM 2118. Third century BCE.)

RIGHT fig.158 A farmer sowing seed illustrating the sixth epigram of Michael Maier's *Atalanta fugiens* (1618).

consistently overlooked, including by Festugière, who re-edited the *Final Count*, for to explain alchemy's origins simply in terms of craft recipes and their subsequent Gnostic spiritualization flies in the face of everything Zosimus says there.

DENDARA REVISITED: AN ALCHEMICAL CHRYSALIS

The Dendara temple was probably still functioning when Democritus's *Four Books* were composed, perhaps in the first century CE.[24] And long before the term 'alchemy' had ever been coined, Dendara's priests displayed signs of an alchemical outlook, knowing, like 'astrologers in reverse',[25] the influence of the heavenly bodies deep in the earth, where the seeds, the metals and minerals all lived in the solar circuit. In fact, Carl Jung was astonished to find alchemical rebirth imagery replicated at Dendara, particularly the praise of the sun god as a sacred snake, appearing like a falcon in the midst of the lotus flower, the divine child separating night from day. 'The opus is a repetition of the Creation, it brings light from the darkness', he wrote, without,

however, commenting further on the Egyptian parallels.[26]

The Khoiak 'seed' festival at Dendara *(see Part 2)* has a distinctly alchemical ring about it, since alchemy has abundant 'farming' references celebrating life coming forth from the earth. Thus, in the Greek alchemical treatise *Letter of Isis to her Son Horus*, Isis advises her son to go and observe the farmer Achaab and 'learn from him what is the sowing and what is the reaping, and you will discover that the one who sows wheat reaps wheat, that the one who sows barley reaps barley.'[27] According to Ibn Umail, the gold the alchemist seeks is 'like a plant sprouting in the garden'.[28] He also compares the quest for the healing 'elixir' to the various transformations of grain into nourishing food—a 'baking bread' theme inscribed also in the Khoiak rites *(see page 115)*.[29] So, too, Michael Maier advises alchemists to 'take as a model the work of the farmer, for gold grows like wheat, and has its own life' *(fig.158)*;[30] and, in the Arabic *Epistle of the Secret*, Hermes, when instructing Theosebia in preparing the 'stone of

ABOVE fig.159 View into the House of Gold located off the western staircase in the temple of Hathor at Dendara, where prototypal alchemical recipes are inscribed.

the wise', says that if she succeeds, her 'stone' will be 'clothed in remarkable colours, wonderful blossoms will appear, the crops will ripen and the fruit will be delicious' *(see page 240)*.

Clearly, these alchemists were not only concerned with 'technical' recipes, but also the cultivation of a fruitful life, and hence, just as Dendara's priests had tended their 'seed beds' in the Khoiak rites, they tended their crucibles to observe the secret 'earth' processes bringing transformation and growth.

Alchemical resonances at Dendara also extend to the location of the rooftop 'chamber of the egg' above the perfume 'laboratory' in the main temple beneath *(see page 112)*, thus connecting 'material' and 'spiritual' transformation exactly as in alchemy. The Khoiak rites also needed specific coloured cloths to shroud Osiris's body *(see page 122)*,

which would have required the art of dyeing, practised by a 'cook of colours' as the dyer was called, an art that would surely have had its own lore in ancient Egypt.[31]

But that is not all. For whenever the priests carried Hathor's statue to the roof from the southern crypt, their procession would have wended its way past the enigmatic room called the 'House of Gold', located off the western staircase *(fig.159)*. Here Hathor gives the pharaoh 'all the precious materials of the mountains in order to complete the work of the House of Gold.' Thoth the 'twice-great' is also depicted, and the inscriptions explicitly mention the creation of cult statues, made from materials that have the appearance of precious metals.[32] Clearly, Hathor's priests were no strangers to the art of substituting different materials so they seemed more costly, something which also preoccupied alchemists in their 'technical' recipes. Furthermore, by bringing this statue from the crypt to the roof, they were also honouring 'two suns': the earthly 'palace' sun enshrined in the crypt and the celestial sun, the Aten, the heavenly 'right Eye' illuminating Hathor's statue in the rooftop shrine. Alchemists, too, work with 'two suns', a copper sun and a heavenly sun, and Zosimus of Panopolis even calls the heavenly sun 'the right Eye of the World' *(see chapter 19)*.

Such 'alchemical' knowledge was extremely well hidden, though, in ancient Egypt, and was also expressed mythically, so that, when transformed into a Greek idiom, it became very difficult to recognize. Despite mentioning Osiris, Isis, Horus and other Egyptian deities directly by name, alchemy's riddling language now seems simply too alien for it to have come from the temples. Enigmatic stones, eternal water, exotic birds, dragons and peacocks' tails streak across every page, creating a chimera of reversals and circumambulations, unions and separations, the inside turned outside, as hidden depths rise to the surface in sudden openings of the eyes and heart to reveal 'the essence that never dies'—all of which seems worlds away from the ancient Egyptian temples and cult life.

The same kind of blind spot initially dogged the response to a highly unusual group of late third-century mummies unearthed in Hathor's once glorious temple precincts at Deir el-Bahri, which, when first seen by their excavator, Herbert Winlock, with their masks covering their heads and torsos painted to show them in contemporary elite Roman-style clothing, seemed nothing but 'atrocities of hideousness' from the 'last days of paganism'.[33] Even though Sokar's *Henu*-boat and solar rebirth symbolism were depicted on the masks, they simply looked too strange to be placed within a genuine ancient Egyptian continuum.

Nowadays, though, against the diverse backdrop of third-century Egypt, these unique mummies are considered authentically Egyptian. To be sure, by this time Thebes was no longer the magnetic hub it had once been during the New Kingdom, or even in the Ptolemaic period, and everything had been vastly scaled down, but these mummies were not buried in a cultural vacuum. On the contrary, their traditional symbolism still reverberated with the ancient rebirth knowledge displayed in the earlier royal and private tombs nearby, showing them to

be participants in the same cycle of life and death that brought the blossoming of plants, the rebirth of life and perennial fertility.[34] What these masks also reveal is that Egyptian mythic beliefs were still alive in the Theban region not long before Zosimus's lifetime—that there was an 'active engagement with the past' continuing throughout the third century and beyond.[35] Old beliefs were being reclothed in new colours by a group of individuals clad in Roman-style white, purple and gold garments.

If seen in their ancient temple context, the alchemical Egyptian 'arts' are no stranger than these Deir el-Bahri mummies. They display the same kind of unusual blend, the same move to don 'new clothes' whilst at the same time continue ancient traditions. And, just as the Isis cults eventually spread to every corner of the Roman Empire and on into medieval Europe, reaching far-flung highways and byways far removed from Egypt's 'Black Land', enough evidence survives to show that Hathor's 'copper wisdom' likewise never died out. This wisdom—with all its feeling for nature, all its reverence for the growth and development of life, for the love that moves the sun and stars—mutated, changed forms and acquired new ideas, like the goddess herself, without ever losing sight of its quintessentially Egyptian roots. Far from disappearing with the pharaohs, it continued to inspire alchemists as an undercurrent to the religions ruling on the surface, keeping alive the sacred Feminine when it seemed all but lost.

In short, alchemy's 'mysteries of the queen' are where Hathor's temple voices can still be heard, speaking in new languages, new ways, but still following a path of love, to which the following chapters now turn.[36]

SEVENTEEN

ALCHEMY OF COLOUR:
THE WORK OF A DAY

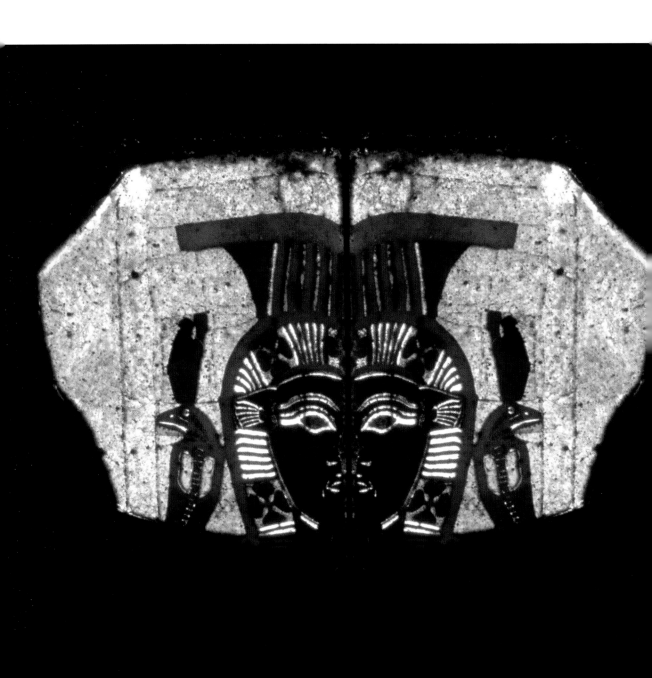

From its earliest beginnings, alchemy has had a language of colour weaving through it, a sequence of black, white, yellow and violet (or purple) which has been retained by Greek, Jewish, Islamic and European alchemists alike—with the exception of 'yellow', which has been subsumed into 'red' under the influence of the Sulphur–Mercury theory of metals, thereby creating an additional sequence of black, white and red.[1] As a sequence, the four colours are often compared with the colours Pliny the Elder says were used by Greek artists,[2] a comparison enhanced by the fact that an early piece of alchemical apparatus, the *kerotakis (fig.161)*, has the same name as the heated palette on which Greek encaustic artists melted and mixed their four basic pigments.[3] Seemingly used in a vaporizing process for softening metals and mixing them with their colouring agents, the alchemical *kerotakis* certainly belonged to the colouring arts, but these four colours also featured in a much wider Greek colour narrative, being applied to the macrocosm and microcosm, the cosmos and the human being.[4]

So, for example, according to Theophrastus, the atomist philosopher Democritus of Abdera (*c*.460–370 BCE) taught there were four colours: white, black, red and yellow (or green). He also observed how gold and copper tones were mixed from white and red, and how iron became redder when placed in fire, all of which suggests that, like later alchemists, he had more than a passing interest in the relationship between colour and metals.[5]

These four colour associations were a 'common possession', observed Peter Dronke, 'and even in antiquity will have been as widespread in popular superstition as in learned science.'[6] To this Greek bias, however, should also be added ancient Egypt, since four basic colour terms similarly tint its language.[7]

By the first century CE Democritus had acquired a legendary status for his wisdom. Confusingly, he was also conflated with a similarly-named early alchemist, an identification which saw Democritus the philosopher and Democritus the alchemist, though historically separated by centuries,

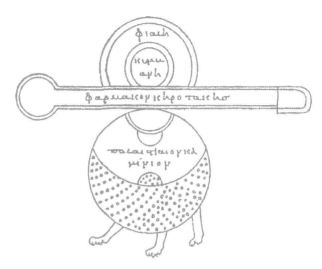

LEFT *fig.160* Alchemy's colours of black, white and red revealed in this Hathor mask inlay. (Mosaic glass, *c*. third–first century BCE. Private collection.)

ABOVE *fig.161* The alchemical *kerotakis*. (Illustration from Ms. Marcianus graecus 299, f.195v, Biblioteca Nazionale Marciana, Venice. Late tenth or early 11th century.)

becoming fused together in later alchemical tradition. It was said that Ostanes, the legendary Persian magus and alchemical authority, had initiated Democritus in the temple of Memphis, taking the opportunity at the same time to initiate the Egyptian priests—a tradition once again disconnecting alchemy from ancient Egypt and fuelling the view it was essentially Persian in origin, a foreign implant in Egypt, with the temples providing little more than an exotic stage backdrop.[8]

CHANGING COLOURS: SYMBOLIZING STATES

What is specific to alchemy, however, is the way each colour defines the intrinsic nature of a particular substance, indicating the qualitative changes of state when it passes through the different phases of a process, the radical transformation through which it is perceived to be truly alive. Thus, 'black', as the foundation of the work, symbolizes the initial dark phase, or turbulent 'death' (called in Greek *melanosis* and in Latin *nigredo*), when a 'body' returns to its primal state, dissolving back into inchoate darkness. Then comes the cleansing, purifying phase of 'whiteness' (in Greek *leukosis* and in Latin *albedo*) associated with silvermaking, which is followed by 'yellowing' (in Greek *xanthosis* and in Latin *citrinitas*), deepening into the 'violet' or 'purple' phase known as *iosis*, though some alchemists, like Olympiodorus, have yellow as the culminating phase *(see below)*.

When this mysterious 'purple' colour, the most elusive of all to produce, is 'fixed' like a true dye, it becomes a permanent heavenly tincture, a luminous power making 'the invisible visible', the 'inner' 'outer'. Suddenly everything glows with a lustre that shines forth from within, a revelation of maturity and fullness as the work comes to completion.[9] Prized as the colour Roman emperors loved to wear, purple marks the complete transmutation of the original substance, a quantum leap which turns this seemingly linear and evolutionary colour sequence into an iridescent circularity. Writing about purple in antiquity, the art historian John Gage noted: 'It was the miracle of purple to incorporate within itself darkness and light and hence the whole world of colour.'[10] To the Islamic alchemist Ibn Umail *(fig,153)*, purple was quite simply the colour of 'eternal life'.[11]

In this alchemical world of colour, to which additional phases known as 'greening' and the 'peacock's tail' were later added, everything is continually moving from initial 'blackness' through cleansing 'whiteness' to 'gold' and 'purple' as bodies change into light and light into bodies, a transformational sequence manifesting in a whole host of elaborate alchemical procedures requiring the agency of fire.

Thus, the *Four Books* ascribed to Democritus the alchemist *(see page 180)* are painted with alchemy's gleaming colour changes. Not only do they contain recipes for imitation purple dyestuffs, cheaper substitutes for the expensive Tyrian dye, but they are also divided into parts devoted to 'Whitening' or 'silvermaking' and 'Yellowing' or 'goldmaking' as well as the arts of dyeing stones. So familiar was Democritus with alchemy's

language of colour, it became woven into the very fabric of his *Four Books* through every aspect of their metallurgical and dyeing recipes.

Implicitly included, too, in the *First Book* is the 'black' or 'death' state, since, immediately after its opening purple-dyeing recipes, Democritus describes how, in the netherworld, he conjured forth his deceased teacher, whose untimely death had left him unable to comprehend alchemy's secrets properly. The only message he is given, however, is: 'The books are in the temple.' Subsequently, whilst banqueting with friends in a temple during a festival, all is revealed when a pillar suddenly opens to disclose the sought-after knowledge, succinctly expressed in the words: 'Nature delights in nature, nature triumphs over nature, nature rules nature.' This phrase encapsulates the dynamic 'sympathies' and 'antipathies' flowing through nature, with which alchemists operate.[12]

This myth, whilst not directly naming an Egyptian temple, implies a connection between 'technical' recipes and sacred knowledge. Whoever the Democritus was who composed these *Four Books*, he evidently sought to link them with temple life, and did so in accordance with an age-old Egyptian tradition of finding ancient wisdom in a secret place.[13]

BLOSSOMING FLOWERS: HEALING DYES

Closely connected with alchemy's colouring arts is medicine, highlighted by the Greek word *pharmakon*, which means not only 'dyestuff', 'paint' and 'colour', but also 'remedy' or 'healing charm', the kind of 'charm' able to alter the nature of a body in both healing and harmful ways, depending on how it is used.[14]

And just as medicines heal, so do dyes. Thus, in the *Dialogue of the Philosophers and*

RIGHT fig.162 Three birds enclosed in the flask of Jupiter, displaying the black, white and red colours of the alchemical work. (13th illustration of Salomon Trismosin's *Splendor Solis*, British Library, London, Harley Ms. 3469, f.24. Late 16th century.)

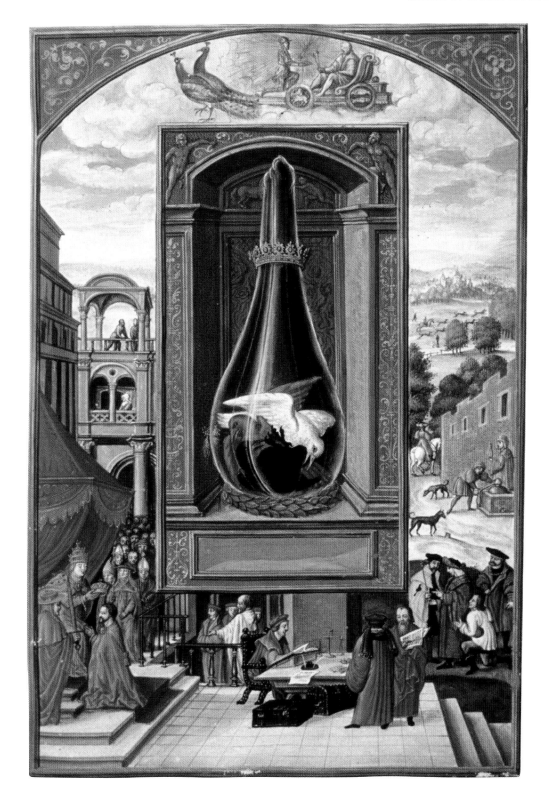

Cleopatra, it is the '*pharmakon* of life' that comes like a remedy to heal the fettered bodies lying in the 'blackness' of Hades, waiting for this medicine's rejuvenating visit and the flow of the rebirth waters so they can reveal their beautiful springtime colours.[15] Whenever the dyes become active, the bodies are healed and the flowers bloom, though, like the seeds in the Khoiak rites, if they are to reveal their true colours, they need to be watered.[16]

As do metals, if we are to believe the ancient Egyptian *Famine Stela*, carved on a rock at Aswan during the Ptolemaic period, which tells how a priest of Imhotep travels to Thoth's ancient city of Hermopolis to seek a cure for the famine and drought plaguing Egypt's land—a healing effected by the life-giving inundation waters that 'bestow stones upon stones' and fruitful plants, washing over them to restore Egypt to life.[17] Hence the listing of the precious ores, stones and plants in the text is surely not simply an inventory of the treasures found in Khnum's southern border region, but rather a testament to the belief in the inundation's power to renew everything in the plant and mineral realm, sacred wisdom guarded by Thoth in Hermopolis.

To explain alchemy's colours simply in terms of their 'technical' application is clearly wide of the mark, for not only are Democritus's craft recipes woven together with a temple revelation *(see page 188)*, but Cleopatra's concern in the *Dialogue* is with the rebirth of the dead.

Alchemy's colouring secrets also found their way into very early Christianity, influencing the non-canonical *Gospel of Philip*, in which dyeing and colour changes are completely blended with Christian spiritual life. Discovered in the famous Nag Hammadi cache of secret writings from Upper Egypt, and thought to have originally been composed in the Syrian region sometime between the end of the first and late third century, this gospel describes baptism as a plunge into a turbulent 'dyeing' vat, in which all 'blackness' is alchemically cast off when an initiate is reclothed in the 'whiteness' of a baptismal garment, then spiritually transformed through fiery anointing.[18] Or, as the Christian alchemist Morienus the monk succinctly says, in words quoted in the Latin treatise *Aurora consurgens*:

Whoever elevates the soul shall see its colours.[19]

These colours, streaming in light, carry energy that dynamically sets an alchemist's inner life in motion and transforms the soul.

ALCHEMY'S COSMOS: PAINTING THE DAY

It is, though, Olympiodorus, in his commentary on a now-lost work by Zosimus, who has most to say about the 'Egyptianness' of alchemy's colours and who explicitly connects them to the ancient Egyptian solar cycle.[20] Strangely, modern-day commentators rarely mention his treatise, and, even when it is quoted, it is not usually for its ancient Egyptian themes—perhaps because these are scattered amidst discussions of Greek cosmological ideas, Presocratic philosophy, metalworking, oracles, extracts from Zosimus and sayings of Hermes.

Exactly when Olympiodorus lived is unclear, though it must have been after Zosimus in the early fourth century, for he calls him the 'crown of the philosophers' and quotes him extensively, including his statement about the Egyptian 'arts' *(see page 181)*.[21] He himself was clearly someone who cherished ancient Egyptian wisdom, referring to 'the writings of the Egyptians' and their quest for 'the divine water'.[22] He also knew about the ritual nature of Egyptian goldmaking. 'Not only have they described thousands of procedures for making gold', he says, 'they have ritualized these things.'[23] What concerns Olympiodorus, moreover, is not simply alchemy as a technique, though he certainly refers to metalworking processes, but alchemy as a living 'art', practised within a cosmos gestating into existence. He alludes to 'the egg' and to the tail-biting Agathodaimon serpent, the 'image of the world' as he calls it,[24] and also perceives humankind within nature, alive with the life-blood of all living things, maturing and growing to fullness and ripeness.

Superficially, it would be all too easy to dismiss Olympiodorus's treatise as yet another alchemist's deluded homage to ancient Egypt, but this would be a mistake, for immediately after mentioning Egyptian goldmaking, he describes the solar cycle in terms of alchemy's colours. First, he says, the Egyptians began their operations at dawn 'when the sun rises on earth', though, true to the spirit of the *Book of Day*'s first hour, he also, quite rightly, states that the numinous moment, and beginning of the work, is just before sunrise, preceding the 'rising' and 'Whitening'.[25] Then, after specifically equating sunrise with 'Whitening' and sunset with 'Yellowing', he allusively refers to 'lead' (the metal that later in the treatise he identifies with Osiris), thus incorporating 'blackness' into this Egyptian colour sequence.

Another Greek version of his treatise elaborates further on the solar cycle. After noting how the Egyptians oriented their temple entrances and exits to the cardinal points, it goes on to state:

> They have attributed the Great Bear (north) to Blackness, the 'Whitening' to rising, the colour of Purple to midday and the Yellowing to setting. They have attributed the white substance to rising, namely silver, and to setting yellow, namely gold.[26]

In light of the *Book of Day*'s reference to the Great Bear when the evening sun god enters the West *(see page 32)*, the 'Egyptianness' of this alchemical 'blackness' needs no further elaboration. Again, 'white' is equated with 'rising' and 'silver', here juxtaposed with the 'yellow' of sunset and gold. Significantly, 'purple' is introduced as the midday colour, thus changing the sequence into a fourfold division completely in tune with the *Book of Day*'s own fourfold division of the solar circuit.[27]

In short, what Olympiodorus transmits is an accurate version of the ancient Egyptian solar cycle, completely expressed in colour. The cycle was evidently still known to him in late antiquity, as it was to Iamblichus, the great Syrian philosopher, theurgist and exponent of the ancient Egyptian mysteries (d. *c.*320–25). In his defence of Egyptian religion known as *On the Mysteries*, he refers to the 'symbolic mystical doctrine' of the ancient Egyptians, in which the divine being 'is changed, according to the zodiac, every hour', and goes on to say that the Egyptians addressed prayers to the sun in accordance with this doctrine.[28] Clearly, in the early fourth century, familiarity with the old solar cycle was alive and well.

COLOUR AND CHARACTER: DEEDS OF LIGHT

Yet long before Olympiodorus, the ancient Egyptians themselves had contemplated the solar cycle's world of heavenly colour, the intimate relationship between gleaming colours and transforming solar life.

The word for 'dawn' in Egyptian is *hedj-ta*, meaning 'whitening of the land', and representing this dawn region in the *Book of Nut* is the Upper Egyptian crown goddess Nekhbet, depicted close to the legs of the sky goddess in the Southeast, symbolically placed in the 'whiteness' of the rising dawn *(fig.120)*. Known as the 'White one of Nekheb', she is the tutelary goddess of Upper Egypt, the guardian of the 'White Crown', the *Hedjet*, or 'White One'.[29]

This dawn 'whitening' is also associated with 'silver' and 'purification'. 'The Two Lands dawn in the colour of silver', eulogizes an Egyptian sun hymn,[30] hence a 'mirror of silver' is offered to Hathor in her dawn 'House of the *Menit*' in Dendara's 'secret palace'.[31] From Dendara, too, comes a beautiful scarab, its body made of lapis lazuli, its wings all silver, emerging from nocturnal 'lapis' darkness into dawn 'whiteness' in the cycle of 'Becoming'.[32] An early word for 'silver' is 'white gold' *(nbw ḥd)*. Later *hedj*, with a gold determinative, is simply used, which is illustrative of how the Egyptians, like later alchemists, differentiated between the two metals simply on the basis of colour, ranging from the cool, silvery-white of silver, the colour of the moon, to the warm tones of gold, the colour of the sun.[33]

From the dawning of divine light until the close of day, the Egyptian sun god journeys through a shimmering play of astral colours, travelling in the *Book of Day* between green, gold and red. Thus, as day dawns in the first hour, travellers in the sun boat follow 'the way of the green bird' *(see page 20)*, in the ninth hour Osiris gloriously manifests his golden divinized nature in the Field of Reeds *(see page 28)*, and then, as the sun boat turns towards the North, the realm of the Red Crown, a deity appears called 'He who makes Redness' *(see page 30)*. In this reddening tenth hour, Re manifests as 'Gold of the Stars', completely encircled by the uraeus's fiery flames, his divine golden essence ringed in serpent redness, radiating that red-gold colour so prized by New Kingdom Egyptians.[34]

Already in the *Pyramid Texts* the celestial mother goddess displays green and red in her perpetual cycle of nocturnal conception, pregnancy, morning birth and regeneration.[35] These colours also pervade incantations to the beautiful but dangerous serpent Eye goddess, whose oscillating nature manifests in 'abundant colours' as she becomes now the 'Green One', now the 'White One', now 'Red'.[36] Colour combinations of 'black, green and white' or 'white and red' or 'green, white and red' stream through these incantations, juxtaposed with the sun (red) and moon (white). They also refer to Egypt (black), to its dual aspects of North (green, red) and South (white, red), and to the combination of land (black), water (green) and fire (red). The two serpent Eye goddesses are the 'Black One' and the 'White One', who turn red with rage, slaughtering the god's enemies, and when this is accomplished, they return, propitiated, as the 'White One'. To the Egyptians, colour reveals the essence of a deity or person, expressed in a word, *iwn*, meaning both 'colour' and 'character', and here the serpent Eye's character certainly manifests in these colour combinations, her 'radiance' and 'rage' constantly changing in a solar world of light.

Just as colour shines through the sun god's journey of life, it also defines Tutankhamun's transforming solar lifetime, which is depicted on a cartouche-shaped box discovered in his tomb at Thebes. On the front, the king is twice shown as a child enclosed in a cartouche, wearing the characteristic sidelock of youth and holding symbols of royal power *(fig.163)*. In the two cartouches on the back of the box, in contrast, he appears once as a mature ruler at the height of his power, wearing the Blue Crown, and once with his face 'black', symbolizing his Osirian 'nocturnal' state as he waits to be reborn again as a child on the front of the box *(fig.164)*. Each image also forms part of the king's prenomen, 'Re is the Lord of Transformations [*kheperu*]', the basket on which he sits representing the word 'Lord' and the sun disk above 'Re'. Surmounting this beautiful box are the

characteristic two plumes and solar disk of Sokar, the metalworking god of Memphis, as if to make clear this oval box is to be understood as his. And just as in the Khoiak rites at Dendara Osiris manifests his power to grow, develop and transform in Sokar's 'chamber of the egg', a solar birth which brings the colours *(see page 112)*, here Tutankhamun symbolically circulates around this Sokarian receptacle, the vessel of life itself, manifesting the changing colours of his solar divinity and transforming in a creation born from the egg.

Hence, though the colours are much more standardized in later alchemy, Olympiodorus was not arbitrarily imposing colour onto the ancient solar cycle, since the Egyptians themselves perceived shimmering colours welling up in solar light. What is astonishing, though, is that a Greek alchemist should still be deeply concerned with the Egyptian solar cycle in late antiquity, ascribing particular colours to each phase and characterizing it as 'the accomplishment of the entire day'.[37] In fact, alchemists more generally refer to their opus as the 'work of a day', though, as Ibn Umail wryly remarks, it would be a mistake to take this literally:

> When ignorant people hear this statement, they think that the work is accomplished in one day and night. They do not know the explanation of Hermes, who said: 'You will still be tired as long as your stone is black'.[38]

In quoting Hermes here, Ibn Umail is warning aspiring alchemists not to be confused by alchemy's riddling language, since there is no quick way to its secrets. This 'work of a day' is a lifetime's work, its treasure is hard to attain, and much patience is needed to transform the stone's initial 'black' state.

Ibn Umail knows his wisdom comes from the Egyptian temples *(see page 180)*; he also knows alchemy as the 'work of a day', and he knows the colours, though nowhere does he connect them explicitly with the ancient Egyptian solar cycle. Indeed, without Olympiodorus's precious commentary on Zosimus's lost work, this connection would be utterly masked, for what Olympiodorus sets out (presumably reflecting Zosimus's own teaching) are the alchemical correspondences with the Egyptian cycle of 'one day'.

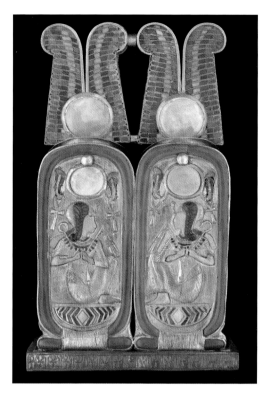

ABOVE LEFT *fig.163* and RIGHT *fig.164* The four transformations of Tutankhamun's solar lifetime, here symbolized by different colours on a double cartouche-shaped box from his tomb. (Egyptian Museum, Cairo, JE61496. 18th Dynasty.)

And just as the ancient Egyptians lived in tune with the sun god's heavenly journey, whenever alchemists see substances changing colours in their fiery vessels, flasks and crucibles, they see them transforming within a living, solar cosmos, uniting 'above' and 'below'. For earth is, as Olympiodorus says, the 'image of the cosmos'. Here he is referring not only to Zosimus's authority, but also to the astrological understanding of the human body as a world in miniature, a mirror of the whole zodiac, from the astrological sign of the Ram at the head down to the fishes of Pisces at the feet, as already mentioned by Manilius in his Latin *Astronomica*, written in verse, probably in the early first century CE.[39]

This is no dualist Gnosticism Olympiodorus is transmitting here, rather an alchemy in which the human body is in the cosmos and the cosmos in the human body, though where once a mother goddess carried darkness into light and light into darkness in the Egyptian solar circuit, there is now simply an iridescent play of colour illuminating this

movement of life from dawn to dusk and dusk to dawn. These transformational colours of black, white, purple and yellow are a cosmos in themselves, a brightly coloured texture of the world manifesting in shimmering light, woven together in alchemy's sacred art. To be sure, Olympiodorus could simply have correlated these colours with phases of the solar circuit without reference to ancient Egypt, but, true to the spirit of Zosimus on whose work he is commenting, what he is giving out here are the principles on which alchemy is founded, the 'divine art', as Zosimus calls it in the *Final Count* *(see page 181)*, and, as both these influential alchemists know, theirs is a wisdom inherited from the temples, an ancient Egyptian 'work of a day'.

EIGHTEEN

MOVEMENT ALCHEMY: COPPER MATTERS

Cosmology alone, though, does not create alchemy, as Mircea Eliade observed when he wrote: 'It is the conception of a *complex and dramatic Life of Matter* which constitutes the originality of alchemy.' He went on to compare this alchemical 'conception' of matter with initiation into a deity's passion, death and resurrection in the ancient mysteries, a 'dramatic spectacle' which could be experienced alchemically whenever matter passed through the four colour phases.[1]

Olympiodorus scatters clues enough in his commentary on Zosimus's work as to which 'life of matter' is key, mentioning various alchemical techniques for 'separating' and 'uniting' copper and gold—washing gold sands to obtain the precious grains and 'separate' them from any unwanted material, then 'uniting' these by means of 'chrysocolla', a blue-green copper carbonate *(fig.173)* widely used in antiquity for soldering gold, including in ancient Egypt.[2] He also refers to different methods of colouring copper, as well as to the different kinds of fire needed to create an enduring, stable body.[3] These are not random, arbitrarily chosen, alchemical operations requiring fire and water.[4] They belong in a treatise about the Egyptian 'divine art'; and in fact Olympiodorus quotes at length a version of Zosimus's statement about this 'divine art' *(see page 181)*. What his commentary also makes clear, as does Zosimus's revelatory teaching in the *Final Count (see chapter 19)*, is that this secret Egyptian tradition involves copper.

Long before the Sulphur–Mercury theory of the generation of metals was developed, particularly by the elusive Jābir ibn-Ḥayyāan and his followers, and took hold in Islamic alchemy, it was copper that fascinated Graeco-Egyptian alchemists[5]—the metal that eventually became associated with Venus, as planet or love goddess, and was known to Islamic alchemists as 'the peacock of the Egyptian temple'.[6] To Olympiodorus, however, living closer to ancient traditions, copper was the metal of 'the Egyptian with the tresses of gold' (Hathor) and 'the Cyprian', the Greek love goddess Aphrodite, whose island home of Cyprus was rich in copper in antiquity, though judging by a bronze statuette from the 12th century BCE, there had evidently been a much older 'copper-woman' revered there. Depicted as a nude Astarte-like figure *(fig.165)*, with her arms curved round beneath her breasts, she stands on a copper oxhide ingot, perhaps to symbolize the fertility of the Cyprian copper mines, on an island where copper-working had long been a sacred activity.[7] Hathor was worshipped there, too, and, reciprocally, during the Late Period, Aphrodite was worshipped in Egypt.[8] 'Direct your way to the South', instructs the 16th-century European alchemist Gerhard Dorn, 'so shall you obtain your desire in Cyprus, of which nothing more may be said.'[9] Clearly, copper held the key to his alchemical work.

CHAMELEON CHANGES: HUMANIZED COPPER

Of all the metals known to early alchemists, copper must have perfectly mirrored their quest to 'induce the nature which is hidden in the interior' to come forth, as Maria the Jewess says, to make actual what is potential and reveal, once all the 'obstacles' have been removed, the treasure longing to be known.[10] For who would ever guess that lumps of uncut malachite *(fig.131)*, the carbonate from which copper was usually obtained in antiquity, with its 'warty excrescences' and, according to a Chinese painting manual, 'the colour of a frog's back', might eventually produce a flowing red substance that was even able to colour gold?[11] To do so, however, the basic ore first had to be pounded and broken down, and then subjected to the fiery smelting process, though, even after this 'death-dealing' preliminary smelting, the metal obtained still contained impurities, and a further heat treatment was required in order to produce shining copper, a beautiful, purified substance of an entirely different order from its original source.

LEFT fig.165 Statuette from Cyprus of the goddess Astarte standing on a copper oxhide ingot. Cyprus was a major exporter of copper during the Bronze Age, and here the fertility goddess, who was also widely worshipped in Syria and Palestine, presides over this trade. Cyprian copper later became associated with the Greek love goddess Aphrodite, including in alchemy. (Ashmolean Museum, Oxford, AN1971.888. *c.*1250–1100 BCE.)

195

What particularly fascinates Graeco-Egyptian alchemists is copper's capacity to change colour in this fiery process. Thus, when writing on the 'arts of the Egyptians' in *Book 5 of Democritus to Leucippus*, a work dealing exclusively with copper, Democritus says 'copper changes colour like a chameleon',[12] a creature famous for its astonishing ability to change colour quickly in order to blend with its surroundings and control its body temperature. Zosimus, too, quoting the authority of Agathodaimon, describes the process of refining copper in terms of alchemy's colour sequence, culminating in a gold-coloured metal:

> After the refining of copper, its attenuation and its Blackening, and thereafter its Whitening, there will occur a solid Yellowing.[13]

Seemingly, the ancient Egyptians valued copper's golden sheen, for, according to the New Kingdom *Harris Papyrus*, when vast quantities of copper ingots brought from Atika were placed beneath Ramesses III's window, they shone with the colour of 'gold of the third quality'.[14] So, too, a statue of Pepi II was said to be made of 'Asiatic copper coloured like gold' *(see page 156)*. Like alchemists and African copper-workers, the Egyptians were not averse to tinting copper to obtain the desired colour; they certainly used copper to redden gold, producing the much-cherished 'red gold' used in New Kingdom jewellery.[15]

To the ancients, the valuable properties of stones and metals derived from their colours. Indeed, without the creative activity of a colouring spirit, or *pneuma*, as Greek alchemists called it, which could be removed from one substance and infused into another, a non-volatile 'body' *(soma)* remained utterly lifeless, a 'body of lead', as it was sometimes called.[16]

'Seek a stone that has a spirit [*pneuma*],' says Zosimus, a stone, therefore, in possession of life—and not just any kind of life. For this stone, he says, is to be found in the waters of the Nile, and to find it is truly to discover the secrets of heart-opened existence:

> Go to the waters of the Nile and there you will find a stone that has a spirit [*pneuma*]. Take this, divide it, place your hand in the interior and draw out its heart. For its soul is in its heart.[17]

Just as there is an invisible 'colouring' spirit within this stone's interior that needs to be brought forth, a sacred power hidden in the heart of matter, so there is in malachite. For the miracle of this marvellous rock is that it has an inner life, one that is made manifest whenever copper is obtained.

Not only does copper change colour, its transformations also mirror human development, since, like humans, copper has to suffer, die and enter the 'black' state if it is to be transformed. Only in this way can its life-giving *pneuma* be obtained. That is the message of the late-antique alchemist Stephanus of Alexandria:

> Copper is like a human being and has a soul and body. It is necessary to make the matter of its body orphaned, so that what we have is its spirit [*pneuma*], namely its dyeing element.[18]

He goes on to say that bodies must be crushed and dissolved in order to extract their dyeing spirit—completely broken down and pulverized to obtain the precious, colouring tincture. The implication is that humans, too, must suffer if their souls are to be transformed and regenerated.

It is this mysterious relationship between copper and humans that Maria the Jewess has in mind when she compares copper's colouring power to human digestion:

> Just as we [*humans*] are nourished by means of solid and liquid food, and as we are coloured only by their proper quality, so copper behaves in the same way.[19]

Revered by Zosimus for her alchemical knowledge, which he often quotes, Maria is particularly renowned for her invention of ovens and retorts. A 'kitchen alchemist' deeply interested in cooking, digestion and nutrition in all its applications, she also knows that copper, like humans, needs to be tended and nurtured if it is to acquire a good colour, grow and thrive, and in turn become a nourishing, 'tinting' substance, since it is only by being nourished itself that it can become a life-giving source. 'Like to like' is an alchemical motto, and not only does copper give life, it also receives life, being able to absorb substances and in turn be absorbed by them. 'Copper does not tint

RIGHT fig.166 Detail from *fig.128* showing metalworkers (almost certainly coppersmiths) using foot-bellows to supply the fire with air, enabling them to regulate the temperature for the metal in the crucible. (Tomb of Rekhmire, Western Thebes. 18th Dynasty.)

BELOW RIGHT fig.167 Detail from *fig.10* showing the air god Shu raising his arms aloft as if to support Hathor nurturing the foetal sun god nestling in the warmth of her womb-vessel. Just as copper in the crucible needs air *(see fig.166)*, so Shu creates air for Re's human birth in the copper cosmos, and such scenes suggest the ancient Egyptians, like alchemists, associated copper-working and human life. (Tomb of Ramesses VI, Western Thebes. 20th Dynasty.)

but is tinted', says Maria in the same treatise, and when 'it has been tinted, then it tints, when it has been nourished, so it nourishes.' Only by being cared for with abundant water, air and heat, says Zosimus, will copper grow and bear 'fragrant flowers and fruits' *(fig.168)*.[20]

Maria ascribes nourishing qualities to copper in the same way as traditional African smiths do.[21] Quite possibly, given Zosimus's statement that she transmitted knowledge about the ancient Egyptian 'furnaces' *(see page 182)*, her teachings about copper —its 'humanness', its 'nurturing' and 'tincturing' qualities—also reflect genuine ancient Egyptian beliefs about copper-working, offering a window through which to glimpse some of the Egyptians' lost copper secrets, ones never set down in writing, but passed on orally through the generations.

Interestingly, in his treatise *Apparatus and Furnaces* Zosimus describes a damaged furnace that he saw in the ancient sanctuary of Memphis, then immediately turns to the furnaces and other apparatus Maria describes, as if ancient Egyptian metallurgy and Maria's work are merged together in his mind.[22] Indeed, in their own way, her copper teachings are simply restating plainly what the Egyptians had already implied when copper created

Horus as a winged bird, empowering him to move, transform and grow *(see chapter 14)*, and when they explained how the gestating sun god came 'forth in copper' during the *Book of Day*'s first hour, precisely the hour when Hathor and the serpent goddesses 'nurtured' his *Ka*-life and the air god Shu brought the breath of life *(fig 167)*.

197

TINCTURING POWER:
MOVING BODIES

Just why this relationship between copper and humans is so important is made clear by Stephanus of Alexandria when he says copper acts 'like a ferment' in the body due to its three active qualities —'movement, sensation and passion'.[23] Its tincturing power is an inner force, a volatile energy, vitalizing solid bodies, an invisible 'ferment' making them rise. Interestingly, Stephanus also applies a geometrical metaphor to illustrate copper's hidden power in the soul, comparing it to a 'plane surface in a solid body'.[24] To perceive copper's activity is to see beyond the surfaces of three-dimensional solid bodies to the 'nature hidden in the interior of things', the hidden reality of life itself.

'Movement is passion', declares Hermes in Book 12 of the *Corpus Hermeticum*, before asking his pupil Tat: 'What is the energy of life? Is it not motion?'[25] It is moving, passionate energy that shows everything within the cosmos to be truly alive—and alchemists experience this whenever they watch copper changing in love's fiery crucible, seeing the very same colours Olympiodorus associates with the ancient Egyptian 'work of a day', as the living metal transforms and moves in harmony with time.

Probably the copper-loving Ibn Umail has the same transformation in mind when describing soul qualities in his treatise the *Silvery Water and the Starry Earth*:

> The soul is not seen in the body but it reveals three things: strength, movement and desire.[26]

Hence, when listing a concatenation of names for the sought-after 'stone' at the beginning of his *Book of the Explanation of the Symbols*, he gives pride of place to 'lead-copper':

> Know that the first of the names of the sage's stone is lead-copper [abār nuhas].[27]

So too, when seeking to discover the mysteries of 'vivifying bodies' and obtaining 'life from lead', it is to the Egyptian sanctuary of Venus that the alchemist Krates wends his way in the Arabic *Book of Krates*. There he is transported into the divine presence of the copper goddess herself and encounters the full force of her loving-destructive tincturing power.[28] Long after the Egyptian temples had closed, Islamic alchemists still know that vivifying copper secrets are hidden in the dwellings of the shapeshifting goddess. Her name may have changed, but, true to the spirit of ancient Egypt, hers is the fiery life circulating in 'bodies' and making their lustrous beauty shine (see page 42). For copper is love and love is alchemy.

And life is movement. 'If you see a movement, you know he is alive', sings the 13th-century Islamic poet, teacher and Sufi mystic Jalāluddīn Rūmī in his alchemically-inspired *Masnavi* ('Spiritual Verses'). He, too, knows that copper causes this life and movement, if somewhat capriciously, since he then goes on to say that knowledge is needed to create 'regulated movements' and turn 'the motion of copper into gold':

> If you see a movement, you know he is alive, but this you do not know, that he is full of Intellect until regulated movements appear, and he, by means of knowledge, turns the motion of copper into gold.[29]

Evidently, by the 13th century, copper alchemy had percolated through to Rūmī's Anatolian home in Konya. As the scholar of Islamic mysticism Annemarie Schimmel wrote: 'Persian poets love to speak of the copper which becomes gold when it is touched by the alchemy of Love or by the hand of the beloved, and of the necessity of suffering in the crucible for the sake of purification.'[30]

Like Thoth's transformation of the fleeing Sun Eye in ancient Egypt, Rūmī knows copper's volatile movements need stabilizing through 'knowledge', though his deprecation of copper in the *Masnavi*—

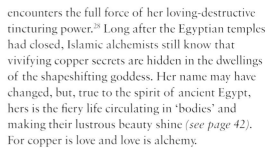

RIGHT *fig.168* Here a fruitful tree symbolizes the copper phase in the alchemical death and regeneration cycle. Reinforcing the connection with Venus is the small scene beneath (often omitted in present-day reproductions), which shows a king and courtiers watching four naked bathing women. Four is the number of copper in alchemy. Copper is also the metal of Venus, and here the allusion is to the biblical King David who fell in love with Bathsheba (2 Samuel 11), whom he spied bathing whilst walking on his palace roof. (Sixth illustration of Salomon Trismosin's *Splendor Solis*, British Library, London, Harley Ms. 3469, f.15. Late 16th century.)

ABOVE *fig.169* Detail from *fig.81* showing a solar boat at the base of two copper naos sistra, here symbolic of the sun's eternal movement in a Hathorian copper cosmos. (South wall of the 'House of the Menit', Crypt South 1, Dendara temple. First century BCE.)

where he often associates it negatively with the 'flesh', the feminine and the sensual, material world, the lower soul or *nafs*, dominated by the animal soul's appetites and passions—creates a split which would have been utterly foreign to the ancient Egyptian perception of copper and gold.[31] The ancient Egyptians knew the copper goddess had to be propitiated, but they honoured an embodied sun god ruling 'humans and gods' together, intimately connected with her and the world of flesh *(see chapter 14)*.

The link between copper and 'movement' is not exclusive to ancient Egypt and alchemy, however, for there is a striking parallel in a creation myth from the Dogon culture of West Africa, telling how the sun is surrounded by eight luminous copper spirals which energize its daily movement.[32] The Dogon also perceive the sun's rays to be 'water of copper'.[33] Meanwhile, according to the

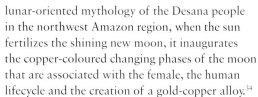

lunar-oriented mythology of the Desana people in the northwest Amazon region, when the sun fertilizes the shining new moon, it inaugurates the copper-coloured changing phases of the moon that are associated with the female, the human lifecycle and the creation of a gold-copper alloy.[34]

Wherever it is found, this metal is alive, and wherever it is worked, something arises between the transforming copper in the crucible and the craft-worker—a magical resonance weaves them together until the very nature of both is mysteriously altered and changed. What affects the metal affects the watcher, for both are woven together in this unified fluid world filled with metallurgical movement and desire, and streaming with life.

WHIRLING COPPER: FOUR AS ONE

There is, though, another reason why alchemists associate copper with movement and change—its 'mixed' or 'composite' nature as an alloy, which Hermes, again in Book 12 of the *Corpus Hermeticum*, explains philosophically, telling his pupil Tat that the very 'identity' of 'composite bodies' causes one body to change into another. It is in their very nature to change. Unusually, he then also draws on the metallurgical metaphor of an alloy to illustrate how nothing dies in this eternally moving cosmos that is a 'plenitude of life', being merely dissolved in order to become new *(see page 129)*.[35]

Correspondingly, in Graeco-Egyptian alchemy, copper belongs to a fourfold 'composite' body called the *tetrasomia*, or 'four in one', formed from the four metals copper, tin, iron and lead, later identified with the four planets Venus (copper), Jupiter (tin), Mars (iron) and Saturn (lead).[36] In fact, virtually all copper in antiquity incorporated these metals, having amounts of lead in it, either naturally, or through having it added in order to make the copper more malleable, as alchemists must have known. Iron, too, was a natural component, and tin was added to harden copper and make bronze.[37] Hence 'four' is the alchemical number of copper.

There is a 'roundness which turns copper into four that consists of one thing', states the alchemical classic known by its Latin title *Turba philosophorum*, not only describing copper as 'composite' and 'four', but also as 'a roundness' and hence a unity.[38] Olympiodorus says the same in his commentary on Zosimus's lost work, if opaquely, when he explains that the *tetrasomia* is initially a 'composite' body in an earthbound 'poor state', which is clearly in need of the alchemical 'art' if it is to display the intended colours of 'black', 'white' and 'yellow' and transform through the human lifecycle from birth to old age:

> The unfortunate one, fallen and chained in the body of the four elements, experiences forthwith the intended colours through the one who subjugates it by means of the art: namely the colours black, white and yellow. Then, having obtained these colours, and having gradually attained adolescence, it reaches old age and finishes in the body of the four elements [the explanatory gloss adds: 'That which signifies copper, iron, tin, and lead'].[39]

Once again, as in Maria's teachings (*see pages 196–7*), copper displays human qualities, for what Olympiodorus means here is that this *tetrasomia*, in its material form and confined to the earthly realm, passes, during the alchemical process, through the colours and human lifecycle: being born, becoming an adolescent and reaching old age, until, at the end of its lifetime, it returns to its original fourfold components.[40] In other words, it experiences the very same lifecycle the ancient Egyptian sun god goes through in the *Book of Day*.

This, though, is not the whole story, since Olympiodorus immediately goes on to describe the 'operation of *iosis*' taking place within the 'circular apparatus', a second process which ultimately brings this substance to its unified, perfected state:

> Completion is reached with them in the operation of *iosis*, as if destroyed by these, and especially no longer being able to escape [the gloss states: 'That is, interwoven with them and no longer being able to escape']. Once more it returns with them within the circular apparatus, keeping at bay the one who pursues it from outside. Now what is the circular apparatus, if not the fire and the cause of evaporation without end, operated in the spherical vessel.[41]

All this might seem complicated and yet another example of alchemy's riddling language (as the text's Byzantine copyist evidently thought, hence the explanatory glosses to help unravel the meaning), but in fact here Olympiodorus goes to the very heart of copper alchemy. For what he is saying is that it is not simply a question of the copper *tetrasomia*'s earthly transformations through the colours black, white and yellow and an evolutionary human lifetime from birth to old age. To be sure, by the very fact of its 'mixed' composition, as Hermes also teaches in Book 12, this *tetrasomia* is indeed subject to change, dissolving back into its original components at the end of life and 'finishing in the body of the four elements'. But Olympiodorus is not interested simply in evolutionary change, in linear development, within a lifetime lived through the ineluctable modality of time past, time present and time future. For, paradoxically, this alchemical copper process is both evolutionary and circular, and what has to happen is that the initial 'linear' process must be repeated again on an increasingly subtle level, which turns the metal back upon itself in order to create a circularity or 'roundness' in the 'purple' phase.

Thus, the *tetrasomia* is transmuted from one level of life to another, qualitatively different, another modality of being, in which time is stretched into another dimension, earth converted into the aerial. The result is that this fourfold substance becomes a stable, indivisible whole, recombined into a higher unity in the purple phase of *iosis*, a synthesis accomplished by turning in an eternally circular motion within the vaporous 'circular apparatus'.[42]

No longer is this volatile substance 'composite' and fourfold. Rather, through this subtle process, it becomes unified and stable—that is, if the preying enemy outside the circular apparatus is resisted and held at bay.[43] 'It is completed, it becomes a whirl', says Zosimus in one of his famous 'visions'.[44]

Unless the *tetrasomia* undergoes this second, much more difficult, 'whirling' iosis phase, as Maria the Jewess makes perfectly clear, 'our true copper' never materializes. (Interestingly, her unusual expression here is distinctly reminiscent

201

of Hathor's epithet 'true copper' at Dendara.) It is precisely this circular motion that ultimately induces 'the nature hidden in the interior' to manifest its beautiful purple colour:

> In thousands of works it is taught how copper is whitened and yellowed properly, but it will not become alloyed properly by doubling [*diplosis*] unless it is changed into *ios* … it is made into a proper alloy only in one single way, by becoming our true copper.[45]

'True copper' is not simply to be understood as a metal here, since coming forth from this erstwhile 'mixed' substance is a unified, shining and living soul, united with its heavenly origin, revealing its eternal nature, and dyed through and through with purple during this circular process (*see chapter 19*).

Immediately after describing the *tetrasomia*'s travails and transmutations, Olympiodorus then briefly mentions the heart and blood, saying how illness or corrupted blood is remedied by forming 'new blood', just as gold manifests in silver like 'fawn-coloured blood'.[46] Something of this 'blood' significance doubtless underlies the purple-red phase of *iosis* reached by the perfected copper *tetrasomia*.[47] No longer are these the 'sick' afflicted metals of a 'composite' earthly body, functioning separately and inharmoniously, for the alchemical 'art' has returned them to a unified state in the circular vessel, like new blood circulating through a healthy, revitalized body, or the healed heart passions radiating life and vitality in the eternal

human soul. For the deep purpose of this copper alchemy is ultimately the reparation of the human heart, the transformation of those afflicted passions that obstruct the flowering of love, as Zosimus indeed reveals to Theosebia in the *Final Count* (*see chapter 19*).

Later images symbolize this 'circular' phase (the *circulatio*, as it is called in Latin alchemy) by a vessel named the Pelican, which is used in the long process of distillation and sublimation, in which a substance is gently subjected to repeated 'rising' and 'falling' within the vaporous flask, ascending and descending until eventually the purified essence is extracted and then condensed. Sometimes a pelican bird is depicted, pecking its breast so that nourishing blood flows to feed its young, which graphically conveys the contact with heart-opened life (*fig.170*).

The circularity is perfectly expressed, too, by the tail-biting ouroboros, and by the circular motion of the 'sun of the year' moving through the signs of the zodiac. As Carl Jung observed, 'The transforming substance is an analogy of the revolving universe, of the macrocosm, or a reflection of it imprinted in the heart of matter.'[48]

The number 'three', the symbol of unity, likewise expresses this circularity, as in the threefold unification of the three elements 'water, air and fire'.

Ibn Umail imagines it in the form of the cosmic egg:

> The egg is like that: without anything strange entering into it, it transforms from one state to another and is changed from one thing into another. Then it turns into a flying bird, like that which was its origin and beginning.[49]

Within this circular egg, where the beginning is the end and the end is the beginning, life moves in perpetual motion, hatching the eternal 'flying bird'—as the ancient Egyptians also celebrated in their midheaven mysteries of life birthed 'from the egg' (*see chapters 2 and 4*).

'Walk to the well [and] turn as the earth and the moon, circling what they love', says Jalāluddīn Rūmī, for 'whatever circles comes from the centre.'[50] And as the dervishes of his Mevlevi Sufi order throw off their black outer robes to whirl in their white garments in their ecstatic dance of 'turning' *(fig.171)*, watched over by their sheikh upon his red sheepskin rug, their revolving around this magnetic centre transports them into the rhythm of cosmic movement, into the mystery of heart-opened eternal life. Drawing deeply from the well of wisdom and its transforming secrets, this Sufi order weaves alchemy's colours of black, white and red into the very fabric of its ecstatic whirling dance.[51]

It is this little-known Egyptian 'copper' tradition, both metallurgical and cosmological in its worldview, that ultimately turns Graeco-Egyptian alchemy into a 'divine art'—something that could never have happened on the basis of the 'technical' recipes alone, the 'noble arts' of Democritus, for many cultures have their metallurgical recipes and craft secrets without necessarily practising alchemy.[52] What creates Egyptian alchemy, on the other hand, are its ancient transformational secrets, cultivated by copper-loving adepts who are familiar with Hellenistic philosophical theories to be sure, but who also know the metal of the love goddess holds the key to knowledge nurtured in the temples of the Black Land.

'Alchemy cannot be reduced to a protochemistry...Everywhere we find alchemy, it is always related to a "mystical" tradition', wrote Mircea Eliade, though, like many others, he assumed it was Gnosticism that provided Greek alchemy's 'mystical' tradition.[53] But Gnostics exiled the love goddess *(see page 212)*, and if Zosimus and Olympiodorus are to be believed, then a much broader canvas is surely needed than Gnosticism, one painted with the copper colours of ancient Egypt.

Indeed, Zosimus himself has yet more to say about copper, taking the trail even further into the Egyptian temples and Hathor's mysteries.

ABOVE LEFT fig.170 Detail from *fig.189* of a pelican pecking its breast to feed its young.

RIGHT fig.171 Mevlevi dervishes from Konya whirling during the opening ceremony in 2002 for the newly-restored 19th-century Sufi sama'khana in the medieval area of Cairo.

NINETEEN

TWO SUNS ALCHEMY:
THE COPPER WOMAN

What is at stake in this copper alchemy is perfectly summarized by Zosimus in his classic statement about 'two suns' in the *Book of Sophe*, whose very title connects it with ancient Egypt, Sophe being a Greek rendering of the renowned Old Kingdom ruler Khufu, said to be the builder of the Great Pyramid at Giza.[1] Interestingly, too, despite all the hostility to Egypt in biblical and Gnostic sources, the treatise's full title, *The True Book of Sophe the Egyptian and the God of the Hebrews, Lord of Powers, Sabaoth* (or at least its title as recorded by later copyists), shows this 'mystical book of Zosimus the Theban' to be wisdom shared by Egyptians and Jews alike.[2]

For some reason, though, the copyists divided the *Book of Sophe* into two separate parts, inserting between them three other treatises about the Egyptian 'arts', including the *Final Count*, perhaps to indicate their shared Egyptian heritage.[3] The first part deals exclusively with the 'practical' treatment of gold and copper, but then, in the second part, before discoursing on the validity of different tincturing methods, Zosimus launches into an extraordinary statement about 'two suns', implying this to be the cosmological wisdom on which alchemy—or *chemeia*, as he calls it—is based.[4] He begins by stating:

> The true book of Sophe the Egyptian and the God of the Hebrews, Lord of Powers, Sabaoth—for there are two sciences and two wisdoms, that of the Egyptians and that of the Hebrews—is more solid than divine justice. In effect, this science and wisdom of things most excellent comes from remote times.[5]

Clearly, this 'science' is ancient knowledge, and Zosimus proceeds to explain it in terms of an incorporeal wisdom that 'no master has produced' since it is:

> Immaterial, seeking none of the bodies plunged into matter and entirely perishable, for it acts without subjecting itself to any change. Now you possess it as a free gift.

So this 'changeless' wisdom is a 'free gift', but then immediately Zosimus goes on to describe a 'fleshly' wisdom which must be actively saved and purified in a 'world-creating' process. To clarify further, he also refers to two different suns: there is a heavenly sun, the 'right Eye of the World' and 'Flower of Fire', and there is an earthly sun, associated with copper, which, if purified, becomes a flowering 'King on the Earth':

> In effect, for those who save and purify the divine soul chained in the elements, or rather the divine *pneuma* mingled with the dough of the flesh, the symbol of *chemeia* is drawn from world-making. For example, just as the sun, the Flower of Fire, is the celestial sun and right Eye of the World, so also copper, when purification makes it flower, is an earthly sun, a King on the Earth like the sun in heaven.[6]

Here in a nutshell this famous Egyptian alchemist goes to the heart of copper alchemy and the cosmological framework within which it is embedded. For what he says here, in a distinctly ancient Egyptian turn of phrase, is that the 'right Eye of the World', the macrocosmic sun, manifests in the earthly realm as a copper king, needing purification in order to grow and blossom, being 'the divine *pneuma* mingled with the dough of the flesh', the very symbol of *chemeia*. Divine in origin this earth-related sun king may be, but through his association with copper, the metal of the love goddess, he acquires a 'flesh and blood' existence, in keeping with the Greek belief that 'flesh' derives from the female.[7]

'Join the male and the female, and you will find what is sought', instructs the copper-loving Maria the Jewess, knowing their union contains the whole secret of alchemy[8]—a union, though, that means, like all earthly life, this copper-gold king possesses a body subject to decay, change, growth and transformation, to all the 'creations, destructions, changes and restorations from one to another' to

LEFT fig.172 Coat of arms emblazoned with 'two suns' in *Splendor Solis*. In contrast to the heavenly sun shining above, the lower sun, displaying the three faces of Saturn, seems unbalanced and hence in need of regeneration. (Detail from the first illustration of Salomon Trismosin's *Splendor Solis*, British Library, London, Harley Ms. 3469, f.2. Late 16th century.)

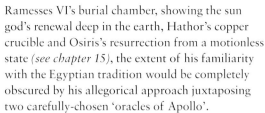

which, according to Stephanus of Alexandria, copper and the elements are also subject. He enigmatically terms the restorations *anakampsis*, literally meaning 'a bending back'. Yet when copper has 'been burnt and restored with oil of roses' many times, he says, it becomes 'without stain, better than gold'.[9]

Amazingly, here is a Greek alchemist concerned, like Zosimus, not primarily with 'making gold', or *chrysopoeia*, as goldmaking is called, but rather with 'purifying copper', a work producing a metal 'better than gold'.[10] Centuries later, the *Rosarium philosophorum*'s author, quoting Ibn Umail, likewise understands the 'copper of the philosophers' to be the sought-after 'gold':

> You should know, therefore, that the copper of the philosophers is their gold. Senior has said: 'Our gold is not the usual gold'.[11]

Indeed, by means of 'our skill', the author says, copper's 'greenness' will be transformed into 'our true gold'.

Manifestly, alchemy cannot be boxed into a simple definition of 'goldmaking'.[12] Nor is this copper alchemy Gnosticism, as Festugière had maintained *(see page 180)*, since it honours the relationship with the Feminine, a 'two suns' alchemy bringing forth a flowering sun king, uniting heaven and earth in a transformational act of 'world-making'. Ramesses III showed 'two suns', identified with the heavenly Aten and the terrestrial sun king ruling in Heliopolis *(fig.151)*, in his tomb at Thebes; now 'two suns' reappear here in Zosimus's *Book of Sophe* as wisdom coming from 'remote times', and they will subsequently be depicted on the stone seen by Ibn Umail during his visit to the Egyptian temple *(fig.153)*.

SOLDERING GOLD: A COPPER UNION

It might be expected Olympiodorus would also have something to say about 'two suns' in his commentary on Zosimus's lost alchemical text, and he does. That he knows the earthly sun's relationship, not just with copper, but also with Osiris, is startling to say the least. Though without the backdrop of the 'two suns' mystery depicted in

Ramesses VI's burial chamber, showing the sun god's renewal deep in the earth, Hathor's copper crucible and Osiris's resurrection from a motionless state *(see chapter 15)*, the extent of his familiarity with the Egyptian tradition would be completely obscured by his allegorical approach juxtaposing two carefully-chosen 'oracles of Apollo'.

The first oracle graphically describes Osiris lying motionless in his tomb, heavily bandaged with only his face visible as he 'fetters and binds the all of lead'.[13] Then, having established this foundation in the 'black' state symbolized by lead, the metal of Osiris and also Saturn, in the second oracle Olympiodorus turns to a love union involving copper. Here an 'alloyed' male, identified with earthly gold, the 'male of the chrysocolla', unites with 'the woman of vapour', who is none other than both the 'Egyptian with the tresses of gold' (Hathor) and the 'Cyprian' (Aphrodite), in other words the two goddesses presiding over copper in antiquity:

> Take the chrysolite, the one which is called 'the male of the chrysocolla', that is to say the man who has been alloyed [together]. These are the drops that create the gold of the Ethiopian earth. There a species of ant extracts the gold, carrying it into the light and rejoicing. Place him with the woman of vapour, until he has been transformed. This is the divine water, bitter and styptic, the one which is called 'the liquid of Cyprus' and 'the liquid of the Egyptian with the tresses of gold'. Coat the leaves of the luminous goddess with this, those of the Cyprian, the Red One, and then thicken it until it coagulates into gold.[14]

This seems another perfect example of Greek alchemy's riddling language. Indeed, it is an utterance from an alchemist dismissed by some as a 'muddle-headed charlatan', not worth taking seriously.[15] What logical thread could possibly connect his bizarre references to the 'male of the chrysocolla', 'ants', 'Ethiopian earth', 'ancient copper goddesses', 'divine water' and 'gold', or link them with Osiris's tomb?

A great deal in fact, given this love union is totally in the spirit of Zosimus's 'two suns' alchemy and steeped also in the ancient Egyptian knowledge of copper's 'tying together' power. For these two oracles belong in a long sub-section tellingly entitled 'On soldering Gold', and, on one level, Olympiodorus clearly has in mind a craft process for joining together gold granules, in which chrysocolla, a copper-based substance (fig.173), is used as the solder.[16]

Much earlier in the treatise he had dealt at length with the 'technical' aspects of treating and washing gold grains extracted from the earth and the work of reuniting them by means of a moderate fire and chrysocolla (see page 195). Now, however, this 'technical' knowledge has become a symbolic love union between Hathor-Aphrodite and the 'male of the chrysocolla', evidently an 'alloyed' male not only needing transformation, but also the 'bitter and healing' liquid of the copper goddesses. And just to make absolutely clear he means 'terrestrial' gold, not a heavenly sun, Olympiodorus weaves into his oracle a legend, well known in antiquity, of gold-digging ants bringing gold to the surface of the earth in Ethiopia, before finally instructing the reader to 'coat the leaves' with the soldering glutinous substance, which will ultimately turn copper into gold, the metal of the sun.

Crucially, this union is founded on Osiris's tomb, where 'lead' is the metal defining the god's death-like state—that is, until copper begins

stirring everything into motion, bringing everything to life, returning everything to the light, 'even from Hades' as Plutarch says (see page 209). Centuries before, in Ramesses VI's tomb, Hathor's 'tying together' copper activity had ensured Osiris's regeneration in the depths of earth, transforming him from a motionless state to copper to gold, reborn with the regenerated sun god (see chapter 15). Now, in these two allegorical 'lead-copper' oracles, Olympiodorus, albeit allegorically, transmits this same 'uniting' Egyptian wisdom, interpreting Hathor's 'tying together' power here in terms of the soldering art, as perhaps the ancient Egyptians also did, since they too used chrysocolla mixed with glue as a solder.[17]

His intent is clear. As an alchemist, steeped in Zosimus's 'two suns' lore, he knows the Egyptian 'divine art' included earthly gold and copper. He also knows the ancient Egyptians dealt with 'technical' operations and 'ritualized these things' (see page 190), and he evidently also knows Hathor's male partner needs 'tying together' for transformation. And through these carefully-chosen oracles, one about Osiris, the other about copper and gold, he welds together the 'technical' and the 'sacred' to teach alchemists about this ancient 'divine art', the origin of their own work, as portrayed all those centuries before in Ramesses VI's tomb and now expressed in the allegorical language of the time to make it understandable. For in this alchemy, copper is the vanquisher of inertia, the mover of life, changing motionless Osiris from lead to gold and 'tying' everything together in a cosmos of love.

GREENING NATURE: THE WAY OF ALL FLESH

Nor is this 'two suns' wisdom confined to Graeco-Egyptian alchemy, for it evidently reached the shores of Europe, to be perfectly encapsulated by the 16th-century Paracelsian alchemist Penotus:

> …we see how the heavenly sun gives of his splendour to all other bodies, and the earthly or mineral sun will do likewise, when he is set in his own heaven, which is named the "Queen of Sheba", who came from the ends of the earth to behold the glory of Solomon…[18]

ABOVE fig.173 Chrysocolla, a copper-based mineral used for soldering gold in antiquity.

Just as the ancient Egyptian queen is a 'heaven on earth' for King Senwosret I in the *Story of Sinuhe (see page 13)*, so alchemy's earthly king is completely encompassed by the female, here identified with the biblical Queen of Sheba. The medieval treatise *Aurora consurgens* calls her the 'Queen of the South'. She is the auroral one, the dawn of Wisdom, a manifestation of the wise female celebrated in the earthy and sensual biblical Wisdom literature attributed to King Solomon, now come to guide the alchemist on her path of love *(see Part 5)*.

One of *Aurora*'s illustrations shows an upper sun shining down onto another sun engraved on a shield—clearly an earthly, 'metallic' sun. Surrounding this lower sun are stylized leaves, crowned at their centre with a grey metallic helmet, the characteristic colour of 'tin' and the planetary metal of Jupiter, hence heralding this lower sun's 'composite', and therefore 'changeable', nature within a greening, leafy world. When a version of these two suns appears in the very first illustration of the 16th-century treatise *Splendor Solis*, moreover, this lower sun seems decidedly unbalanced *(fig.172)*, its suffering expression enhanced by the three faces replacing its eyes and mouth, suggestive of some defect needing to be healed before it can be restored to its original state.[19] As Adam McLean observed, it is a 'picture of the alchemical work of incarnating the spiritual in material form; it brings the Macrocosmic Sun into the lower world (symbolized here by the emblazoning of the image of the Sun on the shield).'[20] And, as the accompanying text makes perfectly clear, this is a 'greening' sun:

> This Stone of the Wise is achieved through the way of greening nature.[21]

This terrestrial sun, this 'Stone of the Wise', shines in everything that grows green and lives, the alchemical 'blessed greenness [*viriditas*]', the colour of the copper goddess Venus, the colour of chrysocolla, the colour of the heart. 'Green is pure yellow love', declares the medieval German *Book of the Holy Trinity*,[22] a work inspired by an alchemical vision of the Virgin Mary as Venus, who 'shines red in the sun' and 'turns green in all fruit'.[23] 'What is not green?' asks the text associated with her 'green shield'. For, as the sun brings 'all things in the heights to gold', so 'everything is green, earth, air, water, fire, sun, moon, stars, people, all creatures…', everything that flares with Venus's beauty is drenched in green, to the extent that the Trinitarian mystery of faith is reflected in this whole world of greening nature.

The same 'greening power' streams through the visions of Hildegard of Bingen, the extraordinary 12th-century German abbess, theologian, doctor and musician, whose notion of *viriditas* Peter Dronke eloquently describes as 'the earthly expression of the celestial sunlight'. 'Greenness', he writes, 'is the condition in which earthly beings experience a fulfilment that is both physical and divine; greenness is the blithe overcoming of the dualism between earthly and heavenly.'[24] He could have been writing about 'two suns' alchemy here, for, as in alchemy, with which she was certainly not unfamiliar, Hildegard seeks a 'fulfilment that is both physical and divine'.[25] 'The soul flows through the body like sap through a tree', she declares.[26] Hers is no life-denying, ascetic work, in which the soul is at war with the body, but rather spirituality utterly grounded in the natural world, where the soul shines in the body like the sun.

The same sunlight dawns at the close of *Splendor Solis*, its final illustration portraying the rising sun, now healed, renewed and restored, coming forth from darkness to bring greenness and renewed life to the shrivelled, bare-branched trees shown in the foreground.

In the *Book of Sophe* Zosimus told how 'two suns' copper alchemy created a flowering 'King on the Earth', while Ibn Umail, whom the author of *Aurora consurgens* quotes extensively, saw 'two suns' inscribed on the stone tablet held by the sage in the Egyptian temple *(fig.153)*. Now this 'greening wisdom' underlies some of the most important and prestigious treatises of European alchemy, having been taken up in a Christian culture imbued with the mystery of the Incarnation, the union of the human and divine. It has certainly travelled a long way from ancient Egypt, yet Zosimus was clear that this 'science' came from 'remote times' *(see page 205)*.

UNITING WORLDS: APHRODITE'S SMILE

What defines this Egyptian alchemy, and sets it so apart from Gnosticism, is this desire to unite the heavenly and the earthly, to affirm the material world, as indeed it must, coming from the Egyptian temples. Plutarch, in his *Dialogue on Love*, composed in the late first century, knows this to be the Egyptian way, for, whilst he makes no mention of alchemy, his concern, nevertheless, is with an Egyptian 'greening' wisdom which, he says, has to be pieced together from 'tiny scraps of evidence'. He tells how the Egyptians reverence not only a 'heavenly' and 'earthly' love, but also the sun as Eros associated with Aphrodite as moon, a 'third love' as Plutarch calls it, a nourishing love that does not approach souls in isolation but specifically 'through the body'.[27] This love mediates where 'the immortal is blended with the mortal', giving 'nourishment, light, and the power of growth to the body' through the radiance emanating from the sun, whilst 'the gleaming ray from love does the same for souls'. It is this 'third love', Plutarch says, that 'graciously appears to lift us out of the depths and escort us upward, like a mystic guide beside us at our initiation', for lovers 'are able to return to the light even from Hades.'[28]

Indeed, if the fiery 'raging element' is excluded, with the soul retaining only 'light, radiance and warmth', then this 'nourishing' love produces:

> a marvellous and fruitful circulation of sap, as in a plant that sprouts and grows, a circulation opening the way to acquiescence and affection.[29]

Intimately united to the body, this Egyptian love makes everything flourish—though it also requires a gentle fire, exactly the kind of refining fire copper-loving alchemists seek in their circulating phase of *iosis (see page 201)*. Plutarch could not have put the matter more plainly—perhaps too plainly for alchemists or Dendara's mystical priests, who celebrated such a nourishing union between Hathor and the sun god in the kiosk on the roof *(fig.74)*, in a temple where Hathor manifested as the 'moon', and 'two suns' were repeatedly portrayed on the temple walls. Plutarch knew this union between the sun and Aphrodite to be a

transforming Egyptian love mystery, mediating between heaven and earth, bringing fruitful growth and life, and even returning lovers from Hades.

Zosimus celebrates this same 'flowering' sun in his *Book of Sophe (see page 205)*, as does Hermes in the Latin *Asclepius*, when he tells Asclepius about the mystery of love and sexual union, which 'you may rightly call Cupid or Venus', a divinizing, fecund love known to those who have 'wisdom and knowledge'.[30] In the famous *Emerald Tablet* Hermes Trismegistus also teaches a wisdom of 'uniting together the power of things above and below' through 'the work of the sun'.[31] Cleopatra, too, in the Greek *Dialogue of the Philosophers and Cleopatra (see page 113)*, delights in the earth's beauty and teaches a harmonious integration with nature. To her, the soul's deification leads not to the rejection of the body and sensuality, but to a fruitful unification of body, soul and spirit, to the resurrection of a divine, living 'statue' where 'everything is united in love'. She is completely immersed in wisdom with passion and heart—wisdom of birth and growth, death and resurrection—and a mystical relationship with the natural world.

This alchemy of happiness also brings joy and laughter. For, after discussing the 'hidden mystery' of copper's transforming colours and 'Aphrodite walking through a cloud' *(see below)*, Stephanus enigmatically refers to the fertile womb as 'the aphrodisiac symbol of joy and love, which is laughter', and goes on to say how 'the melters of gold, understanding what they say, say this, "they laughed"', as they work according to the 'Egyptian method'.'[32]

Just as joyful devotees of Aphrodite encounter her famous smile in the Greek love mysteries, and the Egyptian sun god laughs when Ptah forges his copper firmament *(see page 153)*, so alchemists experience the same smiling revelation through their gold-copper work. Aphrodite's smile is called in mortal language 'beauty'. It is 'the revelation of the soul's essence', wrote the psychologist James Hillman, and he went on:

> All things as they display their innate nature present Aphrodite's goldenness; they shine forth …
> it is the way in which the gods touch our senses, reach the heart and attract us into life.[33]

Whenever Aphrodite smiles, whenever things 'display their innate nature' and make the hidden manifest, 'the gods touch our senses, reach the heart and attract us into life'. This love goddess lives where the invisible and the visible meet, the human and the divine, heaven and earth. She unites the 'immortal and mortal' in her beautiful copper realm—as Plutarch realized when piecing together the mystery of Egypt's 'third love' from 'tiny scraps of evidence'.

MELTING COPPER: THE REFINER'S FIRE

All of which makes the ending of the *Final Count* so perplexing. It seems to strike a blow to the very heart of the Egyptian 'divine art', the secrecy of which Zosimus had carefully described to Theosebia at the beginning of the treatise *(see page 181)*. Later in the text he elaborates further, telling her that Hermes engraved the 'genuine and natural tinctures' in 'symbolic characters' on stelae hidden away in Egyptian temples, making them impossible to decipher 'without a key'. Clearly, he still has in mind the Egyptian temple tradition, but something else concerns him too—copper. For he then cryptically associates these Hermetic 'tinctures' inscribed on stelae with a multicoloured 'melting' operation:

> The other genre of propitious tinctures, the genuine and natural tinctures Hermes has inscribed on stelae: 'Melt the only thing which will be greenish yellow, red, the colour of the sun, pale green, yellow ochre and green, being extracted from the black and residue' ...[34]

Which substance, or 'only thing', is to be 'melted' here Zosimus leaves unsaid, understandably so in light of his previous remarks about Egyptian craft secrecy. Not wishing to betray the Egyptian tradition he reveres, perhaps he feels hesitant even to give out this little morsel, especially as he tells Theosebia that only an initiated person, able to repel the daemons, 'will obtain the desired result'. But the substance he means must surely be purified copper, given the 'only thing' alchemically exhibiting this sun-like spectrum of vivid green, yellow and red colours 'extracted from the black and residue' is this gleaming metal.[35]

Indeed, a Syriac treatise entitled *On the Working of Copper* is far more explicit, referring to the 'books of Hermes' kept in the Egyptian temples in the very same context as copper-working.[36]

Yet purifying copper is never simply a 'technical' process in alchemy. According to Stephanus of Alexandria, to see the 'hidden mystery' of its changing colours is to experience 'Aphrodite walking through a cloud'.[37] It is to discover the Hermetic 'multicoloured etesian stone' and 'cultivate the Muse and things of beauty'.[38] Similarly, here in the *Final Count* it is a revelatory experience, for immediately Zosimus advises Theosebia that in order to attain 'the genuine and natural tinctures' and divinization, she should 'calm' her passions and purify her soul:

> Do not roam about searching for God, but sit calmly at home, and God, who is everywhere, and not confined in the smallest place like the daemons, will come to you. And, being calm in body, calm also your passions, desire, and pleasure and anger and grief and the twelve portions of death. In this way, taking control of yourself, you will summon the divine to you, and truly it will come, that which is everywhere and nowhere. And, without being told, offer sacrifices to the daemons, but not offerings, nor [the sacrifices] that nourish and entice them, but rather the sacrifices that repel and destroy them ... so doing, you will attain the genuine and natural [tinctures] that are appropriate to certain times. Perform these things until your soul is perfected. [39]

This text is an awkward stumbling-block for those who would like to position alchemy solely 'within the long history of science' as a forerunner of modern chemistry, since here is a key alchemist teaching about an inner work of soul-making,

RIGHT fig.174 Illustration of Zosimus of Panopolis and Theosebia, standing beside a distillation oven. They were never portrayed in Graeco-Egyptian alchemy, and this scene comes from an Arabic manuscript copied in Cairo in 1270, which contains the earliest-known pictorial sequence of the alchemical process. (*Book of Pictures [Muṣḥaf aṣ-ṣuwar]*, Istanbul Arkeoloji Müzeleri Kütüphanesi, Ms. 1574. 13th century.)

a purifying process based on copper, in a treatise about the temple 'arts' of the ancient Egyptians.

Clearly, Zosimus understands alchemy as self-transformation, though, at the same time, not a process divorced from material operations.[40] For what Theosebia seeks are the very same 'genuine and natural tinctures' he previously associated with a 'melting' process—tinctures also inscribed on stelae in the Egyptian temples and displaying the colours of purified copper. In short, purifying copper in the crucible mirrors the perfecting of Theosebia's soul, the revelation of her innate, divinized nature, made manifest ultimately when she appears as the 'purple-adorned woman', clothed in the *iosis* colour symbolic of eternal life.[41]

Yet, utterly out of the blue, Zosimus tells Theosebia, when she has been 'perfected', to 'spit on matter', an instruction to which no ancient Egyptian would have subscribed, and one that is utterly perplexing given all he has previously said about the Egyptian temples:

> When you realize that you have been perfected, then, having found the natural [tinctures], spit on matter, and hastening towards Poimenandres [sic], and receiving baptism in the mixing-bowl, ascend quickly to your own people.[42]

Certainly 'spitting' could have a healing, creative function in the ancient world, but Zosimus's advice, often quoted in isolation to illustrate the spiritualizing tendencies of his alchemy, seems tinged with a dualist, Gnostic-inspired worldview setting flesh against spirit.[43]

Even in the *Asclepius*, which perhaps of all the Hermetic writings is most overtly ancient Egyptian, Hermes displays this same disdain for the natural world when teaching Asclepius about the god-like powers of the human being. Despite all the marvels this 'twofold' human can perform as the steward of earth, still union with divine and heavenly life means despising 'the part of him that is human nature'.[44]

The same split runs through the *Poimandres*, the first book of the *Corpus Hermeticum*. 'Desire is the cause of death', says Hermes. Desire binds the soul ultimately to sensual, earthly passions, the cycle of generation and time-bound existence, and is a

hindrance, therefore, to attaining the sphere of the transcendent God.[45]

Taken to its extreme, such dualism spread rampant hostility towards the female who enveloped the divine *pneuma* in flesh. Her unclean garments needed to be discarded in order to attain divine union. Seemingly this outlook infected even alchemists, with some medieval practitioners directing vituperative tirades against copper Venus, whose womb they regarded as the source of all ills.[46]

How to understand the soul's immortality preoccupied late antique thought, yet all this is worlds away from the *Book of Sophe*'s 'two suns' Egyptian wisdom—and from the kind of Egyptian love union between the sun and Aphrodite described by Plutarch in his *Dialogue on Love*. All too easily, in the hands of male alchemists, Egyptian copper wisdom can veer towards a mortification of the passions rather than their harmonization, towards 'perfecting' the purified soul rather than revealing its 'ripeness' and 'fullness', towards an ascetic denial of the world rather than a celebration of the union between 'above' and 'below'. And ultimately, in the *Final Count*, Theosebia's soul work is subordinate to the incorporeal, transcendent world of Poimandres, a dualist vision essentially distorting the ancient Egyptian tradition on which its copper alchemy is based. Indeed, it is this Hermetic-Gnostic thread of Zosimus's work that has so mesmerized modern-day commentators, who are seemingly unaware of a quite different Zosimus, steeped in a greening, unifying 'two suns' wisdom, an Egyptian 'world-creating', not a 'world-rejecting', philosophy.

As elusive and mercurial as the alchemical figure of Mercury himself, Zosimus's alchemy is, in the end, impossible to pin down within some all-encompassing philosophy. Much of what is known about him comes from Greek treatises bequeathed by later Byzantine compilers, who chose to copy, or simply summarize, what most interested them from his considerable writings.[47] Intriguingly, the Zosimus who features in Syriac treatises seems much closer to religious cult life, and interested in talismans, as well as the creation and colouring of statues.[48] He also associates copper-working techniques with the Egyptian priests.[49] Yet despite such glimpses, Syriac sources, like the Greek ones, are silent about where he and Theosebia practised their alchemy, and even whether they did so together or separately. Such biographical details were of little concern, since what mattered was belonging to a tradition passed on through the generations, to the extent that alchemists had no qualms about attributing their writings to mythical or ancient figures.

In short, alchemy has many tributaries. Moreover, once the Sulphur–Mercury theory of metals became well established in Islamic alchemy, adepts could differ considerably in the substance they chose to work with as the first principle. Some favoured mercury, others copper, though they could also weave quite happily at times between them to create a rich, symbolic brew.[50] Which is why Zosimus's statement in the *Book of Sophe* is so hugely important, for it articulates a copper 'science' from the Egyptian temples that eventually flowed into some of European alchemy's most important treatises.

Though strangely, it is not Zosimus who holds the key to this transmission, but his beloved 'sister', Theosebia, to whom the final part of this book now turns, perhaps inevitably, given Hathor's alchemy honours the mysteries of the queen, which are also closely linked with Zosimus's hometown at Akhmim/Panopolis.

RIGHT fig.175 Colossal limestone statue of Meritamun in its original location at Akhmim/Panopolis. She was the daughter of Queen Nefertari and Ramesses II, and her statue belonged to a temple dedicated by Ramesses to the city god Min. Akhmim was an important centre of alchemy, evident already in the work of Zosimus of Panopolis in the early fourth century. (19th Dynasty.)

213

PART 5

ALCHEMY'S QUEEN: THE WESTERN TRADITION

LEFT fig.176 View of the entrance to the tomb of Dhū'l-Nūn in the Southern Cemetery, Cairo. Born in Akhmim in the eighth century he is revered as the 'head of the Sufis' and transmitter of the ancient Egyptian secrets.

TWENTY

THE SECRET EPISTLE:
A MISSING LINK

It has been said Theosebia was 'among the most influential exponents of alchemy in her day',[1] yet her notoriety stems mainly from the Greek letters Zosimus addresses to her in which he discourses on alchemy. Syriac sources are a little more forthcoming, painting a picture of a woman who was a priestess, teaching alchemy in small groups she founded, their members bound by secrecy and an oath of allegiance, much to the chagrin of Zosimus, who obviously disliked some of her companions.[2] In the *Final Count* he refers to one as a 'pseudo-prophet',[3] and elsewhere he calls them 'uninstructed' and names two of them, a female alchemist called Paphnutia and her companion Nilus, whose ineffective alchemy he derides.[4]

Clearly, there were rival approaches to the 'art', and while in later times alchemy may have become 'a science engaged in by men … by men without women', as Gaston Bachelard claimed, this was not how it was practised in antiquity.[5] Zosimus greatly admired Maria the Jewess, the 'divine Maria' as he called her, quoting extensively from her work, which is sadly only preserved in the writings of later alchemists. Similarly, it was to the alchemist Cleopatra that a gathering of 'philosophers' turned for enlightenment, marvelling at her amazing degree of insight into the mysteries of unifying 'above' and 'below'. Women were role-players in antiquity, not acolytes, and were instrumental in transmitting alchemical knowledge.

The anonymous author of the Arabic *Epistle of the Secret (Risālat as-Sirr)* knows Theosebia as the 'daughter of Ašnūs and mother of the priest [or gnostic] Hūn'.[6] What catches the eye in this particular treatise, however, are not Theosebia's biographical details, but rather a letter she has written to Hermes, asking him if he will reveal:

The secrets of the art and its hidden knowledge that the ancients represented in riddles, its names they concealed and its nature, which they hid from the ignorant ones.[7]

To find an 'Arabic Hermes' being asked to reveal the knowledge 'the ancients represented in riddles' is certainly intriguing.[8] That his ensuing revelation is based on a union between a 'king and four females' is even more surprising *(see chapter 21)*. A king associated with four? Four is the number of Hathor at Dendara. It is also the alchemical number of copper *(see page 200)*, the purified metal Zosimus says creates a microcosmic sun, a king on earth *(see chapter 19)*.

According to its introduction, this little-known Arabic treatise was discovered in Akhmim during a visit to Egypt by Caliph al-Ma'mūn (which he made in 832). He was particularly noted for his interest in antiquity and established the 'House of Wisdom' in Baghdad for the retrieval of ancient wisdom, which spawned a huge translation movement of Greek and other texts.[9] Akhmim, the capital of the ninth Upper Egyptian nome, was known to the ancient Egyptians as Khent-Min, after its chief god, Min, to the Copts as Shmin, and the Greeks as Panopolis. It was Zosimus's home town, and the *Epistle of the Secret* was allegedly found in a tomb there, written in a 'foreign' script on a golden tablet which had been placed beneath the head of a beautiful lady, 'robed in seven garments tied together by a single golden knot'.[10]

To discover arcane wisdom in a secret place is standard alchemical narrative, though it is unusual that the keeper of these secrets is a lady lying in her tomb. Surely this is none other than Zosimus's alchemical 'sister' herself, given that Hermes says his revelation must be placed close to Theosebia's body after her death.[11] She is said to be surrounded by 'small beds on which lay the dead with the appearance of youths', a peculiar detail suggestive of an author with a keen 'archaeological' eye for Roman-period burial customs in the Akhmim region, typified by a well-known group of cartonnage mummy cases made mostly for women and adolescent boys. These depict the women with arms outstretched alongside their bodies *(fig.178)*,

Gnostic salvation teaching.[15] Perhaps entrenched opinion that early alchemy was a Gnostic creation influenced him. Perhaps he was affected by his view that Egyptian priests were never connected with metallurgical operations, and metallurgy had no cult significance in ancient Egypt.[16] Perhaps too much was at stake should this Arabic text's indebtedness to ancient Egypt be recognized, thereby contradicting the notion that Islamic alchemists were responsible for transforming the 'inchoate' Greek 'science' into its 'mature' and definitive form.[17] But whatever the reason, whilst his publication rescued this key text from oblivion, unlike so many Arabic treatises languishing unnoticed in manuscript collections worldwide, his conclusions effectively severed the *Epistle of the Secret* from its ancient Egyptian roots and also reinforced the well-rehearsed assumptions about alchemy's 'Gnostic' origins.

Not available to Vereno at the time, however, was the publication of the temple of Repit at Athribis, located on the west bank of the river Nile across from Akhmim/Panopolis. Built at the end of the first century BCE by Ptolemy XII, who also contributed to Hathor's temple at Dendara, this temple must once have matched Dendara in its magnificence, though today many of its stones have long since been pillaged, leaving behind just a tantalizing glimpse of the deities once worshipped there—the fierce leonine goddess, Repit, her fertile consort, Min-Re, and the child god Kolanthes.[18] With its huge Hathor columns once standing in the pronoas *(fig.179)*, there would have been no mistaking this was a temple dedicated to the fiery Sun Eye, closely tied to Dendara's fourfold goddess, radiating her loving-destructive power in the Akhmim region. In fact, the king is shown propitiating Repit, shaking his sistrum for the 'mistress of music', a fleeing Eye goddess, whose return to her consort, Min-Re, heals his blindness.[19] She is also a deity for whom priests bring beautiful cloths from Akhmim across on the east bank, wending their way in joyful procession surrounded by sweet-smelling perfumes. These cloths, so the inscription says, took 121 days to produce, indicative of Akhmim's famous unbroken tradition of dyeing and weaving textiles.[20]

wearing long, close-fitting dresses, brightly coloured in pink and white, and knotted at the breast, the adolescent boys being shown wearing a tunic, cap and mantle. Presumably, when the anonymous author wrote this treatise, these mummy cases still lay in their tombs, not yet plundered and scattered in museums worldwide.[12]

Intriguingly, Ingolf Vereno, in his publication of the *Epistle of the Secret*, repeatedly drew attention to its strong sense of ancient Egyptian cult life. To him, Theosebia seemed like an incarnation of Isis,[13] and her king like the solarized Horus-Re at Edfu, whose companion was Hathor of Dendara.[14] Vereno was certainly on the scent, yet surprisingly, in the end he concluded Hermes's revelation was a

Inscribed in the temple, too, are the names of the day and night hours;[21] and clearly, as at Dendara, the birth and suckling of the divine child, 'the seed of the sovereign', are hugely important. The temple building itself is described as a 'work' of the nurturing goddesses, 'who hold out their arms bearing the child'.[22]

As lord of the Eastern Desert, close to the valuable gold mines, Repit's companion, Min-Re, is a metallurgical god 'who sees gold'. He is, too, a god possessing self-regenerative power, a great bringer of fertile life.[23] Indeed, when his favourite lettuces are offered to him, he tells the king, 'I cause the nursing women to rejoice at your sight', his words being amplified by those of the unidentified god alongside: 'I cause your phallus to impregnate the women.'[24]

Just like the *Epistle*'s anonymous king, Repit's companion is a heated deity, brimming with virility, and here in this temple there is evidence enough to root the *Epistle of the Secret* in the kind of ancient Egyptian wisdom cultivated much earlier at Akhmim/Panopolis in the Graeco-Roman era. In fact, judging by an early fourth-century letter from Panopolis referring to a priestess attached to Repit's temple, it must have been still functioning during the lifetime of Zosimus and Theosebia.[25]

LEFT *fig.178* Roman-period mummy from Akhmim, typical of the distinctive group of mummies from the region, characterized by their pink and white colours, and by the women with their arms outstretched along their bodies. They are accurately described in the Arabic *Epistle of the Secret*, which was allegedly discovered in a tomb in Akhmim. (Allard Pierson Museum, Amsterdam, APM 00723.)

RIGHT *fig.179* Hathor capital in the Ptolemaic temple at Athribis dedicated to the leonine goddess Repit (Triphis) and her consort, the fertile god Min-Re, who were both worshipped in the Akhmim region. The temple is located on the west bank across from Akhmim/Panopolis and dates from the reign of Ptolemy XII, who also built at Dendara. Repit was closely identified with Hathor, Eye of Re, though, unlike the Dendara temple, the Athribis temple is not well preserved. What evidence remains, however, indicates its cults were in tune with the 'ancient knowledge' Hermes teaches in the Arabic *Epistle of the Secret*. (First century BCE.)

CROSSING CONTINENTS: THEOSEBIA'S LETTER

Not only does this Hermetic letter provide a bridge back to ancient Egypt, it also connects forwards to some of the most influential European alchemical treatises, not least *Aurora consurgens, Donum Dei* and the *Rosarium philosophorum*, which it has long been suspected are indebted to Islamic alchemy, without the necessary proof being available to pinpoint this influence. Indeed, they replicate the *Epistle*'s themes with such accuracy, both in image and word, that this little-known Arabic treatise must surely be regarded as a key work in the transmission to the West. To be sure, the known Arabic manuscripts preserving the *Epistle* all postdate *Aurora consurgens*, but already in the tenth century Ibn Umail fleetingly refers to its union of the male with four females *(fig.180)*, so the later versions must have been based on a considerably older text, and one sufficiently important to warrant a mention by the most-quoted Islamic alchemist in *Aurora consurgens*.[26] The consensus is startling, the resemblances unmistakable, though behind them is a vast unmapped terrain with few signposts pointing to the complex avenues through which alchemy reached Latin Christendom.[27] Or, indeed, to how Islamic alchemists themselves came into contact with much older ancient Egyptian knowledge.[28]

It is well known, however, that, as Islam began to spread during the seventh and eighth centuries, Muslims sought out ancient knowledge wherever they settled. Greek philosophy, all the natural sciences, medicine, alchemy and Hermetic wisdom more generally, as well as astronomy and astrology, aroused intense interest in the nascent Islamic world, even if, at times, the Hermetic 'sciences' were forbidden, due to suspect links with esoteric gnosis and mysticism.

Similarly, when this knowledge reached medieval Europe through Islam, due to their increasing contact, it created a period of intense interest in the workings of nature and humanity's place within the natural world. This was a time when people were seeking a personal experience of divinity, needing to know what was happening inside their own souls, and so manuals began to appear giving guidance about the soul's journey. People were looking beyond the limits imposed by traditional ecclesiastical authority and exploring new ways of synthesizing ancient philosophies with the Christian religion, fostered by a search for the deeper relationships between the visible cosmos and the wisdom of the invisible creator, the 'sympathies' binding earth to heaven, as reflected also in sacred scripture. Nature was a sacred text, redolent with hidden meaning, and just as the Jewish sapiential books promoted a long tradition of nature as the sphere of female Wisdom, so Hermetic alchemy spoke to these medieval yearnings, directing human experience towards knowledge of the divine not by excluding nature, but by working with her.

The light of intellectual discovery lit up medieval Europe, creating what has been called a '12th-century Renaissance'. The lamp of love shone too, fuelled by veneration of the lady, which swept like wildfire among the Islamic and Christian 'faithful of love'.[29] Troubadours sang praises of their lady and poets honoured their Venus, worshipping her in the guises of Lady Wisdom, the Virgin Mary or Mary Magdalene, all of which provided fertile ground for Egyptian alchemy to flourish. As Ibn Umail said, alchemy can be known by 'people of any religion'.[30] And this love alchemy was certainly in tune with the spirit of the age, indeed helped shape it, flowing, if sometimes disguised, in surprising channels. Not though as dogma, even if at times medieval alchemists seemed simply to be searching for alchemical proofs of their Christian beliefs, demonstrating their validity with recourse to the natural world. For whenever Lady Love calls and the heart responds, her presence starts to have an effect on the human soul, and as her transforming vision captured the imagination of the medieval world, so it brought alchemy to a new time and place.

RIGHT fig.180 Illustration from *Donum Dei* showing four female faces surrounding a mating couple; above them in the neck of the flask is a winged child. This treatise is based on the alchemical principles taught by Hermes in the Arabic *Epistle of the Secret*. (Bibliothèque de l'Arsenal, Paris, Ms. 975, f.13. 17th century.)

SEEKING THE STONE: TWELVE PRINCIPLES

Alchemy has never been territory without a map, but the *Epistle of the Secret* sets out the principles on which it is based as never before, opening up a completely new vista in which to view the European 'marriage' treatises, with their enigmatic content and emblematic illustrations.

The four colours had already indicated there was a definite pattern underlying Graeco-Egyptian alchemy *(see chapter 17)*. Their sequence was also tenaciously retained in later texts and images, so much so that, even without a text, illustrations can sometimes tell their own symbolic story within a complete 'colour' paradigm. Also, Zosimus's classic *Book of Sophe* statement about 'two suns' provided, in its way, a 'theoretical' framework *(see page 205)*, fleshed out a little by Olympiodorus with his allegorical oracle about the union of the 'male of the chrysocolla' and the bitter and healing 'woman of vapour', the copper love goddess Hathor-Aphrodite *(see chapter 19)*. One of the reasons why Olympiodorus wrote his allegorical treatise, so he said, was to dispel the confusion caused by the multiplicity of explanations about the art circulating even in his own day.[31] But neither he nor Zosimus precisely defined the principles, at least not in the discursive language Hermes uses here in the *Epistle of the Secret*, which is nothing short of a full-blown exposition setting out the fundamental 12 'steps' of transformation from beginning to end, and, in guiding Theosebia through them, this expert way-shower uncovers alchemy's hidden tracks.

True to the unifying spirit of the *Emerald Tablet*, Hermes tells Theosebia the 'art' ultimately comes from 'one thing'. However, though its ultimate goal is unity, it is also founded on two 'operations', four 'unions' and 12 'divisions', and, in words echoing the two divisions seen by Ibn Umail on the sage's stone in the Egyptian temple *(see page 179)*, he then proceeds to define the two 'operations'.[32] The first he calls the 'Whitening', which is associated with the moon and 'without shadow'; the second is solar, red and also 'without shadow', being the 'work of the Sun' (in other words the 'Reddening'). The four 'unions' Hermes defines as 'burning',

'nourishment and rusting', 'colouring' and 'separation', though he also says some ancients identify these with the four 'keys' of 'mixing', 'putrefaction', 'birth' and 'colouring'. Like Hermes himself, alchemical language is slippery, never completely fixed; there are nuances and variations, and evidently alchemists could vary in their terminology—which is a delightful reminder not to box their work too tightly into a rigid conceptual framework.

Clearly, though, Hermes here draws on an extensive Egyptian tradition. For in *Book 5 of Democritus to Leucippus* on the 'arts of the Egyptians', Democritus likewise divides 'the work' into a two-colour operation, albeit with 'Yellowing' defining the second operation rather than the characteristic 'Reddening' of Islamic alchemy *(see page 187)*. Thus, he tells the doctor Leucippus:

> The work comprises the Whitening and the Yellowing, as well as the softening and roasting of copper ore.[33]

Interestingly, Democritus is specifically relating these operations to copper, and in fact throughout the treatise he is keen to explain the various 'technical' processes of preparing copper for tincturing purposes, comparing the alchemical recipes to the way doctors make their remedies.

According to Ibn Umail, 'reddened copper' is the goal of the alchemical work, the 'stone' able to withstand the heat of the fire, and whilst Hermes refrains from mentioning the metal directly, he must surely also have copper in mind here in the *Epistle* when defining the two 'operations'.[34] Indeed, in the much-copied European treatise *Donum Dei*, the text beneath the flask depicting the 'philosophical putrefaction' quotes Hermes and explicitly instructs:

> …therefore smelt our copper [aes] with a slow dry fire…
> Care for it as a nurse until its body is constituted and its tincture extracted.
> But do not extract it all at once, rather little by little it will come forth, until at length it is all completed at the right time.
> Hermes, father of the philosophers.
> I am black, white and red, white and yellow.[35]

Here, within this putrefying, all-encompassing blackness, copper's tincture is to be extracted at the right moment, it has its own timing, beyond the reach of any precipitous action. The closing words, 'I am black, white and red, white and yellow', herald the power to move, transform and grow through the whole spectrum of alchemy's gleaming colours, working with a metal particularly valued in the Middle Ages for the life and light it radiated.[36]

In fact, 'putrefaction' is the second of the 12 divisions Hermes defines in the *Epistle of the Secret* after he has described the two 'operations' and four 'unions'. These divisions, he says, are correlated with the zodiac, harmonized, therefore, with the sun's annual passage through the zodiacal signs.[37] Hermes thus perpetuates the astrological outlook widely cultivated at Akhmim during the Graeco-Roman era, evident in the zodiacs adorning the ceilings of tombs in the region. He also divides these 12 divisions into nine 'fundamental' steps and three 'subsidiary' ones, setting out the nine fundamental ones as follows:

'Marriage', 'Putrefaction', 'Setting in Motion', 'Ascent', 'Return', 'Mixing', 'Watering', 'Solidifying the Stone' and 'Roasting'.

These are then completed by three subsidiary phases:

'Conception', 'Birth' and 'Rearing'.

Clearly, this 'ancient knowledge' Hermes is teaching is a birthing mystery, the reactivation of the cycle of life through successive phases, initiated by a procreative 'marriage' or 'union', which then leads to the inchoate, black phase of 'putrefaction', followed by 'setting in motion' and 'ascent'. Then comes the second, or 'Reddening', operation, incorporating the phases of 'return', 'mixing' and 'watering', resulting in a perfectly cooked and digested 'stone', and ultimately completed in fiery 'roasting'—that is, if all the obstacles have been successfully surmounted.

This is the map of Hermetic 'marriage' alchemy, but it is important not to confuse it with territory, since, like any map, it needs to be experienced if its real meaning is to unfold. It is also deeply embryological, having 'conception, childbirth and

rearing' at its core—as had the sun god's life with Hathor in the *Book of Day*, Osiris's rebirth in the Khoiak rites, the Ptolemaic cult for Repit in her temple at Athribis *(see page 219)* and Cleopatra's teaching in the Greek *Dialogue of the Philosophers and Cleopatra (see page 113)*. Whether flowers in the fields, stones in the earth, substances in the flask, a child in the womb or the alchemist's soul itself, to encourage everything towards 'maturation', towards revelation of its own innate beauty and life, is the endeavour of Egyptian love alchemy.

Importantly, too, whilst 'putrefaction' defines the second phase, it belongs within a very different alchemical context than when it occurs in a regenerative 'death' process, as, for example, in Zosimus's Greek 'visions' *(see page 169)*, where the instruction is given to build a temple of 'one stone', having neither beginning nor end and a source of pure water within. To find its entrance requires 'taking the sword in your hand', and once inside, the priest, the 'Man of Copper', will be seen changing from his natural colour into a 'Man of Silver' and then, shortly afterwards, a 'Man of Gold'.[38]

Not 'marriage' or 'union', but rather 'wielding a sword' initiates this death and regeneration drama, as the introductory scenes in the 16th-century treatise *Splendor Solis* perfectly illustrate. The first illustration depicts a temple entrance *(fig.172)*, the second one an alchemist holding a flask, and then, in the third, a martial knight is pictured, sword in hand, standing astride twin fountains spouting red and white water.[39]

Splendor Solis is often grouped with the 'marriage' treatises and, insofar as they represent complementary aspects of wholeness, they are linked. But its Osiris-like death and regeneration sequence is far closer in spirit to Zosimus's 'visions' than to 'birthing' alchemy. To be sure, the imagery can overlap, including the 'two suns' coat of arms depicted in both *Aurora consurgens* and *Splendor Solis (fig.172)*, but, like the ancient Egyptians, alchemists worked with both the 'daytime' and 'night-time' cycles, which can be confusing, yet are essential to differentiate.[40] What concerns Hermes here in the *Epistle*, though, is the daytime cycle.

RECASTING WISDOM:
THE GODDESS RETURNS

There is no alienation from nature in this birthing process, no dualistic division of matter and spirit as in Gnosticism, no rejection of the material world as an illusion which has to be abandoned in a hierarchical 'vertical' mode of ascent. Time is not a hindrance, rather it is a means of bringing into manifestation what is already there in 'potential' at the very beginning, an alchemy attuned to the movement of the sun through the zodiac.

Dendara's astrology-loving priests had built the southern crypt around the same 'birthing' principles ruled by the four Hathors *(see chapter 7)*, and clearly their wisdom had not been lost. For at a time when religious life in the Graeco-Byzantine world had increasingly marginalized women's roles, bolstered by a Platonic metaphysics that devalued 'biological' birthing' in favour of 'spiritual' procreation, it was alchemy that helped keep alive Hathor's 'birth' mysteries, which Islamic alchemists were now rediscovering and making them their own.

In fact, the Arabic *Epistle of the Secret* captures the mythic spirit of ancient temple life in a way quite unlike anything in the Greek alchemical writings. To anyone familiar with the Crypt South 1 at Dendara, where Pepi I is shown honouring four Hathors *(see page 97)*, the *Epistle's* union of a king with four females has a completely authentic ring about it. Both Zosimus and Olympiodorus stated their wisdom came from the Egyptian temples, but such specific 'temple' details were veiled in their Greek writings—understandably so, since it is well-known that Egyptian mythological symbolism was considerably 'pared down' by writers influenced by Greek philosophy and thought.[41] Interpreting it allegorically or philosophically, they sought to 'render it plausible' for a Greek audience who might have had difficulties relating to 'the plethora of arcane divinities encountered in the ancient texts'.[42] The *Epistle's* author, however, seems untroubled by this mythic dimension.

Certainly, Islamic alchemists still knew the Dendara temple, and according to the Arabic *Epistle of Hermes of Dendara*, in which Syrian

elements have been detected,[43] this particular letter was discovered beneath a statue of Hathor *quadrifrons* (called Artemis) in the temple. All of which suggests a much more complex process of transmission than a straightforward copying, or reworking, of Greek alchemical sources, though how this happened is shrouded in mystery.[44] What matters, though, is that it did happen, and once Hathor's secrets touched the heart of the *Epistle's* anonymous author, they began to exert their powerful influence again, transformed into something quite new in the world of Islam, yet still very recognizably ancient Egyptian.

To be sure, Vereno called the *Epistle* an 'extremely heretical text' from the standpoint of orthodox Islam.[45] Yet, interestingly, the *Qur'ān* itself affirms the developmental cycle of life as an important aspect of creation, specifying the various phases of embryonic development in the womb leading to the birth of an infant, who is then nurtured and brought to maturity. If the Creator can accomplish this miracle of life, says the *Qur'ān*, then surely a person can also be taken through the further phase of resurrection after death.[46] Hence, from this perspective, together with its emphasis on 'unity' *(see page 222)*, the *Epistle's* 'science' is not incompatible with *Qur'ānic* revelation. Indeed it might even enhance a living experience of faith.[47]

Curiously, too, when its Hermetic 'marriage' is reinvested with a sacramental Christian outlook in *Aurora consurgens*, overtly projected onto the daytime cycle of the hours ruled by Lady Wisdom, it comes far closer to the ancient Egyptian 'work of a day' *(see chapters 22, 23)*.[48] Manifesting as the glittering, starry Queen of the South, she comes to call the alchemist to follow her path, a 'science' begetting the alchemical colours of 'yellow and red, midway between the white and black'. Hers is a 'science' belonging to the 'rising dawn', that joyful 'golden hour' called 'Mother of the Sun', a betwixt and between time which is the 'true golden hour at the end of their labours' for those who 'correctly perform the various operations', neither complete light nor complete darkness, a time when the invisible gateway opens for those longing to commune with her.[49] Through a mosaic of quotations, taken especially from the Jewish

RIGHT fig.181 The Queen of the South and her attendant, accompanied by a small child holding a vase aloft, stand before King Solomon. The silver vase is decorated with a star, the symbol of quicksilver and the planet Mercury, and the queen here manifests as Wisdom. (*Aurora consurgens*. Glasgow University Library, Ms. Ferguson 6, f.237r. 17th century.)

Wisdom books, the Psalms, the *Song of Songs* and the New Testament, and interspersed with relevant sayings from various alchemical authors, she brings renewal and new life to those who follow her sevenfold path, which is revealed in a sequence of seven parables. Certainly, its Christian author freely draws on sacred scripture, but the aim is clearly to weave these seven parables into the 'work of a day' that is utterly alchemical— and, though this has gone completely unnoticed, ultimately ancient Egyptian. For they cover one vast cycle encompassing the darkness of night, birth and redemption at dawn, ascent to the zenith, midheaven second birth and finally the mysteries of death and union with the divine beloved, a cycle initiated by Wisdom, 'shining with the rays of 12 stars, prepared as a bride adorned for her husband', the lady of the hours whose transformational way is to be actualized in the devotional life of her followers. Given that many European alchemists belonged to religious orders, they would easily have reconciled *Aurora*'s seven parables with their daily life, regulated by the seven divine offices of the day marked by the canonical hours.[50]

In the end, though, this alchemical 'science' depends on a person's aptitude and capacities, since it is not wisdom leading to a uniform goal. 'Exert your body', Hermes advises Theosebia at the close of the *Epistle of the Secret*, 'and apply your understanding, so that you comprehend what I have revealed to you about it [the stone].'[51] Carefully tended, this alchemy nurtures the soul with an inexhaustible supply of life. It is 'a gift', Hermes says, a 'treasure that never fades, is never depleted', but it also needs to be protected from 'misunderstanding and half-truths' perpetrated by the unworthy.[52] As *Aurora*'s author wryly remarks: 'Lettuce should not be given to asses when cabbage will suffice for them.'[53] Hence Theosebia must undertake never to reveal its secrets, remaining silent and living quietly, 'like a wild dove singing beautifully'.[54] For this ancient knowledge speaks through stillness, through the music of silence, and whoever listens to the songbird, to the stirrings of the heart deep within, will grow in spiritual understanding—following a path of love known to seekers of Egyptian wisdom for centuries, and which Hermes is ready now to reveal.

TWENTY ONE

DIVING DEEP:
THE FISHER KING

Right at the start Hermes instructs Theosebia to create a sacred space for her alchemical operations, and whilst it is very unclear where these actually take place, she is to burn temple incense to cleanse her chamber, uttering also an invocation to 'the Lord of the Highest Building' and 'the Spirit of the Strongest Source of Light' in the four cardinal directions at a favourable time.[1] As Hermes says in the Arabic *Epistle of Hermes of Dendara*, 'Whoever serves continuously the highest light, so what is desired will happen', and likewise, Theosebia invokes this supreme source of illumination.[2] She is then ready to begin.

First she must seek out 'the King of the World', whose personal name Hermes never reveals. Rather, true to the spirit of Islamic mysticism, in which the real name of the beloved is always withheld, he calls the king by a string of epithets, some curiously similar to those of Horus of Edfu and Akhmim's city god Min—he is the 'Wonderfully Coloured One', the 'Pure White One', the 'Sun of the Wise' and 'Beautiful Male amongst the Noblest Minerals'.[3] But whoever he really is, having found him, Theosebia is to:

> Weigh him with the scales of equitable harmony and refine his body until he becomes spiritual.[4]

In accordance with Jābir ibn-Ḥayyān's famous alchemical principle of the 'balance', aimed at creating a perfect equilibrium between substances, her king has to be weighed and measured in order for his body to be refined and spiritualized. Indeed, an illustration in the Arabic *Book of Pictures* (*Muṣḥaf aṣ-ṣuwar*), which was produced in Egypt in 1270, depicts Theosebia as a huge figure holding scales with tiny embracing figures of herself and Zosimus in the left pan and six figures in the other pan (*fig.182*)—a weighing perfectly in tune with

ancient Egyptian goddess wisdom. For Shentayt manifests as lady of the scales in the Khoiak rites at Dendara (*fig.88*), and, similarly, Maat ensures creative equilibrium rules in the sun god's journey through the hours (*see pages 19, 25*). Here in the *Epistle*, though, this perfect balance is about to be sorely tested.

FOUR FISH MAIDENS: AN ELEMENTAL HOUSE

Hermes tells Theosebia she is to unite the king with 'four females' or, as he later says, some 'wise ones' work with seven, corresponding to the number of the planets, who must 'belong to the same species', differentiated only by their 'colour and gender'.[5] Trying to unite dissimilar substances is always a fruitless task in alchemy, hence there must be an innate 'sympathy' drawing them together. These four women are said to be 'white', resembling 'moist, shining, beautiful and round pearls', jewels associated with the alchemical secrets of colouring, which also adorn the head of Venus in her Egyptian temple in the Arabic *Book of Krates*.[6] They belong to the 'fishes of the sea of wisdom', mercurial sea creatures assigned to the astrological 'division of Hermes', in other words the planet Mercury, an easily overlooked but crucial detail.[7] For underlying this figurative language is the all-important alchemical union of 'mercury/quicksilver' and 'copper' (represented by the four females), a union symbolic of the moist, fertile womb and absolutely key to this Hermetic 'birthing' mystery (*see also chapter 23*), though a combination scarcely mentioned in the modern-day alchemical literature mesmerized by the Sulphur–Mercury theory of metals.[8]

True inheritor of Thoth's wisdom that he is, Hermes also says these four fish maidens need to be 'sought with guile'. He knows well their elusive nature, and also that 'nothing can be accomplished' without them, since submerged in their deep-sea depths are the secrets of rebirth—as they are in Hathor's ancient Egyptian watery world, which is wonderfully evoked by the aquatic imagery decorating a tiny, blue-glazed vessel, including the seven Hathors and a large fish (*figs.115, 116*).[9] The birthing power of the *Epistle*'s four fish maidens is

soon to manifest in their mercurial sea of wisdom, and, indeed, their fourfold union with the king reaches deep into ancient Egyptian territory. For not only does Pepi I offer to four Hathors in the southern crypt at Dendara at the close of day *(see page 97)*, the time when the sun god unites with his mother for renewal, but the god Amun is also called the 'Bull of the Four Maidens'.[10] Utterance 205 in the *Pyramid Texts* cryptically mentions a fecundating union between four goddesses and the virile solar king as 'Great Bull', ensuring he possesses solar nourishment.[11] It also brings vision: 'He belongs to the seer and it is he who sees.' Certainly vision is something the *Epistle*'s king will sorely need in his own imminent union with the four females.

Just as Hathor's name means 'House of Horus' *(fig.183)*, these four women surround their king 'like the walls and roof of the house'. They are the sacred space within which he stands at the centre as the 'pillar' (or 'tent-pole').[12] Hermes also compares these four walls with the four elements; and his words are graphically pictured in European alchemy, either geometrically in the form of a

square composed of the four elements, or embodied by four faces, identified with the four elements, shown surrounding a royal couple as they mate in a flask.[13]

'Our science is in every house', declares Ibn Umail, knowing the divine secret can be discovered anywhere, in whichever materials are close at hand, and is available to all who care to seek.[14] And having prepared her own 'house', Theosebia is ready to purify its king and four women with the 'green-gold water of life'.

GENESIS IN THE FLASK: THE EYELESS VESSEL

Theosebia times her purification to coincide with the sun's entry into 'the Ram', the sign of Aries, which inaugurates the astrological year. It is, says Hermes, 'the most favourable for what you intend'. First Theosebia is to blindfold one of the four women, then unite her with the king so that both become completely unified and display the colour of Saturn. Enclosed in the 'Purifying Vessel of the Wise with no Eye', they sink into a lifeless, putrefying state for three weeks, warmed by fire:

As soon as you have purified them, cover the eyes of one of them, lead her to that king and unite them in such a way it makes a later separation impossible, turning them to the Magnesia of the Wise as is appropriate for both bodies. They will disappear into each other, and the colour of Saturn will go forth from them. Then enclose both until they are lifeless, and let them both putrefy in the 'Purifying Vessel of the Wise with no Eye' for three weeks … and when completed then they have dissolved into one another in a marital union and conception has occurred.[15]

Completely true to the 'conception, birth and rearing' principles Hermes has already set out *(see page 223)*, everything becomes 'like sperm' in the fertilizing waters and three weeks are to elapse before 'conception has occurred'.

Then, after being removed from the vessel, the king is to unite successively with the other three women in a glass vessel, undergoing pulverization in the 'Mortar of the Wise' until all traces of blackness are gone.[16] During this period he is said to change 'from black to a dust colour and then to white', culminating in a union with the 'moist and

shining' fourth female in an apparatus called 'the Elephant's Head', heated now by the 'moist fire of the sun' until 'he becomes white' and assumes the nature of Jupiter.[17]

Nothing in the matter-of-fact way this Arabic Hermes instructs Theosebia, however, betrays the conflict, the suffering and pain endured in this visionless, pulverizing initial phase, when the eye is occluded and everything thrown out of balance,[18] though something of its anguish is captured in *Aurora consurgens*'s first parable, called 'Regarding the Black Earth in which the Roots of the Seven Planets are to be found'.[19] The previous five sections have all celebrated Wisdom and the 'science' she brings *(see page 224)*, but now, in this first parable (or sixth section), everything is engulfed by 'a huge cloud darkening the entire earth, which absorbed the earth and covered my soul'. Everything has been split apart in this dark night of despair. The elements are in conflict, a tempestuous storm is raging, flesh has lost its health and, overwhelmed by the 'putrefying' and 'corrupting waters', a voice cries out for deliverance, for a redeemer 'in whose embrace I am made young, to him indeed I will be a father'. What is longed for is an heir 'whose

ABOVE LEFT fig.183 Hieroglyphic writing of Hathor's name meaning 'House of Horus'. (Detail from a wall-painting in the tomb of Nefertari, Valley of the Queens, Western Thebes. 19th Dynasty.)

RIGHT fig.184 The saturnine phase of 'putrefaction' in *Aurora consurgens*, here represented by a blindfolded couple accompanied by a sage, and exactly replicating Hermes's instructions to Theosebia in the Arabic *Epistle of the Secret*. (Glasgow University Library, Ms. Ferguson 6, f.236r. 17th century.)

beauty even the sun and moon admire', whose appearance will bring the cleansing of the seven 'until they appear as pearls' in the time of the 'Whitening'.

The illustrator of a version of *Aurora consurgens* now in the Glasgow University Library seems to have been very familiar with the *Epistle*'s alchemy, for here a couple, both blindfolded, face each other in the presence of a wise sage and small child holding a banner *(fig.184)*. Surrounding the scene is a quotation from the opening chapter of the New Testament Gospel of John, 'And the light entered the darkness and the darkness comprehended it not', here expressing the alchemical 'black' state.

Gratheus seems to have known it, too, when composing his highly unusual vernacular verse text somewhere in the Lower Rhine region in the second half of the 14th century.[20] Four glass flasks, he says, are needed to begin the work, and subsequently he includes a tiny miniature depicting a king and queen in a flask, followed by miniatures showing their union and a child in a flask.[21] His are the earliest known depictions of the 'chymical' marriage in medieval Europe, here undoubtedly Christianized, and associated by Gratheus with Christ's death and resurrection.[22]

Much more explicit is the *Rosarium philosophorum*'s famous series portraying a royal couple floating in a watery landscape, naked and locked in an erotic embrace, a sun and moon at their feet. Then this dream-like landscape changes abruptly into a sarcophagus-like vessel, plunging the couple into the second phase of 'putrefaction', and they are depicted within it four times, clearly in a generative union, since a small child is seen hovering above them in a cloud.[23]

The much-copied *Donum Dei* shows the mating couple enclosed in a sequence of womb-flasks, 12 in number, and a 17th-century version, now in Paris, includes four female heads surrounding the couple as they 'dissolve' in the waters of the third flask, called 'Perfect Solution' *(fig.180)*.[24] Above them, in the neck of the vessel, is a winged child, with nine poppy seed heads sprouting from the top of the vessel, which is placed in a verdant green-gold landscape.[25] Three united purple flowers and a single red flower grow on either side of the flask,

their lustrous colours symbolic of the purple *iosis* phase and final 'reddening', for even though everything is still embryonic, not yet born, this is a living dissolution, bearing within it the possibility of a final maturation in glorious purple-red.

Right from the very beginning there needs to be a vision of the goal to inspire this work, especially as the very next flask in the *Donum Dei* sequence is the 'Putrefaction', the burning of copper, when all might seem lost and hopeless. This is the putrefying, black phase all alchemists must experience, ruled by the planet Saturn, the leaden, dark god who must revert to the original black state. 'No generation without corruption', alchemists ubiquitously teach, since there can be no new life without a willingness to be dipped into oblivion, to be dissolved in a union with four females. The ancient Egyptians honoured this feminine, death-dealing power; likewise Zosimus tells Theosebia in the Arabic *Book of Pictures* she is responsible for his 'blackness' and 'death' *(see page 175)*, and these four females 'dissolve' their unnamed king in 'putrefying' blackness, their power now at work in European alchemy.

All of which also helps to explain the 'baffling and paradoxical representations' of 'Christ crucified by the Virtues' that first appear in the Rhineland around 1250 and seem to have been particularly favoured by German nuns. Here, too, a fierce, spearing female is depicted, as, for example, in a Psalter illustration showing 'Love' *(Caritas)* spearing Christ whilst three other Virtues, named as 'Humility', 'Obedience' and 'Mercy', nail him to the cross, their death-dealing power manifesting on Good Friday.[26] In another example, an early 14th-century stained-glass window from the

---

RIGHT fig.185 Here sun and moon, male and female, do battle at the beginning of the alchemical work, though the union of their opposing powers is indicated by the shields. Originally the disk on the woman's head would have been silver but has tarnished over time, and her twofold nature is symbolized by the griffin she rides. (*Aurora consurgens*, Zentralbibliothek, Zurich, Ms. Rhenoviensis 172, fol.10v. Early 15th century.)

Cistercian convent of Wienhausen in Lower Saxony, 'Mercy', 'Truth', 'Justice' and 'Peace' surround Lady Love as she stabs Christ in the heart.[27]

The emergence of these 'Four Daughters of God' as personifications of the Virtues is well documented; less easily explained is why they should be shown in such incongruous, death-dealing roles, and why they appear both at the Annunciation and the Crucifixion.[28] Their link with alchemy's four females has gone unnoticed, as the medieval artists no doubt wished, since alchemy could sometimes court extreme controversy in the Middle Ages.[29]

SETTING IN MOTION: SEPARATION AND UNION

Changing from Saturnine 'blackness' to Jupiter's 'whiteness' (the only two planets specifically named in the *Epistle of the Secret*), the *Epistle*'s king moves from darkness to light, now able to change colour, and hence transform, during this initial phase. Probably, though, it is *Aurora consurgens*'s

second parable, entitled 'On the Flood and on Death, which the Woman both brought in and put to Flight', that best encapsulates the vitalizing energy here 'setting everything in motion' in the phase following 'putrefaction' *(see page 223)*.[30] For this unnamed woman manifests both as life-giver and death-wielder. 'The arrows of my quiver are drunk with blood', she declares, but she is, too, a bearer of abundant new life. 'My belly has swelled from the touch of my beloved', she exults, her pregnant state holding out the promise of 'a light coming forth from the darkness' and the appearance of the 'Sun of Justice' from heaven. Quite clearly, too, this Jupiter-like 'Sun of Justice' is to be understood here as a terrestrial king, an earthly sun, identified with Christ as the newborn babe 'to whom the wise men from the East brought three precious gifts'. Knowing well the woman's 'separating' and 'uniting' powers, the author, referring to the New Testament parable of the lost sheep, also confirms the stray '100th sheep' has been found and returned to the flock. Everything

is now reunited—as it was, too, in ancient Egypt when the sun god's 'Eyes' opened in the rising dawn (*see chapter 1*).[31]

One of *Aurora*'s illustrations superbly evokes the woman's attracting-repelling power (*fig.185*). It shows the male riding on a lion and the female appropriately on a griffin, a creature with features of both a lion and an eagle, and generally noted for its twofold nature in medieval art. Perfectly poised though they are, this couple are also locked in battle, each thrusting a spear towards the other in a death-dealing gesture. Yet there is an 'attracting' power at work here, for the shield of the lunar female is emblazoned with the sun and the solar male's shield with the triple moon, symbolizing their union in this 'teeth and claw' realm.

This second parable might seem randomly introduced, but in fact there is a very deliberate author at work here, steeped in biblical and alchemical knowledge and intent on 'Christianizing' alchemy's 'separating' and 'uniting' copper woman, with her power to 'set everything in motion' and bring forth new life—though as long as this beautifully illustrated medieval text remains cut off from its Islamic and ancient Egyptian roots, it is doomed to remain 'incomprehensible'.[32]

Probably its author also had in mind here the 'first Eve', whose separation from Adam, according to the early *Gospel of Philip*, brought death into the world and whose reunion with him brought life, an identification reinforced by the spearing naked female on the griffin (see above), reminiscent of Eve's naked state in the biblical Garden of Eden.[33] Certainly this identification with Eve is how the death-dealing female is viewed in the *Book of the Holy Trinity*, which first appeared in southern Germany in the early 15th century, though here Eve watches as a female, depicted as a melusine, all human above, all fish (or serpent)-tailed from the waist down, fatally wounds Adam with her piercing lance.[34] Just like the *Epistle*'s four females, she belongs to the fishes of the sea, her Venus-copper nature symbolized by the green colour of her coiling, lower body, tinged, though, according to its Christian author, with 'eternal impurity'.[35]

Rooted in 'two suns' copper alchemy, the *Book of the Holy Trinity* also shows the crucified Christ

eagle plunged into human suffering, earthly death and mortification as the 'black sun', his suffering state identified with Saturnine lead and the 'torment of metals' undergone by the fourfold leprous *tetrasomia* (*see page 201*). 'Calcine copper' instructs the text inscribed in a semicircle beneath the eagle, thus reinforcing Christ as an earthly king, though evidently alchemy's separating-uniting female has strayed far from ancient Egypt, being now integrated with Judaeo-Christian ideas of original sin and, through her identification with Eve, even responsible for humanity's 'Fall'.[36]

Clearly, by the Late Middle Ages 'copper' alchemy had taken hold in the German lands, or even earlier, since there are remarkable affinities between *Aurora consurgens* and a work by Hildegard of Bingen (1098–1179) called *Scivias* ('Know the Ways'), in which she gave a wide-ranging commentary on her famous visions, not just their meaning for the individual soul, but also their theological and social significance within the Christian faith. Already her love of 'greening' power has been compared to 'two suns' alchemy (*see page 208*), but in *Scivias*'s second vision of the first part the parallels with *Aurora* are even more striking—as when she discourses on Adam's fall into blackness, enveloped by a dark cloud, with all the elements of the world in turmoil. She refers to the Sun of Justice, to the rescue of the lost sheep and the precious 'pearl' lost in the filth that is purified like gold in the crucible, before turning, in her fourth vision of the foetal child nestling in the mother's womb (*fig.186*), to the infant's capacity to move with 'vital motion'. In fact, shortly before this embryological discourse, she also cites the biblical passage, 'I will kill, and I will make live; I will strike, and I will heal; and there is none who can deliver out of my hand' (Deuteronomy 32:39), without, however, in contrast to *Aurora*'s second parable, explicitly identifying this twofold 'animating' power with the female.[37] Hildegard never mentions the word 'alchemy' specifically, but these are remarkably similar themes, and *Aurora*'s author may well have looked to *Scivias* for inspiration, especially in light of the further similarities between the two texts (*see chapters 22 and 23*).

ABOVE *fig.186* The heavenly soul develops in the maternal womb in Hildegard of Bingen's *Scivias*. Here the foetus's umbilical cord is attached to a fiery golden kite flying in a starry sky. Also depicted are figures carrying various vessels containing fatty cheeses symbolizing male seed; to the right are the various afflictions hindering the soul's attainment of the golden tent shown at the top. (Illumination from the lost Rupertsberg manuscript *c.*1175, facsimile 1927–33. Abbey of St. Hildegard, Rüdesheim/ Eibingen.)

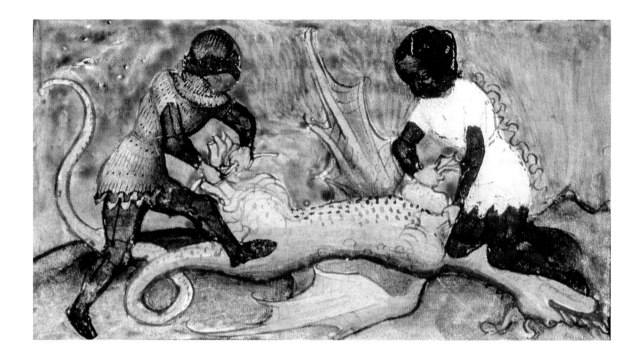

WOMANSPIRIT RISING: ENDANGERED ASCENT

Now able to change from black to white, from darkness to light, the *Epistle*'s king has clearly moved from the lifeless state in the vessel 'without an Eye' to a 'set-in-motion' body, ready therefore for 'ascent'. Though Hermes is extremely brief about this next phase, simply saying the four females are to be 'washed', 'purified' and 'coloured' so that:

> The spirits rise, the beings transform,
> the quicksilver appears.[38]

Yet this 'ascending' phase is extremely important, and is certainly amplified in *Aurora*'s third parable, called 'On the Gate of Copper and Iron of the Babylonian Captivity', hence a gate made from the metals of Venus and Mars respectively, which, together with tin (Jupiter) and lead (Saturn), form the *tetrasomia*.[39] Sadly, though, Hermes's instruction to 'wash' and 'purify' the four females is all the ammunition this Christian author needs to launch into an extraordinary attack, completely in tune with the misogynous attitudes displayed by some medieval theologians.

The parable begins innocuously enough, expressing the hope for release from 'the prison of darkness' and food for 'my ravenous soul', but it soon longs for purification 'from the greatest sin' and from the assaults of 'the noonday demon'. Clearly, it is timed for noon, when light is increasing and heat intensifying, though, rather than the ancient Egyptian blissful union of Re and Hathor-Maat at the zenith *(see page 25)*, *Aurora*'s author prefers to dwell on the shadow side of this expansive phase, not least the dangers of spiritual 'pride', which, according to monastic tradition, particularly afflict the carnal soul in the noonday hour. Just as the Apophis snake attacks the Egyptian sun boat during the heat of ascent, so the dreaded demon of *acedia* (as it is called in the Christian tradition) seeks to lure its victims away from the river of life.[40]

Indeed, it is the 'high-necked' daughters of Zion who are particularly susceptible to this demon's wiles, as, though their instruments are silent during the 70 years of Jewish exile by the waters of Babylon, they 'flirt with their eyes'.[41] Driven by their sexual desires, these unclean daughters display the shadow side of Venus 'copper', their haughty

bearing typifying pride's love-denying passion, which, in the *Book of the Holy Trinity*, is 'against justice' and identified with the red crown encircling the sword of Mars (iron) wielded by the androgynous Antichrist.[42] In short, held captive by the imprisoning gate of Babylonian exile, these Mars–Venus passions potentially obstruct any further growth and development.[43]

Clearly, *Aurora*'s author knows this noonday phase involves musical 'daughters', but whereas in ancient Egypt they procure zenithal union, shaking their sistra to soften the sun god's potential anger and pride and using their charm and music to call forth his magnanimity and compassion towards others *(see chapter 3)*, here the Lord is implored to 'wash away the filth of the daughters of Zion with the spirit of wisdom and understanding.' Certainly, the ancient Eye goddesses required propitiation— cool water needed to flow during noonday ascent— but these 'daughters' safeguarded life during this testing noonday time *(see chapter 3)*, and such a blaming, shaming attitude towards them was completely absent. And yet, and yet, despite *Aurora*'s complete reinterpretation, replicated here is the same pattern of ancient Egyptian ascent, mirrored also in its illustration of the brother–sister pair fettering the dragon *(fig.187)*.

Not that this vengeful tone sounds in all European 'marriage' treatises. For example, the *Rosarium philosophorum* concludes this 'Whitening' phase with the crowning of the hermaphroditic king and queen beside a moon (or silver) tree, accompanied by a riddling text worthy of the ancient Egyptians:

Here is born the noble Empress rich,
The masters say she is like her daughter.
She multiplies, producing innumerable children.
They are immortally pure and without nourishment…
I became a Mother and yet remain a Maid,
and was in my essence lain with.
That my son became my father
As God has decreed in essential way,
The Mother who gave birth to me
Through me will be born on earth.[44]

Here in zenithal 'Whiteness', the united royal couple, shown standing on a moon and holding aloft four rearing serpents *(fig.188)*, three placed in the king's chalice and one coiled around the queen's arm, triumphantly hold together the 'opposites'. So, too, this ambivalent 'noble Empress' is a transforming queen—according to the text she is 'like her daughter', who 'becomes a mother and yet remains a maid', a contradiction expressed too by her son 'who became her father'. Just as the Hathorian queen grows ever younger during ascent to the zenith *(see page 39)*, so the *Rosarium*'s shapeshifting 'noble Empress' likewise changes from mother to daughter, moving through the human lifecycle towards increasing vitality and power, however not yet able to provide nourishment for her children, as she rightly says, since this nourishing phase is yet to be achieved *(see chapter 22)*.

Despite now belonging in the world of Christian Europe, where a very different view of the female and human sexuality prevails, this 'marriage' alchemy really does begin to sound like a re-formed ancient Egyptian 'work of a day', much more overtly tied to the 'hours' in fact than the Arabic *Epistle of the Secret*, even though, paradoxically, it was through Islam that this knowledge reached the West. To be sure, all these womb-like flasks, putrefying unions, poundings in mortars and graphic depictions of the mating couple might seem worlds away from ancient temple life, yet in the regenerative dance rites performed during Amenhotep III's first Sed Festival at Thebes, celebrating the king's nocturnal union with Hathor of Dendara, a song inscribed above the acrobatic dancers tells of the 'pounding' and 'roasting' of seeds, when vision is impaired and eyes become 'bloodshot', during the nocturnal pain of transformation. Incarnating as Hathor's musical

child Ihy, the king is a dancer in these rites, identified with a child god whom the *Coffin Texts* graphically describe as an 'inert one' lying motionless in the waters of Nun, his body blackening and decaying in stench and foulness, before eventually rising forth, fragrant as his mother, to new life at dawn.[45] As always, the Egyptians give but the subtlest of hints, yet here is an Egyptian king in a 'dissolving' nocturnal ritual, seeking renewal with the propitiated, glittering goddess 'Gold', who is implored to bring him through the darkness into the rising dawn—the time of 'Whitening' as it is called *(see page 191)*. And no child of Hathor, according to an inscription at Dendara, escapes 'pulverization', for, like the circular movement of the grinding stone used to pound the ingredients for her sacred drink, Hathor journeys across the heavenly circuit 'with her son within her'.[46] Manifestly, to grind this grain is to revolve around the heavens in her matrix of life.

The names have changed, the rites have altered, and a whole new way of experiencing Hathor's goddess mysteries has evolved. Yet the same fiery female energy pounds away, the same volatile rage and radiance transform the alchemical king, who rises, exactly like all the earlier Horus rulers, from the darkness of night, soaring heavenwards with a starry goddess who 'sets in motion' the entire solar circuit. From Hathor's temples in ancient Egypt to medieval Europe there is a reciprocal dialogue spanning centuries—a movement backwards to the roots of the tradition and forwards to new contexts and new discoveries, as time's arrow shoots both ways until the lost stone is found.

RIGHT fig.189 Alchemy's second crowning. (*Rosarium philosophorum*, Stadtbibliothek St. Gallen, Switzerland, Vadiana Collection, VadSlg Ms. 394a, f.92r. 16th century.)

SEEKING THE STONE:
THE SOUL'S CODE

Ascend from earth to heaven, and descend again from heaven to earth, and unite together the power of things above and below', declares Hermes Trismegistus in the famous *Emerald Tablet*, and here in the *Epistle of the Secret* the limits of 'ascent' have been reached. Hence Theosebia must begin the 'return' phase, crossing the threshold into the second 'operation'— the 'Reddening' work of the sun.

'As soon as ascent is complete', Hermes tells her, 'there is a return to what is below, so that, after the spirits, souls also ascend.' Attention now needs to be given to 'what is beneath' when weighed in the 'scales of wisdom'.[1] No longer is time to move ever forward in a linear direction, a one-way progression. Now it is to turn back on itself, like the ouroboros-snake with its tail in its mouth, like the roundness of the egg, or the roundness of purified copper in the purple *iosis* phase *(see page 202)*, circling in the great wheel of life, travelling in a circular motion.

Throughout this second operation, Hermes uses highly coded language, weaving with mercurial alacrity between the 'technical' laboratory and 'soul' work. Sometimes Theosebia seems to be in an alchemical kitchen, measuring, preparing and cooking substances according to his recipes, sometimes deeply immersed in an illumination mystery. In the *Final Count*, Zosimus had already intertwined 'melting copper' and the purification of Theosebia's soul *(see chapter 19)*, so this Arabic Hermes is following a well-trodden path. Indeed, in light of Maria the Jewess's statement that 'when copper is refined by melting it diminishes by a third of its weight', his easily-missed remark that the work 'will be decreased by a third', not only suggests the number three, but also indicates refining copper is all-important in this second operation.[2] Though trying to tie his instructions too precisely to 'technical' procedures would be to fall into the kind of literal trap alchemists repeatedly warn against. To be sure, Hermes's teaching here has a metallurgical basis, but what also concerns him is soul work, and he is far more interested in moving between different modes of experience because of their symbolic correspondences than in instructing Theosebia

simply about 'technical' copper-refining procedures. Thus, his 'recipe' at the start of this second operation, namely to divide the substance into silver and gold, then take a little gold stone and heat it, whilst relevant to the 'Whitening' and 'Yellowing' Zosimas says occurs when copper is refined *(see page 196)*, might equally well apply to the 'sage's egg' coloured 'white' (silver) and 'yellow' (gold)—the embryological solar egg of rebirth intrinsic to this second operation.[3]

GROWING A SOUL: COOKING THE EGG

For the first time in his teaching, Hermes now mentions the 'stone', instructing Theosebia to cook her 'lead-like stone' until it becomes black:

> Maintain a medium heat, cooking for 40 days so it will become a lead-like stone which will be covered with the blackness of the highest planet [Saturn] and its colour.[4]

Once again there is a 'back to black', a repetition seemingly of the first operation *(see page 229)*, but this is a very different blackness, since Saturn has now become the 'highest planet'.[5] Just as the divinized Osiris manifests in the *Book of Day*'s midheaven ninth hour, when Re 'comes forth from the egg' in his glorious 'second birth' *(see page 28)*, so this saturnine planet, which alchemists associate with Osiris, is in the heights now. It is, though, Hermes's instruction to 'cook for 40 days' with a medium heat that completely unmasks the embryological nature of Theosebia's work here, for among Greek writers, including Hippocrates, 40 days is a decisive period for embryonic development and survival in the womb.[6]

Ibn Umail says the same when telling how the soul manifests in the womb on the 40th day, being then nourished and strengthened by maternal blood:

> On the 40th day the soul comes into being, and appears in it. Then, from the 40th day on, blood appears and flows in the embryo through its navel, becoming its nutrition. And the soul becomes strong, developing little by little, and becomes stronger.[7]

Ibn Umail is not simply claiming medical knowledge here; he is a copper-loving alchemist, and, as the Arabic *Book of Pictures* makes perfectly clear, after 40 days, when the sperm has entered the womb and 'God turns it into a creation', this is like washing and heating copper.[8] In fact, in a Greek text attributed to Zosimus, after mentioning there are 1,000 ways of heating copper to prepare it for tincturing, a warning is given to 'take care that the time of gestation does not last less than nine months, without which there will be a miscarriage.'[9] Elsewhere, Zosimus associates the 'cooking' of the foetus in the 'womb of fire' with dyeing processes within the alembic.[10]

Ibn Umail elaborates further, saying this heavenly foetus requires 'water' 'air' and 'fire' and will develop through these elements in a three times three rhythm of nine months to maturity:

> Know that water serves the embryo in the womb for the first three months, then air serves it for the second three months, then fire for the third three months cooks it and perfects it.[11]

Indeed, Cleopatra specifically identifies the 'womb of fire' and the 'womb of air' with red copper, and the 'womb of water' with mercury, in the *Dialogue of the Philosophers and Cleopatra*, thereby, as in the *Epistle of the Secret*, indicating copper and mercury together define the womb's 'fertile state' *(see page 227)*.[12]

Noticeably excluded from this 'work of the three', however, is the fourth element, 'earth', just as it is in ancient Egypt when the Theban god Amun-Re manifests in 'light', 'air' and 'water', his body being composed of the three elements that sustain 'life' in the cosmos.[13] Clearly, they sustain alchemy's heavenly foetus, too, the reason for which Marsilio Ficino succinctly explains when he says universal 'life' flourishes 'much more above the earth in subtler bodies, which are nearer to soul'. It is by experiencing the movements of these three elements, as well as those of the heavenly bodies, Ficino says, that human beings 'receive the motion of the Life of the world', doing so in seven steps drawing down the life of heaven.[14]

After all, in the *Corpus Hermeticum*, Hermes teaches his pupil Tat if he wishes to comprehend the order of the cosmos, how 'the motionless is set in motion' and the 'invisible made visible', he should look to 'how the human being is crafted in the womb'.[15] To the Greek Hermes, this developing foetus provides a wonderful unseen image of eternal 'Becoming', an imaginative vision of participation in a moving 'life' world born of the hidden creator. Now, in the *Epistle*, his Arabic counterpart teaches the same mystery, adding purified copper to the mix to reveal this heavenly womb's secrets.

Still, though, Theosebia must cook for a further '80 days', which, together with the previous 40 days, brings her to 120 days, an important time period in Islamic alchemy.[16] Furthermore, if related to the astrological sign of Aries (March–April), which inaugurated the first operation, it brings Theosebia's work to the solar sign of Leo (July–August), the fifth sign of the zodiac, the sign of the lion, which is associated with the heart, energy, blood and feeling, courage and passion, the time when the sun is hottest and the 'son of the year' is born, both in alchemy and ancient Egypt.[17] Cooked for 80 days, this lead-like stone miraculously changes from black to white, becoming a 'white shining spirit', beautiful like the 'whiteness of the Eye':

> This is the splendid alabaster stone the ancients have always sought.[18]

The Eye occluded in saturnine, first darkness, sealed in the 'Purifying Vessel of the Wise with no Eye' in the first operation *(see page 229)*, now shines dazzling 'white', like the longed-for 'splendid alabaster stone'—a code name also for the alchemical 'egg'.[19]

'Make a little movement like the foetus', says Rūmī, drawing deeply on this alchemy, 'that they may give you light-perceiving senses', for this Hermetic revelation needs subtle perception, vision unclouded by turbulence, seeing from soul to soul in the heavenly womb world.[20]

Yet this work cannot be hurried. Theosebia's 'cooking' still needs care and constancy, for whilst her 'alabaster stone' is able to change colour now from black to white, hence able to move and transform, it still lacks something—it needs further nourishment. It needs 'strength', which is obtained now by adding the heavenly 'sulphur of truth', while making sure, as the heat is increased, not to leave the precious stone alone in case 'its blossoms are destroyed'.[21]

Theosebia's gestating 'stone' also 'yearns', as it must if it is to manifest the 'desire' that, together with 'strength' and 'movement', enlivens the whole body *(see page 198)*. Strength, though, so easily turns to aggression, to the need for domination—fiery sulphur readily burns and destroys. Likewise, desire can flare up in destructive anger, creating a harshness of soul. So this fiery sulphurous process needs careful handling if it is to become fruitful life, a peaceful 'circulation of sap', as Plutarch says *(see page 209)*. 'Take care that you do not put the king and his consort to flight with too much smoke', warns *Aurora consurgens*.[22] There are perils to be avoided beneath the Leo sun's heat; great care is needed if everything is not to dry up.

As if realizing the difficulties and seeking to encourage Theosebia, Hermes adds, 'There is hope.' He tells her she is not alone—there are unseen companions anxiously watching her work, 'wise ones who throughout its whole long operations have concern and who know about it.' Perseverance is needed, but if she succeeds, they 'will rejoice'.[23]

Her stone, though, is thirsty—it needs 'watering', drenching in the moisture Hermes calls 'the shining rain which the wise know'.[24] Then, if returned to the cooking vessel, and the temperature raised still higher, it will be:

> Clothed in remarkable colours, wonderful blossoms will appear, the crops will ripen and the fruit will be delicious.[25]

Beautiful colours now start to shine in all their rainbow-radiant diversity, fanning out like a peacock's tail, the *cauda pavonis*, as European alchemists call this sudden, colourful eruption that is associated with copper's purification and Venus *(see page 195)*. This is a sign that the sought-after 'purple' is imminent. Not quite yet, though—first the stone needs to be returned to the oven again and moistened with the 'the sea-water of Pontus', the everlasting water from which 'whoever drinks will never thirst'.[26]

Then comes the miraculous moment when the eternal colour manifests, when the stone, 'clothed in purple', has 'power far stronger than in the previous phase', and by strengthening it still further in the 'heat of the sun', Theosebia will be gifted with heavenly illumination:

> The clouds will be dispelled, the heavenly region will become clear and you will forever attain the fruit of the wise.[27]

Centuries before, a similar 'cloudless' illumination had inspired the festival rites of Horus at Edfu, when the divine statue was taken to the temple roof to be enlivened by the sun's rays:

> The sun appears, the clouds are hidden, the sky is clear ... Horus of Edfu has come in procession to unite with Re ... whose rays have entered his body.[28]

So, too, in the New Year rites at Dendara, Hathor's cult image had left the crypt's secret palace in the eighth hour and been taken in procession to the roof for her union with the heavenly Aten. Now Theosebia is blessed with this same heavenly illumination as Hermes guides her into 'ancient knowledge'.

But still she must moisten her stone further with 'pyrite water', cooking until it is coloured in 'eternal, enduring purple'.[29] Then she must put it in 'crowned vessels known only to the priests', allowing air to enter through a vent at intervals of a third, a 30th and a 60th (all multiples of three) to enable the stone to breathe and remove all traces of 'moistness', the 'corrupting humidity' as *Aurora consurgens* calls it, that might blemish fruitful life and prevent her attaining the 'eternal colouring of truth the wise have always sought'.

Surrounded by a beautiful scent, she is then to place the stone in golden vessels and raise them in her temple's prayer niche *(mihrab)*, and if she has understood all this, Hermes says, she will be:

> Queen of the World, the heiress to the knowledge of the wise, the fortress of this knowledge, and source of the greatest riches, from which the souls of the seers never have enough, and which can never be understood by the ratio of things.[30]

Possessing 'knowledge of the wise', seeing beyond 'the ratio of things', Theosebia's inner beauty and truth shine forth: she has become 'Queen of the World'.

HUNTING THE LION:
THE VANISHING KING

Where, though, it might be asked, is her partner, the 'King of the World'? For throughout this second operation Hermes makes no mention of him; indeed, as soon as Theosebia starts 'cooking', he seems to disappear completely beneath the Leo sun.

Curiously, the story of the famous lion hunt of King Marqūnis provides the necessary clue, as told by the king to his mother and narrated by Ibn Umail in his treatise the *Silvery Water and the Starry Earth*.[31]

Marqūnis tells his mother how a lion has fallen into a pit he has dug, above which he has placed a glass retort so he can observe its fate—and his own, since the lion is here a thinly disguised symbol of Marqūnis himself, who is both the hunted and the hunter in this story.[32] What has lured the lion into the pit is a sweet-scented 'stone', alias a woman, which Marqūnis has placed at the bottom of it, lighting a gentle fire beneath it using charcoal, which moves 'like the creeping of black scarabs'. No lion can resist pursuing prey exuding attracting odours, and, trapped by the female stone, this one is swallowed by it. One by one his paws are amputated, then his head, and all are cooked in heavenly water 'falling from the sky when it rains' and obtained from the hearts of white and yellow statues. Dismembered, cooked and distilled, the lion eventually coagulates into a luminous stone, shining like a sapphire to 'light up the temple'. Trickery was needed to entice him initially, but when cooked in heavenly water within this female matrix, all his compulsive, leonine passions—the greedy paws, the roaring head—are transformed into a radiant stone of light.[33]

This, too, seems to be the fate of Theosebia's disappearing king when she cooks her stone, returning him to the transforming maternal vessel for his heavenly rebirth. For the Arabic *Book of Pictures* depicts her cradling Zosimus's inanimate body in her left arm, whilst tethering a winged man

ABOVE fig.190 Theosebia cradles Zosimus in her left arm, whilst tethering his purple winged soul with the rope she holds. *(Book of Pictures [Muṣḥaf aṣ-ṣuwar]*, Istanbul Arkeoloji Müzeleri Kütüphanesi, Ms. 1574. 13th century.)

clothed in a purple garment with the rope she holds in her right hand *(fig.190)*. Above them is a green-gold stone, from which, according to the caption, heavenly water flows, 'the colour of the sky'.[34] Just as heavenly dew falls like rain in the maternal vessel to transform Marqūnis into a shining stone, so it falls here to enliven Zosimus's lifeless body, transforming him into a heavenly light-filled soul.[35] To the right of this group, he is

241

RIGHT fig.191 Here the parturient mother's waters flow from her as she holds an astrolabe towards the sign of Leo in the encircling zodiac, indicating this birth belongs to the midheaven phase and is a 'work of the sun'. (*Aurora consurgens*, Zentralbibliothek, Zurich, Ms. Rhenoviensis 172, f.11r. Early 15th century.)

shown again, clothed in a purple robe and standing between Theosebia and another woman, unnamed in the caption *(fig.177)*.

Just as purified copper enlivens inanimate bodies with life-giving *pneuma*, so Theosebia empowers his lifeless body with a flowing, growing, light-filled soul; in fact the accompanying text explicitly refers to copper's 'humanness' and its regeneration.[36] Without the beloved, as Rūmī says, 'my soul becomes just like a lifeless corpse.'[37] And like Marqūnis in the lion hunt, who is both the observer of the lion's fate and the lion himself, paradoxically Zosimus is here both 'the watcher' and 'the one who is transformed', pointing towards Theosebia with a gesture that is unequivocal.[38]

Such a heavenly triad also graces Queen Nefertari's midheaven chamber of 'life' honouring 'the three' at Abu Simbel *(fig.44)*. Here Ramesses II stands between Nefertari as 'daughter' and the enthroned mother goddess, Taweret, 'who gives birth to the gods', the female powers sustaining 'life' in this maternal heart realm where everything

is constantly growing, flowing and transforming, perpetually changing from one state to another in the eternal cycle of 'Becoming'. Here, too, Hathor and Isis make their beautiful *Ka*-like gesture of 'justification', blessing Nefertari as Egypt's exalted queen, her crowning attuned to the returning inundation waters in mid-July and the sun god's birth from the egg, when the solar circuit is both rising and setting as a turning sphere. That 'ancient knowledge' sought by Theosebia in the *Epistle of the Secret* is enshrined right here at Abu Simbel, and it is hers, too, when she manifests as 'Queen of the World'.

Hence it is not surprising to find Ibn Umail calling the sought-after stone 'the Mother of Gods', though to find its transformational secrets requires all the skill and watchfulness of a hunter stalking prey.[39]

RIGHT fig.192 Seven planets encircle the alchemical flask in *Aurora consurgens*. At the top is Saturn, reflecting his position as the 'highest planet' in this phase. (Glasgow University Library, Ms. Ferguson 6, f.237v. 17th century.)

RAISING THE WATER: THE DISTILLER'S ART

All this also explains why *Aurora consurgens*'s fourth parable, entitled 'On the Philosophic Faith that is based on the Three', the middle of the seven parables, towards which the action soars and away from which it flows, is unexpectedly introduced after the noonday third parable. 'All perfection consists in the number three', it rightly states.[40] And, indeed, after initially discoursing on the Christian Trinity and the promise of enthronement on the 'throne of David', the author then gives a startling twist to the Holy Spirit's fiery activity, associating it with a threefold baptism of 'water, blood and fire' so that earthly things become heavenly and a 'living soul' created, a threefoldness that is then extended to foetal life in the womb, which is sustained by 'water, air and fire'—Ibn Umail's embryological statement being quoted almost verbatim.[41]

At first sight such an embryological and threefold theme, coming immediately after 'ascent', might seem utterly disjointed, opaque and extraneous, whereas in fact it is completely in tune with the unfolding 'work of a day'. One of *Aurora*'s illustrations graphically depicts a parturient mother clad in a white head cloth as she squats amidst the amniotic fluid gushing forth from her body *(fig.191)*. Completely encircled by the zodiac, she tellingly points with her astrolabe towards the sign of Leo, positioned just after the zenith, so there can be no mistaking hers is a midheaven child, the 'son of the year' attuned to the great circling wheel of life, maturing under the hot Leo sun.[42]

This embryological fourth parable also connects the Holy Spirit's sevenfold gifts with a sevenfold distillation process. 'Distil seven times and you will have separated it from the corrupting moisture' are its concluding words, thus mirroring Hermes's instruction to Theosebia in the *Epistle of the Secret* that she must strengthen the heat 'seven times' during the refining process.[43] The Arabic *Book of Pictures* shows Theosebia standing with Zosimus beside a distillation oven, depicted with seven openings in its design, evidently for these seven steps of the distilling process, and sometimes identified also with the seven planets *(fig.174)*.[44] The medieval alchemist Gratheus is even more explicit, comparing the clay vessel with seven holes to the pregnant female matrix.[45] Rūmī, too, beautifully expresses this sevenfold activity in his *Masnavi*:

> O soul, there is a time the seven planets
> make every foetus turn around in service.
> And at the time the foetus gains the soul,
> just then the sun becomes defender of it.
> The sun will stir this foetus into movement;
> the sun will suddenly give life to it.[46]

Warmed by the sun's rays, the hidden foetus stirs into life, turning in harmony with the seven

planets, and, drawing still further on alchemy, Rūmī rhapsodically then praises the heavenly sun's transforming power, its hidden ways 'remote from human senses'. For, he sings, 'gold gets nourishment from it', 'rocks are turned to gems by it', it 'makes the ruby red', 'the horseshoe flash with sparks', it causes fruits 'to come to ripeness' and even consoles 'the bewildered soul'.[47] Whether they are stones, gems, fruits or the human soul, all, like the gestating embryo in the womb, are brought to maturity through the sun's fiery power.

A common alchemical term for distillation is 'raising the water', already mentioned by the Greek alchemist Stephanus in the context of purifying copper.[48] 'They named their copper—I mean their stone—a raincloud', says Ibn Umail.[49] Watching their precious substance being carefully heated in the distilling womb vessel, and seeing the vapours repeatedly rising and falling in a circulatory process then eventually flowing through a tube at the top of the vessel into a cool glass to condense the precious

essence, alchemists are transported into the hidden mysteries of heavenly soul life. Like rain falling from the sky, these vapours water the substance in the vessel, a process perfectly capturing the spirit of foetal transformation in the nourishing, cosmic matrix.

It is this purifying distilling vessel, the 'true philosophical Pelican', with its unifying circular process, that turns 'the outside to the inside, the inside to the outside, likewise the lower and the upper', according to the *Tractatus aureus*, hence the prominence of a pelican in the *Rosarium philosophorum*'s scene of the second crowning (*fig.189*).[50] Here the hermaphroditic winged couple, clothed in black, white, red and gold, are shown triumphantly standing on a coiled triple-headed serpent, proclaiming their sovereignty in this 'circular' realm of 'the three'. In the background is a peaceful 'Leo' lion, while close by a pelican pecks its breast to obtain blood to feed its three hungry chicks in the nest, emblematic of the distillation

LEFT *fig.193* Wisdom nurtures the philosophers in her treasure-house in *Aurora consurgens*. (Zentralbibliothek, Zurich, Ms. Rhenoviensis 172, f.13v. Early 15th century.)

RIGHT *fig.194* Ibn Umail holds the key to Wisdom's treasure-house. (*Aurora consurgens*, Zentralbibliothek, Zurich, Ms. Rhenoviensis 172, f.27r. Early 15th century.)

process. Appropriately, too, sprouting from the tree to the right of the couple are gold blossoms, symbolic of the sun and illumination. All is precision—every detail is carefully chosen to illuminate this 'work of the sun' in the heavenly realm of 'the three'.

Maternal nourishment streams forth from the *Rosarium*'s heart centre of life, as it does in *Aurora*'s fifth parable, called 'The Treasure-House that Wisdom built on a Rock', a female house of blessings filled with intoxicating life and pouring forth abundant food and drink *(fig.193)*:

> Wisdom has built a house and whoever shall enter in shall be saved and find sustenance … they will be drunk from the plenty of your house… If anyone hears my voice and opens the door, I will come in to him and he to me, and I shall be satisfied with him and he with me. How great is the multitude of sweetness which you have hidden for those entering this house … it has been founded on a firm rock…[51]

Here is a 'living fount of water which makes young', a spring of eternal renewal. No wonder then that *Aurora*'s illustrator depicts Ibn Umail at the entrance to this treasure-house *(fig.194)*, one of the 24 elders possessing the key to its door, with a tree of life behind him, its trunk encircled by a golden crown. Or that he is shown with his friends excitedly pointing to the sacred vessel placed in the centre of the Egyptian temple/church where the sage dwells *(fig.154)*.

This journey, like any healing journey, has moved from the first parable's inchoate state to an experience of heavenly second birth, though only the 'justified' find their way to Wisdom's house to behold 'the brightness of the sun and moon'—those able to endure the fourth parable's fiery

crucible separating 'the pure from the impure by removing all accidental things from the soul that are vapours … those things that are unlike and bringing together those things that are like.' For Wisdom's nourishing house is founded on '14 virtues', or 'cornerstones', an unusual number for the Christian virtues, but, nevertheless, the strengths residing in the soul, providing the firm foundation for this heavenly house of truth built on a stone.[52]

'You have tried me by fire and no wrong-doing was found in me' is the plea of justification, as 'what is sown is reaped' and the fruits of previous actions harvested.[53] And for the 'children' able to unlock Wisdom's door through the 'science', a gleaming alchemy of happiness awaits, a fruitful harmony between body and soul, if all the elements flow harmoniously. For, as the fifth parable states, 'As long as the elements are in balance, the soul delights in the body, but when they are in discord within it, the soul declines to dwell therein.'[54]

EMBRYONIC ENLIGHTENMENT: DANTE'S NEW LIFE

Curiously, the *Divine Comedy* composed by Dante (1265–1321) throws some interesting light on *Aurora*'s soul-making, particularly his visionary experience on the summit of Mount Purgatory in the Earthly Paradise, which occurs from two o'clock in the afternoon onwards, precisely the time of second birth in Egyptian alchemy.[55]

Quite unexpectedly, canto 25 introduces the mysteries of generation, including an embryological account of the aerial soul's development, transforming through water, air and the sun's heat, created with powers both 'human and divine'.[56]

Subsequently, in canto 27, Dante hints at his own fearful entry into the fiery 'womb', which concludes with Virgil giving him a mitre and a crown, the regalia of kingship.[57] An elaborate pageant then follows, including 24 elders heralding the arrival of Dante's beloved Beatrice, clad in a green mantle and crowned with Minerva's leaves. Turning her fierce eyes upon him, however, Dante's wisdom-imparting lady initially confronts him with his infidelities and his forgetfulness of her, and wounds his heart with her gaze. He is filled with remorse. Her judgement of him is harsh, yet she is, too, a beautiful incarnation of Aphrodite, implored by her handmaidens to unveil the radiance of her 'smile', which she does at the end of *Purgatorio*, when Dante joyfully emerges from the waters of Eunoe, into which he has been plunged, 'born again', in the concluding words of canto 33, 'even as new trees renewed with new foliage, pure and ready to mount to the stars'.[58]

The correspondences with *Aurora*'s fourth and fifth parables are startling, telling the same initiatory story of heavenly second birth, albeit one alchemically, the other poetically. In fact, William Anderson, without going into detail, suspected Dante might have been 'influenced by the mystical ideas underlying alchemy, in his search for the inner transmutation of the soul.'[59]

Just as *Aurora* instructs, 'Distil seven times' (*see page 243*), so from canto 27, when Dante fearfully enters the fiery womb, to canto 33, when he concludes his journey through *Purgatorio*, there are seven 'purifying' cantos, 'seven steps through which something from on high can be attracted to the lower things', as Marsilio Ficino says.[60] Clearly, too, this new life into which Dante is born sustains his poetic life, inspiring him to bring forth new 'forms' contrasting with the 'old style' of poetry composed by his predecessors, an association between love and poetry germane to the innovating Florentine poetic tradition of the 'faithful of love', to which he belonged.[61]

LEFT fig.195 Zosimus and a companion (Hermes?) stand beside a tree with three branches coloured purple, white and red. His companion points towards the new leaves growing at the tree's base, as if to indicate life's eternal renewal. (*Book of Pictures [Muṣḥaf aṣ-ṣuwar]*, Istanbul Arkeoloji Müzeleri Kütüphanesi, Ms. 1574. 13th century.)

To be sure, nature's womb as the source of all living forms, constantly in motion, constantly changing and transforming, haunted medieval authors before Dante, but his embryological account, his purification told in seven cantos and his 'judgement' by Beatrice leading to his 'green' renewal are an alchemical brew too distinctive to ignore. In short, they point to a lively exchange between alchemists and Dante's circle of poets, sparked by a shared devotion to the death-dealing and life-bestowing Lady guiding them on her path of love.

Ironically, too, these overlaps reopen the controversy about who wrote *Aurora consurgens*, which circulated under the name of none other than Thomas Aquinas (*c.*1225–74). His influence on Dante is well-known, but so far there are no known manuscripts of *Aurora* earlier than the 14th century, and, given alchemists' fondness for pseudonyms, this authorship is notoriously difficult to prove. That such a respected doctor of the Church might have had something to do with *Aurora*'s composition does seem to stretch the bounds of credibility, and it seems to be more of a case of an anonymous Christian author, albeit one extremely well versed in biblical wisdom, seeking validation for the work.[62] Yet more seems to be behind this attribution than meets the eye, and the authorship conundrum is highlighted by *Aurora*'s final parable *(see page 257).*

Not that Dante was introducing anything new, or *Aurora*'s author, since, in *Scivias*, composed back in the 12th century, Hildegard of Bingen had explicitly linked her vision of a foetal child within the mother's womb with the fiery workings of the Holy Spirit and new birth *(fig.186).* Clearly, hers is a heavenly child, connected to the tail of a four-sided golden kite flying in the starry sky, within which is a tripartite division. Hildegard also associates mother and child with the 'splendour like the dawn' which has the brightness of 'purple lightning' inside it, moving 'with vital motion'. She says this child shows itself to be alive 'by the movements of its body', and compares it to the way 'the earth opens and brings forth the flowers of its use when the dew falls on it.' She also says this child 'knows not only earthly but also heavenly things, since it wisely knows God.' She speaks like an alchemist, and, like alchemists, she sees the soul as the vitalizing power, breathing life into the body 'as the tree from its root gives sap and greenness to all the branches'; she also compares the ripening of fruit to the maturation of the soul. Even more startling in this embryological 'greening' vision is her association of the sun, water and air with this maturation process: 'And how is this fruit matured? … By the Air's tempering. How? The Sun warms it, the Rain waters it, and thus by the tempering of the Air it is perfected.'[63] Immediately, she then associates this threefold work not only with God's grace illuminating a person like the sun, but also with the 'breath of the Holy Spirit', clearly relating this illumination to the judgement occurring when the soul separates from the body.

Hildegard was a doctor as well as an abbess; she wrote on health matters, and her commentary on this unusual vision shows her familiarity with Jewish and Classical embryological ideas about the activity of male seed in the womb.[64] Yet alchemy and medicine have always gone hand in hand, and in fact she ends her embryological vision with the plea to 'seek a physician' to heal the soul's wounds. As mentioned earlier, nowhere does she refer to alchemy by name. Nor, unlike *Aurora*'s author, does she include quotations from alchemical authorities, but why would she, since commenting on the Trinity could be controversial enough in the 12th century, and linking the Holy Spirit with alchemy would be even more so and expose her unnecessarily to the hounds of orthodoxy. Yet her imaginative 'embryological' vision, coinciding with the first wave of alchemy surging towards 12th-century Europe's shores, anticipates *Aurora*'s fourth and fifth parables in every way—and Dante's new birth—though it would take time for this wisdom to be fully absorbed during the late Middle Ages and to reappear, in a full-blown exposition complete with citations from alchemical authorities, in *Aurora consurgens*.

TWENTY THREE

THE FIRE-TRIED STONE:
QUICKSILVER RED

Still there is a final phase when 'purple' must deepen into 'red'. In the *Epistle of the Secret*, Hermes explains this 'roasting' in only a few enigmatic lines,[1] yet the scale of this 'Reddening' work is vast, cosmic, requiring a quantum leap into unity as all opposites come together—above and below, heaven and earth, life and death, inside and outside—in this concluding operation.

Hermes tells Theosebia if she wishes never to repeat her work again and have a peaceful heart, she should take her 'beloved pure stone' and 'mix it with the appropriate noble spirits'.[2] Then she must add 'hidden sea water', and, if this is unfamiliar, she must use

> Mercury of cinnabar, and if you don't know it, then the red-adorned fugitive, and if he is unknown to you, then pure eternal perseverance. Bring them to one another and mix them accordingly, so that no trace of him [the beloved stone] is visible in the mixture. Then, add from the hidden leaven called joy of the soul, as much as a tenth of it, bring this before you in the Sun of the Wise and prepare for him the aludel; or bring him in the glass vessel that you know. Stir until you see him clothed in the colours, beautifully and durably clothed without undergoing changes, please God. Proceed with him like at the first time. So know, Theosebia, when you carry this out as I have described to you, he [the beloved stone] will continually increase and never diminish, even if you have to care for all humankind.[3]

Mixing 'mercury of cinnabar' with soul mysticism, Hermes's instructions seem baffling, yet there are clues. Evidently, the 'stone' requires a further 'soaking'. And 'a tenth' is a fraction rich in meaning, since, in the Hermetic number symbolism, ten symbolizes completion and eternal regeneration.[4] Importantly, too, the cryptic remark that Theosebia should proceed 'like at the first time' suggests a repetition of the first operation, for alchemy is not about transforming substances into something entirely new, but rather making manifest what has always been hidden within them, 'turning the inside out', as alchemists say. 'Wise sunshine' is also needed, and, as Theosebia can only accomplish her work if 'God wills', devotion to the divine is all-important.

Clearly, the transmuting agent Hermes particularly has in mind is 'mercury of cinnabar', or 'the red-adorned fugitive' as he calls it, which is obtained from bright red cinnabar, a compound of mercury and sulphur, and requires roasting in order to release the precious mercury. But whereas fumy sulphur readily manifests in this separating process, the purified, mercurial substance, the all-pervading alchemical remedy said to enter bodies like a poison or dye, is extremely difficult to find. 'The cinnabar is the sulphur and the mercury is what is rarely found', says Ibn Umail enigmatically, knowing elusive mercury holds the key to the final 'unification'.[5] Not surprisingly, since mercury is indefinable, difficult to pin down. It can feel solid and liquid, dry and moist, heavy and light, and is therefore able to contain opposites within itself simultaneously.[6] It also likes to move around and has a predisposition to combine with metals. Hence, by its very nature, this mercurial liquid-solid, called 'silver water' (*hydragyros*) in Greek, and the 'fugitive servant' (*servus fugitivus*), or 'fugitive stag' (*cervus fugitivus*), in European alchemy, completely overturns notions of paired opposites and binary perception. Mercury is the source of union, Ficino says, because he 'is masculine as well as feminine and the father of Hermaphroditus', and, in an alchemy based on the principle of 'like to like', it perfectly encapsulates the unifying power needed in this final 'Reddening' operation (*fig.197*).[7]

CELESTIAL BODY: FIRE IN EARTH

The refining power of 'water, air, and fire' have been operative in the previous phase (*see chapter 22*), but now, in this final 'roasting', earth reasserts its importance once again, though, as *Aurora consurgens*'s sixth parable makes perfectly clear, this is a very different 'earth' than at the beginning

LEFT *fig.196* Ash of Ashes. Here the tree's triple roots are shown growing within the alchemical flask. At the centre of the flames is an astral circle, coloured blue and red, representing the 'water' and 'fire' operative in this final phase when everything returns to 'ashes' and stars enter the earth. (*Donum Dei*, Bibliothèque de l'Arsenal, Paris, Ms. 975, f.23. 17th century.)

LEFT fig.197 The three-legged hermaphrodite held in the talons of a huge blue eagle, the bird of Hermes. Blue symbolizes water and mercury in *Aurora consurgens*, and here the hermaphrodite's feet are immersed in a mass of blue birds, some 'winged', others 'wingless', heavenly creatures now forming the fertile earth of the work's final phase. Both rooted in earth and raised aloft by the eagle, the hermaphrodite experiences the 'unification of opposites' ruled by Hermes-Mercurius. (Zentralbibliothek, Zurich, Ms. Rhenoviensis 172, end-paper. Early 15th century.)

fire, but by a fire containing the opposites, a 'fire which is water', an 'unnatural fire', an 'igneous water' or 'non-burning fire', as alchemists say, flames that cook but not consume, creating a body resistant to heat.[10]

Furthermore, 'rooted' in the very centre of this earth, according to *Aurora*'s sixth parable, are the 'seven planets', leaving their 'virtues' there, 'thus in the earth is water, which causes various buds from the heavens to sprout forth, producing fruit, bread and wine to gladden the heart...'.[11] The title of *Aurora*'s first parable, "Regarding the Black Earth in which the Roots of the Seven Planets are to be found' has already hinted at this planetary 'rooting' *(see page 229)*, but only now, in the sixth parable, are its implications fully realized. For these starry bodies pour their blessings into an utterly transmuted earth, a purified earth, a celestial and fruitful earth, which 'makes the moon in its season and then the rising sun', providing the stable, permanent ground for the 'air, water and fire' of heavenly soul life above. 'For you have founded the earth upon its stable bases, it shall not be moved forever and ever. The deep is its clothing, above it shall stand air, water and fire', declares the sixth parable.[12]

True to the spirit of Hermes's instruction to proceed 'like at the first time' *(see page 249)*, and also the repetition in Nefertari's temple between the hall and the sanctuary at Abu Simbel *(see pages 56–7)*, *Aurora*'s first and sixth parables subtly mirror each other. But whereas in the first parable the father is plunged into the putrefying blackness of the corrupting waters, despairingly calling out for someone to rescue him, here in the sixth

of the work. Entitled 'On Heaven and Earth and the Locations of the Elements', this parable tells how the body 'that is the earth' is returned to the 'ashes' of death before being 'mixed with permanent water which is the ferment of gold',[8] becoming transformed into 'boiling' or 'evaporated water' before congealing into its earthly form again:

> When the heat of that fire reaches the earth itself, it is dissolved, becoming boiling water and steaming, thereafter returning to its own previous earthly form.[9]

Manifestly, this is no solid, physical earth, but rather a fermenting golden earth, rising from heated water, a leavened body, displaying fullness utterly unlike before, warmed not by an ordinary parable, as the earth becomes fiery and vaporized, the longed-for heavenly heir manifests his eternal nature. He is the 'second Adam', the *anthropos*, or 'Philosophical Man', as the sixth parable calls him, the perfected human redeemed through Christ's death and resurrection. Formed from 'pure elements', his composite mortal body shines now as a celestial eternal body, the 'simple and pure essence' that 'remains forever' at the heart of a new creation:

> For as in Adam all die, so also in Christ all shall live … for the first Adam and his sons took their origin from the corruptible elements, and therefore it was necessary for what was conceived to be corrupted, but the second Adam, who is called the Philosophical Man, from pure elements entered into eternity. And because the second Adam is made of a simple and pure essence he shall remain eternal, as Senior [Ibn Umail] said: There is one thing that never dies for it continues by perpetual increase.[13]

Just as the ancient Egyptian sun god's starry golden image manifests in the tenth hour and Ramesses II reveals his 'risen earth' nature at Abu Simbel *(see page 54)*, here in *Aurora consurgens* there is the same, albeit Christianized, revelation of an eternal body, similarly identified with the 'name' and 'divine essence', the creative power of the eternal 'Word'. Made in the divine likeness, no longer subject to change, this heavenly heir is reborn through the image, stamped by the seal with an 'imprint that never alters', as Dante says of Adam's manifestation in the second planetary heaven of Paradise ruled by Mercury *(see page 257)*.[14]

In his Greek treatise *On the Letter Omega*, Zosimus equates Adam with Thoth as the 'giver of names to all corporeal things'. He also associates his external nature with 'virgin earth', or 'blood red earth, fiery or carnal earth', formed from the four elements identified with the cosmos and the four cardinal points, and juxtaposes this with his interior, spiritual nature, the primordial human, whom he calls Phos.[15] Ibn Umail knows this heavenly, cosmic body as 'the human being of the sages'.[16] And here in *Aurora* he manifests as the second Adam living in a regenerated creation, an *apokatastasis*, returning the world to its original

state. 'When you have water from earth, air from water, fire from air, and earth from fire, then you have the Art in all its fullness and perfection', concludes the sixth parable, praising the mercurial power flowing through this paradise regained.[17]

UNIFYING ESSENCE: THE TWO IN ONE

It is upon such a transmuted fiery earth that Theosebia stands in the beautiful closing illustration of the Arabic *Book of Pictures (fig.198)*. Her glorious apotheosis is preceded, however, by a scene of her with Zosimus, flanking a 'Man of Gold' lying on the ground with his head brutally severed. Death by decapitation symbolically defines this 'Reddening' phase, as it did in ancient Egypt *(see page 32)*, and witnessing this scene are three winged figures, with six other figures shown in the background, all firmly tethered to ropes held by Zosimus and Theosebia, seemingly to bind their volatile activity.[18] Correspondingly, *Aurora consurgens* also depicts a couple in a completely red setting, the female clothed in white, the male in red, lying on the ground with their heads drastically severed, with four flowers enclosed in an alchemical flask eerily burning alongside, in a 'calcination' operation. Their slayer is an axe-wielding serpent-tailed demon, his head shaped like a seven-pointed star, his body all blue, the colour of quicksilver, suggesting that he is none other than Hermes-Mercury in his death-dealing cosmic aspect, separating the head from the body in this return to eternal life.[19]

Yet in his mercurial world, where the opposites are united, death is life, and the final scene of Theosebia in the *Book of Pictures* shows two images of her *(fig.198)*, dressed in red and gold, each gesturing towards the other, with the figure on the right having a tinge of purple on her sleeves and so hinting at the *iosis* phase previously experienced *(see page 188)*. She is the 'two in one', making manifest her eternal nature, and, according to the accompanying caption, standing in a 'green place' where plants and 'three trees' grow. The caption also explains that 'hot water, the colour of the sky', flows beneath her, a priceless gloss in light of *Aurora*'s sixth parable about the 'heated' earth.[20]

نطاق عليها ثلث نحوه والوزن حتى جمرة وحضره ولو والسماء

ومكا الفضه وبيد هذه الصوره هذا القول

وقار

Not long before the *Book of Pictures* was produced in Egypt in 1270, the great Egyptian Sufi poet 'Umar ibn al-Fāriḍ (d.1235), wrote his long poem the *Tā'iyya*, or *Poem of the Way*, a profound exploration of spiritual pilgrimage guided by female Wisdom; and he too experienced himself as the 'two in one', accomplishing this by 'unifying his essence':

> I caused myself to behold myself, inasmuch as
> in my beholding there existed none other than
> myself who might decree the intrusion [of duality]
> … since I had unified my essence.[21]

This 'unification of essence', Ibn al-Fāriḍ says, is not for the purpose of becoming an ascetic, or to become burdened with intellectual wisdom, but is rather a 'kinship' and 'heritage of the most sublime gnostic', whose aspiration is to 'produce an effect' in the world for the benefit of others.[22] So, as in alchemy, his aim is not primarily personal illumination, but rather to become a channel of

creative life, a source of blessing for others on Wisdom's path. He also knows that nobody is able to accomplish this unification for another person:

> I saw that he who brought me to behold and led
> me to my self was I.[23]

The 16th-century alchemist Gerhard Dorn says the same: 'You will never make from others the One you seek, except there first be made one thing of yourself.'[24] He knows, too, the difficulties of discovering this innermost 'pure essence' that only a few find, the healing 'Philosophers' Stone' that is a fount of abundant life, the 'incorrupt medicine' so powerful it can work miracles in the world.[25] Beyond the comprehension of ordinary sensory perception, it is a divine gift, transforming the literal into the mystical, the transient into the eternal.

Thus *Aurora*'s author states that 'Philosophers … have left it to the glorious God to reveal it to whoever he will and withhold it from whom he will.' Hermes intimates the same to Theosebia in the *Epistle (see page 249)*.[26] So does Petrus Bonus in his *New Pearl of Great Price*, when he says the 'art is partly natural and partly supernatural, or divine' and that the 'hidden stone' cannot be understood by the senses, but is revealed through inspiration or oral teaching. To comprehend this 'divine and glorious mystery', he says, requires 'understanding through the heart', not 'seeing through the eye', for only the heart can perceive the stone's regenerative power to 'conceive, beget, and bring itself forth' in a virgin birth, as the 'begetter and begotten, old man and boy, father and son, all become one'.[27]

In his *Masnavi*, with his usual incisiveness, Rūmī poetically captures this mystery in a few lines, knowing it requires moving beyond 'comparing this with that'. For then

> A third throat is produced, and for its care
> there is God's drink and His illumination.
> The severed throat will drink the drink, but only
> the throat that's free from 'No' and dead in 'Yes'.[28]

Rūmī knows that dying to the opposites of "no" and 'yes' is needed for divine illumination—as well as a very specific form of symbolic death: 'severing heads from bodies'. He also knows the difficulties, and prescribes alchemy as the remedy for those trapped in the world of sensory perception. 'Try alchemy', he pointedly advises, 'change copper into gold.'[29]

TINCTURING VENUS: MERCURIAL MEDICINE

Mercurial Reddening needs copper, unspoken in Hermes's brief instructions for the *Epistle*'s final operation *(see page 249)*, but obvious, nevertheless, from *Aurora*'s seventh and final parable, entitled 'Regarding the Confabulation of the Lover with the Beloved', though this is best conveyed through alchemical imagery, as Abū al-Qāsim al-'Irāqī well knew when he warned against excessive reliance on increasingly corrupted texts—an 'unstable sea without banks', as he called them. Being adrift on this ocean of words could be extremely hazardous, he observed, reminding alchemists that 'enigma in pictorial form is simpler than enigma in words.'[30]

Certainly, the famous image of the *Rosarium philosophorum*'s green lion biting the sun *(fig.199)*, despite the fact that the precious red substance dripping from his mouth is called 'our mercury' in the text, speaks volumes about copper in this final 'Reddening'.[31] For, whilst this green lion might roar and sulphur burn, what the *Rosarium*'s author is keen to stress is mercury's link with the 'copper of the philosophers', which is 'their elixir, from spirit, body and soul completed and perfected.'[32]

Nowhere, though, is this final 'copper-mercury' union more startlingly visualized than in an illustration for *Aurora*'s seventh parable showing a winged young woman in a brilliant red setting *(fig.200)*. Surrounded by a nimbus of golden light, she stands on a silver moon, though its colour has oxidized in the Zurich manuscript's best-known version and darkened over time, as has the original grey colour of her skin—an intriguing reversal in art of alchemy's changing colours.[33] She parts her close-fitting white tunic to reveal the caduceus enclosed within her mandorla-like womb, a gesture 'turning the inside out' to show the snake-staff wielded by Hermes-Mercury (and the ancient Egyptian god Thoth) as the unifier of opposites.[34] *Aurora*'s author never discloses the woman's true identity, but given the sixth parable's celebration

ABOVE LEFT *fig.198* Theosebia as the 'two in one' in the final illustration of the Arabic *Book of Pictures*. Clothed in red and gold, she stands in a green place where trees grow, and, according to the caption, 'hot water the colour of the sky' flows beneath her. (Istanbul Arkeoloji Müzeleri Kütüphanesi, Ms. 1574. 13th century.)

ABOVE *fig.199* The green lion bites the sun. (*Rosarium philosophorum*, Stadtbibliothek St. Gallen, Switzerland, Vadiana Collection, VadSlg Ms. 394a, f.97r. 16th century.)

of Christ as the heavenly Adam, she may be the second Eve.[35] She also brings to mind the pregnant 'woman clothed with the sun, with the moon beneath her feet' in the biblical Book of Revelation, beautifully depicted, for example, in a 15th-century painting preserved in the Cordoba mosque in Spain *(fig.201)*. According to this Christian apocalyptic text she is in labour, calling out in fear of a malevolent dragon threatening her child, and after giving birth she flees into the wilderness, where she receives eagle's wings to escape from her pursuing

RIGHT fig.200 Surrounded by a nimbus of light, and standing on the full moon (its silver pigment having oxidized over time), Venus 'makes the hidden manifest' as she reveals the unifying snake-staff of Hermes within her womb. Here copper and mercury, Venus and Hermes, combine to bring the alchemical 'work' to completion. (*Aurora consurgens*, Zentralbibliothek, Zurich, Ms. Rhenoviensis 172, f.29v. Early 15th century.)

FAR RIGHT fig.201 Painting of the Book of Revelation's pregnant 'woman clothed with the sun', crowned with 12 stars and with the moon beneath her feet. It was painted by an anonymous artist and probably dates to the 15th century. (Mezquita of Cordoba, Spain.)

enemy, her son having been 'snatched up to God and his throne' (Revelation 12).[36] Here is a Christian 'woman who flees' easily reconcilable with alchemy's volatile 'fleeing female'. In fact, when Wisdom rises at dawn to initiate *Aurora*'s work of a day *(see page 225)*, she is crowned with 12 stars exactly like Revelation's 'woman clothed with the sun' *(fig.201)*.

Aurora's illustrations are not Christianized, though, and the green colour of her wings *(fig.200)* reveals who she really is: she is all copper, all Venus, all love, illuminated by the sun as she shows the caduceus within her.[37]

Alchemically, the union of copper and mercury symbolizes the fertile womb *(see page 227)*. To Zosimus in the Arabic *Book of Pictures* it is a metallurgical dyeing operation, and he explains to Theosebia that when mercury embraces copper, both solidify, and this hard substance is the true dye.[38] Hence texts sometimes name purified copper as the dyeing spirit and others red mercury, the

elusive 'red slave', for, as that great lover of copper Michael Maier says in his *Atalanta fugiens*, it is the union of Venus and Mercury that ultimately produces the alchemical hermaphrodite.[39] 'Join therefore the male, the son of the red slave to his sweet-scented wife', instructs the *Turba philosophorum*, and joined together 'they will generate the Art.'[40] This is followed perfectly here in *Aurora*, when copper Venus 'turns the inside out' to reveal the powerful mercurial medicine enclosed within her womb *(fig.200)*.

At the beginning of the seventh parable, in words echoing those of the lover in the *Song of Songs* attributed to King Solomon, a female voice passionately calls to her beloved from the depths of the waters, asking not to be despised because of her blackness:

> Be turned to me with all your heart, and do not cast me aside because I am black and swarthy, because the sun has discoloured me, the deep waters have covered my face.[41]

And there is good reason not to despise her, for if this work is to come to fruitful completion, the 'black' state must be lived through once again. 'Unless the grain falling into me dies', this female voice later proclaims, 'it will remain alone, but if it dies, it brings forth threefold fruit.'[42]

Once before, in *Aurora*'s second parable, the death-dealing and life-giving woman celebrated her unifying and birthing power *(see page 231)*. Now here, in the final parable, she rejoices again as the cosmic 'mediatrix of the elements', the alpha and omega of existence with her beloved, able to transform the elements into their opposites, to 'slay and make alive', whilst hiding the 'science' within her:

> What more shall I say to my beloved? I am the mediatrix of the elements making one accord with another; that which is warm I make cold and the reverse; that which is dry I moisten and the reverse; that which is hard I soften and the reverse. I am the end and my beloved is the beginning, I am the whole work, and the whole science is hidden in me … I slay and make alive, and there is none able to deliver from my hand.[43]

Once again, as with the first and sixth parables *(see pages 250–51)*, there is a repetitive structure guiding this 'science', returning the end to the beginning, for both the second and seventh parables honour her life-giving and death-dealing power, and the correspondences are heightened by her curious grey flesh. For this is the colour of tin, the metal of the planet Jupiter, thus recalling the second parable's reference to Christ as the light rising from the darkness, and the 'Sun of Justice' appearing from heaven *(see page 231)*. It is also a transitional colour created when black changes to white, so moving to the 'whiteness' manifested by Jupiter when conception occurs and everything is set in motion in the first 'birthing' operation *(see page 229)*. Now, however, in this final union, the cosmic mediatrix bears the caduceus in her fruitful womb, the healing snake-staff of Hermes Mercury, whose very nature is to unify opposites, returning heaven to earth in a repetition spiralling deeper into life's birthing mysteries.

Tellingly, *Aurora*'s seventh parable belongs to the 12th hour, the closing hour of day, when

Wisdom becomes like a fruitful vineyard after 'labourers' have worked there in the first, second, third, sixth and ninth hours, and now receive the 'just' gift in the 12th hour:

> I am that chosen vineyard into which the householder sent his labourers at the first, second, third, sixth and ninth hours, saying: Go also into my vineyard, and at the 12th hour I will give you what is just.[44]

She also knows she will come forth in the rising dawn, 'shining exceedingly, bright as the sun, fair as the moon, besides what is hidden within',[45] dancing into the daylight to give birth to new life, just as Hathor brought forth Ramesses all those centuries before at Abu Simbel *(see chapter 4)*.

Again and again the seventh parable returns to the Jewish Wisdom books to express this union, especially the richly sensual imagery of the *Song of Songs*, long interpreted in the Christian and Jewish mystical tradition in terms of the soul seeking God. 'As the vine I have brought forth a pleasant odour, and my flowers are the fruit of honour and riches', the female lover sings. She will clothe her beloved in a purple robe, adorning him with a 'crown of gold' and 'robe of righteousness', placing golden sandals on his feet, dyeing the gold red in a 'love as strong as death' and resurrecting him through love.[46]

Their voices ring out in this primordial hour, calling to each other, seeking each other and finding each other, as they long for union. 'You are she who enters through the ear, through my domain', her lover declares, for they are in that liminal sphere where 'hearing' alone is the operative faculty, the last to remain for the dying, and the one through which their love is experienced in this 12th hour.[47] And that is why their voices sound so ambivalent, so fluid and elusive.

Sometimes the lovers revel in the fruitfulness of flesh, sometimes they anticipate the yet-to be-born.[48] Sometimes they celebrate their union as life draws to a close, sometimes when it is beginning. They are simultaneously in the 12th hour and the rising dawn, the betwixt and between golden hour, neither night nor day, flowing like quicksilver between the worlds, between heaven and earth, the human and divine, sunset and sunrise, in a unified

realm moved by copper and love.[49] Thus it is that they are depicted against a golden background, naked and hermaphroditic, held in the talons of a huge winged blue eagle, the bird of Hermes, his body the colour of quicksilver, in the eternal flux of life streaming forth in the union of heaven and earth *(fig.197)*.[50]

DANTE'S PARADISE: FOURTH HEAVEN OF THE SUN

And once again the correspondences between *Aurora consurgens* and Hildegard of Bingen's *Scivias* are striking. Her first vision of the third (and final) part opens with an 'alchemical' glimpse of 'the stone in the East', an iron-coloured stone above which the 'Shining One' is royally enthroned, encircled in gold like the dawn. But this is a revelation containing the dark 'mire' of human existence too, for held at this royal person's heart centre is a 'blackness', surrounded with precious stones and pearls, the size of a human heart.

Then Hildegard discourses at length on the heavenly fortified city within which the virtues shine and where, in her ninth vision, she sees Wisdom herself atop the 'house of seven pillars'.

In the tenth vision she urges to 'cultivate the field of your heart' and lead a fruitful life, to live in the 'fruitfulness of the flesh'. Then, as the sun sets in the 11th vision, come the tribulations at the end of time, brought by five beasts in the North facing West. 'For people rise and set like the sun', she says. 'Some are born and some die.' All must return to cosmic dissolution and regeneration at the end of time, and in her 12th vision, dominated by the colour red, she describes how 'Fire and Air and Water burst forth, and the Earth was made to move.'[51] She speaks of the mysteries when temporal time is changed into the eternity of God and how the elements are also changed: 'Fire, without its raging heat, will blaze like the dawn; Air without density will be completely limpid; Water without its power to flood or drown will stand transparent and calm, and Earth without shakiness or roughness will be firm and level, and so all these will be transformed into great calm and beauty.'

'I will give you the "just" gift,' says *Aurora* in the 12th hour when the 'mediatrix of the elements'

manifests. The Egyptian sun god, too enters the West as judge at the close of day *(see page 96)*, and Hildegard articulates the same in this apocalyptic 12th vision in which Christ manifests with 'power of judging terrible to the unjust but gentle to the just'. But, musician that she is, she ends *Scivias* with a joyful musical celebration, opening with praise of Mary as the second Eve, in which 'words are the body' and 'music the spirit', uniting the celestial and the earthly, the human and divine, and culminating in a mystical marriage with Christ as the bridegroom.

All this distinctly echoes *Aurora consurgens*. Indeed, *Scivias* seems to offer a visionary blueprint for *Aurora*—its knowledge of 'the ways' now become an explicit alchemical 'work of a day', minus albeit the extensive theological, and social, reflections strewn throughout *Scivias*. Perhaps then it was this extraordinary 12th-century Benedictine abbess who helped pave the way for alchemy's reception in medieval Europe, a way-shower for *Aurora*'s later author, sowing the seeds for a deeper understanding of alchemy's ancient 'birthing' wisdom to be discovered by Western Christendom.

But there was another influential shaper of Western spirituality who was seemingly drawn to this 'rising dawn' alchemy. For the beautiful symbolism of *Aurora*'s Venus standing on a moon, enclosing the mercurial snake-staff within her, surrounded by a nimbus of golden light *(fig.200)*, strangely echoes Dante's experience in the *Divine Comedy* when he ascends with Beatrice through the first four heavens of Paradise, which are sequentially ruled by the planets Moon, Mercury, Venus and the Sun.

Right at the beginning of their ascent, immediately after his rebirth from the womb on Mount Purgatory *(see page 246)*, Dante gazes at the sun's glowing light and experiences his human nature being indescribably transmuted into the eternal quality of heavenly life, as if, he says in the first canto, God had adorned the heavens with a 'second sun'.[52] No Egyptian-oriented alchemist would have missed the meaning of this 'second sun', and Dante makes sure his soul's alchemical transmutation continues during his ascent with Beatrice through the planetary heavens of Paradise. Indeed, when the fourth

heaven, ruled by the Sun, is reached, in the tenth canto, Solomon, Thomas Aquinas and his famous teacher, Albert the Great, are among the luminaries shining there; and this canto concludes by referring to the clock chiming the dawn hour as the 'bride of God' rises to sing to her 'bridegroom' so he may love her.[53]

Back in Mercury's second heaven Dante had discoursed on the Fall of Adam and Eve, their redemption through Christ's death and resurrection, and Adam as the divine 'image'. In Venus's third heaven, he had celebrated the unifying love where each dwelt in the other as the 'two in one': 'I in you even as you in me.'[54] Now, in the solar fourth heaven, as the harmonious cosmic wheel revolves in eternal joy, this love, bestowed by a 'ray of grace', as Aquinas says, will always multiply.[55] Hermes says the same in the *Epistle of the Secret* when he tells Theosebia her stone will always increase, 'even if she has to care for all humankind'.[56] He repeats it in *Aurora consurgens*, declaring that if someone possesses this 'science', even 'if they live for 1,000 years and every day must feed 7,000 people, there will be no lack.'[57] And now Aquinas celebrates this divine gift.

Dante composed the *Divine Comedy* not so long after Aquinas's death in 1274, and, as already mentioned, *Aurora consurgens* circulated under Aquinas's name *(see page 247)*. It is known that both he and Albert the Great, the Universal Doctor, busied themselves with alchemical matters, Albert writing his own study of minerals.[58] Aquinas also influenced Dante, who here makes him the mouthpiece for 'rising dawn' love, with all its echoes of *Aurora*'s closing parable, perhaps unsurprisingly, in light of the visionary discourse on the *Song of Songs* Aquinas gave to the monks of Fossanova Abbey on his deathbed. Apparently, as the famous Dominican scholar lay dying, the immediacy of love overwhelmed him in an experience far removed from his previous writings, which, towards the end of his life, he had come to regard as 'but straw'. Whatever the truth behind *Aurora*'s authorship (and it might be better to regard Aquinas as the honorary, rather than the actual, author), Dante's ascent through these four heavens is so attuned to *Aurora*'s alchemy that it would be perverse not to recognize this.[59]

To be sure, his further hierarchical passage through Paradise—the three subsequent planetary spheres, the eighth stellar sphere and then the ninth sphere of the angelic hierarchies culminating in the Empyrean Heaven—ultimately seems to part company with the 'two suns' transmuting alchemy on which *Aurora* is based. But the resemblances, so powerfully highlighted by Dante's 'embryological' rebirth on Mount Purgatory *(see page 246)*, all point to alchemy, not as some weird pursuit on the fringes of medieval society, but rather as a leaven within the 'faithful of love' network to which Dante belonged, helping to shape its creed of love and veneration of the Lady.[60]

In fact, the Hermetic art coloured the whole fabric of medieval culture. It was at the very heart of spiritual life, and continued to be so in the Renaissance, including the Rosicrucian movement.[61] Indeed, its sanctification of time, its 'birthing' wisdom, completely chimed with the illuminated *Books of Hours* created in the Late Middle Ages as devotional books for lay people, dedicated to the Virgin Mary and with beautiful scenes depicting Christ's 'conception, birth and rearing' synchronized with the liturgical hours of the monastic day.

TURNING THE INSIDE OUT: THE THREEFOLD WORK

In *Aurora*'s seventh and final parable, three different 'earths' delineate its fruitful landscape, with grain being sown first 'in pearls', then in 'leaves' and finally in 'the best earth, namely of gold'—'threefold fruit' sown as the 'food of life which comes down from heaven'.[62] The lovers also build three 'tabernacles', two for themselves and a third for 'our sons', held together by a threefold cord 'not easily broken'.[63] Correspondingly, three faces of Lady Wisdom inspire these fruitful cultivators in their threefold work.

Perhaps, though, her shapeshifting manifestations are best summarized by the 13th-century Italian Dominican friar, Jacopo de Voragine, in his much-copied *Golden Legend*. For in his etymological preface to Mary Magdalene's life, he plays on the meaning of her name, calling her first the 'bitter sea' *(amarum mare)*, secondly the 'illuminator'

and thirdly the 'illuminated'.[64] Whether intended or not, the contours of 'rising dawn' alchemy are clearly visible in her triple manifestations, applied here to her role as Christ's beloved companion, as she was, of course, in early Christian alchemy.[65] After all, Olympiodorus had called Hathor-Aphrodite's 'divine water', the death-dealing turbulent water flowing in alchemy's initial, chaotic state, 'bitter' and 'styptic' *(see page 206)*, which matches this description of Mary Magdalene's first aspect.[66] Then as 'illuminator' she becomes the 'light-receiver', enlightened by the power of her inward illumination, which she imparts to others and which cannot be taken from her because of its 'continuity'. Thirdly, as the 'illuminated', she is eternal, being gifted with the light of 'perfect knowledge', her transfigured body shining with the light of 'heavenly glory'. Certainly, Mary Magdalene's triple nature here readily corresponds with the traditional threefold path of 'purgation', 'illumination' and 'union' in the contemplative Christian journey, but these aspects also strikingly correspond with *Aurora*'s alchemical Feminine.[67]

That 'ancient knowledge', with its wisdom of growth and birth, its honouring of nature, the Feminine and time's circularity, which seemed to have vanished in the groundswell towards 'linear' time favoured by the early Church Fathers, was resurgent once again in medieval Europe, a 'path of the just' guided by Wisdom, moving in one continuous circle from dawn to dusk. But what Wisdom also brought when she came calling was the 'two suns' copper alchemy Zosimus had taught centuries before—a 'heavenly' sun and a blossoming 'earthly' sun, as pictured on *Aurora*'s coat of arms and on the tablet the wise sage holds in Wisdom's treasure-house *(fig.154)*—and this was ultimately rooted in ancient Egyptian 'two suns' wisdom. Just as the sun god in the *Book of Day* journeys from an earth-related birth in copper in the first hour to the heavenly revelation of his divinity in the tenth hour, so *Aurora*'s journey moves from the human, earthly king at the beginning, identified with the infant Christ worshipped by the three wise men, to the revelation of the divinized second Adam, born from Wisdom's eternal womb when the purified heavenly soul

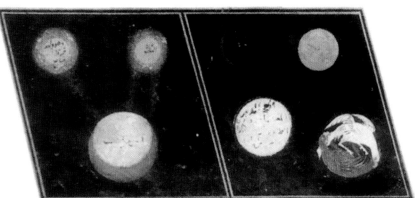

RIGHT fig.202 Detail from *fig.153* of the tablet Ibn Umail saw during his Egyptian temple visit. In the left half are two suns shining down on a moon; the imagery in the right half includes two conjoined birds in the bottom right corner, one winged, the other wingless. (Topkapi Sarayi Müzesi, Istanbul, Ms. A2075. 14th century.)

returns to the glorified body, and this was a revelation utterly dependent on the purification of copper.

These are not unrelated 'suns'; they belong within an indivisible process of unifying heaven and earth, a fruitful unification pouring forth great abundance, as Plutarch well knew when describing the Egyptian mediating 'third love' *(see page 209)*. Indeed, the intentional repetitions between *Aurora*'s opening and closing parables go to the heart of 'rising dawn' alchemy, a distinctly different path from the hierarchical mode of ascent to the One so characteristic of Neoplatonism and the celestial hierarchies of Saint Dionysus the Areopagite, which also influenced medieval Christian spirituality right through into the Renaissance. Egyptian alchemy is about transmutation, its aim being not to supersede previous stages, but rather to reveal the spiritual, hidden life active in the original material substance: to 'make the inner outer'. From the very beginning, the seeds are present and the potential of life is unfolding, changing shape and growing to maturity in a unifying, cosmic birthing that returns the end to the beginning and integrates the human and divine.

'Without the earthly sun, the work is not perfected', says the anonymous author of the *De arte chymica*.[68] Similarly, the *Book of the Holy Trinity* identifies Christ with the circling rainbow as 'the sun above and below', thus bringing together divinity and humanity, the heavenly and the earthly. 'The whole world hangs in the rainbow, the one of the sun above and below. It goes round suspended like a circle in the midst of the airy clouds of the judgement of God's mercy', declares the text accompanying a scene of Christ as the blossoming Sun of Justice. Clothed in a purple-pink robe, with a flowering branch sprouting from his shoulder and a sword in his left hand, he sits within a green, red and gold rainbow, the colours of purified copper *(see page 210)*.[69] Even if at times the author politicized these 'two suns', cleverly applying them to ideas of temporal and spiritual power in the early 15th-century political situation of southern Germany, this work completely relies on the kind of Egyptian alchemy Zosimus set out centuries before in his *Book of Sophe*.[70]

Yet unless these European treatises are understood in the context of their ancient Egyptian heritage, it becomes all too easy to conclude, as some modern-day commentators have done, that they are products of authors with psychotic delusions, containing compelling imagery to be sure, but ultimately incomprehensible.[71] In reality, nothing could be further from the truth. For copper 'two suns' alchemy, the Hermetic art, had become medieval Christendom's own, clothed in a radically new form certainly, and tied now to original sin and the Fall, yet, at the same time, displaying an unfolding transformational pattern utterly in tune with the ancient 'wisdom of the queen', that glorious 'work of a day' enshrined in Hathor's temples.

واسمال الأقلم الى بوعده وصار بقعه با مره وطوعه وطلع المشترى
وهو راكب الحوت وبيد القوس الذى من ضار به لا يموت وعلى
لونه ووجه الأصفر ار لا انه من الكواكب رطب حار وكلا اهلا
لا قلم بلونه وصفته ملك هذا الأقلم صوره المشترى راكب الحوت
وبيد القوس وقدامه فها دايره هلال اصمر وحية بجناحين
وهذه الاحرف فى هذه الموضع

Ironically, contact with Islam had returned medieval Europe to a much earlier tradition of Christian alchemy, one that was already detectable in the non-canonical *Gospel of Philip*, which was discovered in the Nag Hammadi cache of secret writings.[1] In fact, according to Islamic lore, it was a Christian monk named Morienus (or Marianos) who had taught the secrets of alchemy in the late seventh century to the first Muslim alchemist, Khālid ibn-Yazīd, a prince from the Ommayad court in Damascus, and, fittingly enough, Morienus reappeared as the torch-bearer of alchemy when, in 1144, Robert of Ketton provided a Latin version of Morienus's instructions to Khālid in his *Composition of Alchemy (De compositione alchemiae)*, adding that practically nobody in the Latin world knew what alchemy was until then.[2] In doing so, he seamlessly reconnected Christian Europe with its own lost knowledge, which had never really disappeared but simply flowed into other channels, waiting for the ouroboros-snake to close the circle, bite its tail anew and restore this long-forgotten wisdom to the West. Whether this particular Latin text really introduced alchemy to Latin Christendom is doubtful, but the story of Robert of Ketton's translation provides a neat line of transmission, though it scarcely explains how medieval alchemists came into contact with the kind of 'ancient knowledge' Hermes teaches in the *Epistle of the Secret*.

Moreover, the Persian philosopher and Sufi, Shihāb al-Din Suhrawardī (1154–91), who founded the School of Illumination that cherished mystical wisdom obtained through illumination and the soul's purification, gave a very different transmission account. According to him, true wisdom and the 'science of light', which originated with Hermes, 'the father of the philosophers', persisted through the ages, dividing at a certain point into an eastern and a western branch and being known in the East by the great Persian sages and in the West by such figures as Asclepius, Empedocles, Pythagoras and Plato. Ultimately, this 'Pythagorean leaven' passed into Islam and devolved on 'the brother of Akhmim' and 'the traveller from Tustar', the latter referring to the Persian Sufi, Sahl al-Tustarī, the 'brother of Akhmim' to none other than the highly elusive Dhū'l-Nūn al-Miṣrī (d. 860/861).[3]

It might be tempting to overlook Suhrawardī's information. It would be a mistake, however, since not long after Dhū'l-Nūn's death his influence had reached Ibn Masarrah in Andalusia, who was born in 883, and by the early tenth century had founded a Sufi retreat in the hills above Cordoba.[4] Before doing so, he had also visited Medina, where he meditated on the rooftop of Maria the Copt's

LEFT fig.203 An alchemist rides on the back of a huge fish facing towards a fiery dragon, all painted in the green-gold colours of life. Here these creatures of water and fire belong in the final mystery of unification, and the fish associates its rider with Dhū'l-Nūn, 'Lord of the Fish', the great Sufi mystic from Akhmim. (Illustration from al-'Irāqī's *Book of the Seven Climes [Kitāb al-aqālīm al-sab'ah]*, which is in an anthology of Arabic alchemical texts copied in the 18th century. Al-'Irāqī's original 13th/14th-century version is now lost. British Library, London, Add. Ms. 25724, f.30v.)

RIGHT fig.204 Interior of Dhū'l-Nūn's tomb in the Southern Cemetery, Cairo. The tomb was undergoing renovation when this photograph was taken in November 2015.

house, the Egyptian mother of the Prophet Muḥammad's son, Ibrahim, who sadly died in childhood. Curiously, he was also seen to measure the rooms of Maria's house with his hand's breadth, telling his disciples that he intended to model his new retreat in the Cordoba hills exactly on the plans of her sacred dwelling.[5]

What is extraordinary is that here is a western Sufi, not only seeking to establish his Sufi way in an Egyptian feminine house, but also transplanting it to Andalusia long before the translation movement in medieval Europe had started. Even more thought-provoking, though, in light of the king's transformations in a female house in the *Epistle of the Secret*, is Ibn Masarrah's connection with Dhū'l-Nūn, who came from the very same city where the *Epistle* was apparently found *(see page 217)*, and whose lifetime coincided with its reputed discovery date. According to Suhrawardī, Dhū'l-Nūn was the inheritor of ancient knowledge in a western lineage stemming from Hermes; and Theosebia had also turned to Hermes in her own quest for the 'knowledge known to the ancients'. Is all this coincidence? Or is the transmission of the *Epistle*'s alchemy bound up in some way with the mysterious 'brother of Akhmim', who is called 'head of the Sufis' because of his enormous influence in the development of Sufism, and revered by Sufis as the transmitter of the Egyptian secrets?

Furthermore, he is also closely linked with the Sufi Order of Builders, a connection particularly suggestive given that Hermes tells Theosebia to invoke the 'Lord of the Highest Building' and place herself under the protection of the 'Strongest Source of Light' *(see page 227)*. 'Building' connotations continue in the description of the four women surrounding their king like the four walls of a house whilst he functions as the central pillar *(see page 228)*. Such 'building' work would, of course, have been perfectly familiar to the ancient Egyptian

Hermes, alias Thoth, whose scribal companion, Seshat *(fig.205)*, is called 'Lady of Builders' in the Old Kingdom *Pyramid Texts*, and to Dendara's priests in their work of 'rebuilding' Osiris *(see Part 2)*.[6] Horus, too, is a 'builder' in the *Pyramid Texts*; he 'builds and makes the pharaoh live'. Hence, the king can be named 'a pyramid', a 'temple', a 'shrine or 'palace', his name being 'founded' like a building.[7] Just as in Islamic Akhmim, a 'building' tradition weaves through the very fabric of ancient Egyptian spirituality, to the extent that chapter 25 of the *Book of the Dead* says 'becoming a builder' belongs in a creation sustained by the power of sacred words and remembrance of 'Names'.[8]

Whilst Dhū'l-Nūn's father was Nubian, he himself is said to have been born in Akhmim sometime around 771, though there is an Islamic tradition that his actual birthplace was Adwa in Nubia.[9] Interestingly, according to the medieval Egyptian writer al-Qiftī, he used to frequent the temple at Akhmim, doubtless the famous temple dedicated to the fertility god Min, which had strong Hermetic associations in the world of Islam, and was still remarkably well preserved right through into medieval times.[10]

Temple lover that he was, perhaps Dhū'l-Nūn would also have known Repit's temple across on the west bank at Athribis, dedicated to a goddess incarnating the power of Hathor of Dendara, a fierce lioness worshipped with her fertile consort, Min-Re, in a temple celebrating the mysteries of 'conception, birth and rearing' *(see page 219)*.

Indeed, Dhū'l-Nūn's study of the ancient Egyptian signs and images, so it was said, revealed to him the ancient mysteries, a claim scholars are very quick to debunk nowadays, perhaps too hastily in light of his reputation as an alchemist and the evidence connecting him with the *Epistle of the Secret (see below)*.[11] Even his name, meaning 'Man of the Fish'—which is usually compared to the designation of the biblical Old Testament prophet Jonah in the *Qur'ān*, whose sojourn in the belly of a whale before being regurgitated became a potent symbol of death and regeneration in Western alchemy—is curiously reminiscent of the *Epistle*'s king and his marriage with four women 'belonging to the fishes of the sea'.

LEFT fig.205 The scribal goddess Seshat, the companion of Thoth and Lady of Builders. (Temple of Ramesses III at Medinet Habu, Western Thebes. Dynasty 20.)

263

MAN OF THE FISH:
ANCIENT KNOWLEDGE

Dhū'l-Nūn's life had not always been one of great wisdom and spirituality, since he had apparently lived a dissolute life in his youth until jolted one day by a strange experience in the desert. He tells how he fell asleep and when he woke he saw a lark fall from its nest to the ground, which suddenly opened to reveal two bowls in the earth, one made of gold and the other silver, filled with sesame seeds and water from which the lark ate and drank.[12] Seeing this helpless creature finding sustenance deep in the earth somehow revealed his own soul's fallen state, and at that moment he knew he had to return to God and be regenerated.

With its gold and silver bowls deep in the earth, this episode has a distinctly alchemical flavour, though Dhū'l-Nūn's connections with alchemy are certainly controversial and the strange pursuit of 'making gold' seems irreconcilable with his elevated spiritual path.[13] Yet several alchemical treatises are attributed to him.[14] Moreover, one of his companions is known to have been Isrāfīl, and, judging by his name, to which a surname was never appended, he was probably an alchemist.[15] Interestingly, Isrāfīl is also the name of the Islamic archangel who blows the last trumpet on the Day of Judgement, the 'kindler of the dawn' when bodies are resurrected—a hint perhaps that Isrāfīl the alchemist might have had more than a passing interest in 'rising dawn' alchemy and its regeneration mysteries. He was with Dhū'l-Nūn when he died in Giza in 860, after a long life, and subsequently transmitted some of his sayings. According to later sources, he was also Dhū'l-Nūn's teacher, though Dhū'l-Nūn himself said a teacher whose wisdom he particularly revered was a lady he met in Mecca called Fāṭima of Nishapur.[16]

Whatever the truth of this 'Isrāfīl' connection, particularly instructive is the title of a work by the ninth-century Akhmim alchemist 'Uthmān ibn Suwaid, as listed in the catalogue of Ibn an-Nadīm, the tenth-century bookseller in Baghdad, which indicates he defended Dhū'l-Nūn from a 'false charge'.[17] Presumably the 'false charge' related to the accusation of teaching a 'new science' that

jurists made against him towards the end of his life. This resulted in him being brought in chains before the Caliph in Baghdad. Rather than condemning him for heresy, however, so struck was the Caliph by Dhū'l-Nūn's spirituality that he asked him to be his teacher.[18] Alchemists work with the principle of 'like to like' and it is difficult to see why an alchemist from Akhmim would have bothered to defend Dhū'l-Nūn without some shared connection. Moreover, alchemy is called a 'science' in Islam, so the 'new science' Dhū'l-Nūn was accused of spreading may well have involved alchemy.

Most telling, though, is Ibn Umail's brief reference to Dhū'l-Nūn in his treatise the *Silvery Water and the Starry Earth*, composed in the early tenth century—not long, therefore, after Dhū'l-Nūn's death—and prefaced with his famous visit to an Egyptian temple *(fig.153)*. Twice Ibn Umail refers to the 'letter of Hermes' concerning the marriage of 'the male with four wives', who are 'from a moist, beautifully rounded pearl'. He also refers to the 'three marriages' of the male and the female after everything has been whitened, and, in doing so, he cites Dhū'l-Nūn as an authority, quoting one of his alchemical sayings.[19] This cannot be explained away as simply random: Ibn Umail knew this 'marriage', he also knew it was a teaching given by Hermes to Theosebia, and, in his mind at least, quoting Dhū'l-Nūn helped to amplify its meaning.

How much of this early Akhmim alchemy has been lost is obvious from an-Nadīm's catalogue, but, clearly, the *Epistle of the Secret*, with its Akhmim origins and four ladies belonging to the 'fishes of the sea of wisdom', did not come out of the blue. Rather it stemmed from a city indelibly connected with Hermes and Zosimus, and with a 'Man of the Fish' who was honoured as the guardian of 'ancient knowledge' and transmitter of the Egyptian secrets.[20] There is no smoke without fire, and there must surely have been a historical core to all the narratives passed down through successive generations about Dhū'l-Nūn. That Ibn Suwaid wrote a book to defend him not long after his death, and also that Ibn Umail mentioned him in the context of the *Epistle*, are credible pointers to the tradition's authenticity. Yet, in the end, alchemists are not out to prove historical facts, but rather to cultivate connection with Dhū'l-Nūn through the heart's mysteries.

Just how important he was to later medieval alchemists can be gleaned from an illustration towards the end of the Arabic *Book of Pictures* copied in Egypt in 1270, which shows Zosimus and Theosebia holding a red and gold cupola enclosing nine figures, whilst in the foreground is a smaller figure of Theosebia standing in a lake and pointing towards two birds on the shore *(fig.206)*. Within this lake, the mystical sea of unity, swims a tiny whale—a small detail perhaps, though sufficiently important to be mentioned in the caption, and a coded message which, to initiated alchemists,

would surely have resonated with Akhmim's 'Man of the Fish'. In short, whoever illustrated the *Book of Pictures* seemed to have wanted to link it with Dhū'l-Nūn, and an alchemical lineage extending even further back to an Akhmim tradition involving Zosimus and Theosebia.

Interestingly, after a succession of owners in Egypt, the *Book of Pictures* was eventually taken to the Sufi *madrasa* of Inebey in Bursa, in modern-day Turkey, which was established by Niyāzi Miṣrī in the 17th century. He had spent time in Egypt before returning to Bursa, where he founded a new Sufi order, and clearly, as one of the first books marked with the *madrasa's* stamp, the *Book of Pictures* was a treasured possession connecting these Anatolian Sufis to their Egyptian heritage.[21] Whether the

RIGHT fig.206 Zosimus and Theosebia carry a cupola protecting nine winged figures, all tethered with a golden rope. In the foreground Theosebia stands in 'water of silver', in which swim a dog and a whale. Here the whale's inclusion indicates the *Book of Pictures* stems from an 'Akhmim' alchemical lineage honouring Dhū'l-Nūn, 'Lord of the Fish', and going back to Zosimus of Panopolis and ancient Egypt. (Istanbul Arkeoloji Müzeleri Kütüphanesi, Ms. 1574. 13th century.)

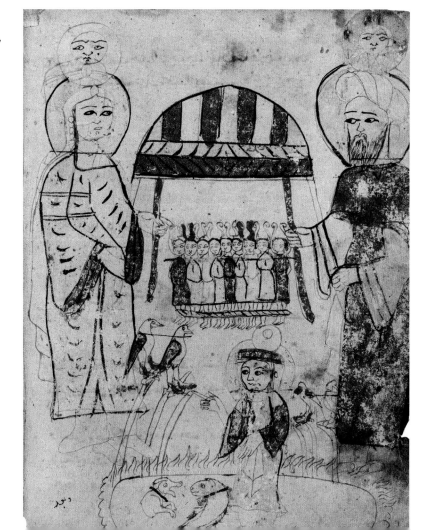

265

book's previous owners in Egypt were Sufis is unknown, but its arrival in Anatolia shows it found a home within a Sufi milieu. It also demonstrates this Akhmim alchemy reached far beyond the banks of the river Nile. In fact, in the 13th century, in the Anatolian city of Konya, Jalāluddīn Rūmī had already drawn considerably on Egyptian love alchemy in his *Masnavi*, creatively reinterpreting it for the purpose of guiding seekers on the transformational Sufi path.[22] And if the geographical span of Egyptian alchemy could encompass Anatolia, it could surely have extended to Europe, especially at a time of such intense cultural exchange between Muslims, Christians and Jews, including in Sicily and southern Spain, where there was a strong Sufi presence.

Around the same time as the *Book of Pictures* was being copied in Egypt, the alchemist Abū al-Qāsim al-Irāqī was similarly adhering to this Akhmim tradition in his richly illustrated *Book of the Seven Climes*. Not only does one illustration include Egyptian hieroglyphs, albeit in a distorted form, but the upper section also shows a winged and a wingless bird, the latter resting on a red flower, and to the left of them, an alchemist working at a distillation oven, engaged in a purifying process *(fig.207)*. In the lower section a black bird perches on the Egyptian royal *serekh* beside a cartouche enclosing the name of the Middle Kingdom ruler Amenemhet II.

The imagery might seem jumbled and meaningless, but there is method in this seeming madness. For al-Irāqī was seeking to connect his alchemy with ancient Egypt, Ibn Umail's 'winged and wingless' birds and Zosimus's 'two suns' wisdom, according to which copper must be purified if the earthly king is to flower and become an image of the heavenly sun *(see page 205)*. Indeed, the bird on the *serekh*, while not a falcon, distinctly echoes the Horus falcon perched on the *serekh*, the rectangular frame which enclosed the pharaoh's Horus name and symbolized his presence in the palace as Egypt's ruler. This is not some fanciful assortment of images—al-Irāqī knew the importance of 'pictures' *(see page 253)* and evidently had looked at Egyptian imagery to convey a rich visual expression of the ancient copper tradition.

ABOVE *fig.207* Illustration from al-Irāqī's *Book of the Seven Climes* influenced by Zosimus's 'two suns' alchemy and ancient Egypt. (British Library, London, Add. Ms. 25724, f.50v. 18th century.)

Also in the same manuscript is a beautiful scene of a man serenely seated astride a huge green and red fish that is rising up to face a large green-gold dragon *(fig.203)*, again surely evoking Dhū'l-Nūn, the 'Man of the Fish', here symbolically uniting the alchemical opposites of fire and water represented by the two creatures.[23]

It was not the factual biography of Dhū'l-Nūn's life within chronological time that concerned al-Irāqī and other alchemists; what mattered to them was to connect with the heritage of the revered 'Man of the Fish', and through him with the ancient temples, as seemingly did the great

Egyptian poet 'Umar ibn al-Fāriḍ (d.1235), who cryptically refers to the 'Man of the Fish' in his Sufi classic, the *Poem of the Way (Tā'iyya)*.[24] 'In this influential poem, composed of some 750 verses addressed to a disciple, Ibn al-Fāriḍ narrates the growth of his soul, inspired by 'ancient knowledge' and his love for female Wisdom. She it is who guides him along the path of love, leading him to the realization of his soul's divinity, which is accomplished by 'unifying his essence' *(see page 252)*. Just as 'mirror-images' of Theosebia close the Arabic *Book of Pictures (fig.198)*, so Ibn al-Fāriḍ knows himself to be both the seeker and the one who is sought, the 'two in one'.[25] Throughout his journey, he is also acutely aware of all the trials and obstacles hindering his spiritual growth, all the illusions preventing the discovery of the divine presence hidden as a pearl in his heart, and he sets them down in detail to assist his disciples following the Sufi path of love.

Like Hildegard of Bingen, nowhere does he refer to alchemy by name, though his statement about cleansing 'the mirror of my essence from the rust of my attributes', sounds distinctly alchemical.[26] What he does state very clearly, however, at the end of his profound poem is his soul's remembrance of 'ancient knowledge':

> In the world of reminiscence the soul has her ancient knowledge—my disciples beg it of me as a blessing.[27]

Amidst the veil of symbols he weaves, both to indicate and, at the same time, hide his secret, Ibn al-Fāriḍ says just enough to reveal the Egyptian spiritual lineage to which he belongs—that 'ancient knowledge' his Persian contemporary Suhrawardī explicitly associates with Dhū'l-Nūn *(see page 261)*. And, like Muhyi'ddīn ibn 'Arabī (d.1240), the Andalusian poet and philosopher known to Sufis as the 'greatest sheikh'—whose spiritual vision is opened at Mecca by a beautiful Persian girl called Nīẓam, 'Eye of the Sun', a manifestation of Wisdom in all her life-giving and fierce power—Ibn al-Fāriḍ's 'way of illumination' is guided by Wisdom. She it is who awakens love in his heart, arousing in him both fear and ecstasy as he seeks union with the divine beloved through a 'kinship of love'.[28]

Love mysticism pervades the soul-making of these influential Sufis. They also lived during a period of renewed spiritual activity in the Islamic world, which saw a whole chain of Sufi centres established. What had once been informal meetings, with a few friends gathering together in private houses, now metamorphosed into the various Sufi orders that exist to the present day, open to all seekers of wisdom, both men and women. Sufis were becoming more organized; they needed to map out the spiritual path, the way of life that aided spiritual realization. They were also keen to establish the spiritual genealogies and 'chains of transmission' relevant to their burgeoning orders. Interestingly, many of their members came from the artisan and craft communities in urban centres, hence alchemy, with its fiery knowledge of transforming one substance into another and its reconstitution of nature into an eternal state, became a particularly powerful aid for teaching the mysteries of regeneration and spiritual rebirth. This was a transformational wisdom (though kept very well-hidden) completely in tune with the Sufi path.

ALCHEMY OF PRAISE: GREEN MESSENGER BIRDS

Slippery as the fish with which he is identified, Dhū'l-Nūn's alchemical connections are, however, extremely hard to unravel, especially as his mystical teachings are found only in what has been transmitted by other writers over the centuries. Nevertheless, in his reverential homage to Dhū'l-Nūn entitled *The Shining Star: On the Spiritual Virtues of Dhū'-l-Nūn the Egyptian*, Ibn 'Arabī scatters some clues.[29] He is known to have stayed with a small group of Sufis in Cairo in 1206, so perhaps this inspired him to gather together copious anecdotes, sayings and spiritual wisdom from various sources associated with the ninth-century 'Man of the Fish' whom he deeply revered. His homage would surely have provided an inspirational source of wisdom for Sufis in Egypt and elsewhere, reminding them of the rich spiritual tradition they had inherited and could continue in their own lives.

In particular, Ibn 'Arabī's homage includes a devotional praise-poem, a 'science of the ways'

composed in 22 verses, which sets out the 'states and stations' Dhū'l-Nūn experiences during his spiritual journey. The very fact that throughout this praise-poem Dhū'l-Nūn calls his path a 'science' suggests these are to be understood as alchemical (*see above*).[30] In fact, elsewhere Ibn 'Arabī refers to alchemy as a 'natural, spiritual and divine' science.[31] What is also particularly striking about these 22 verses is that, whilst the Prophet Muḥammad's nocturnal ascension *(Mi'rāj)*, as mentioned in *Qur'ānic* revelation, is often the prototype for mystical experience in Islam, Dhū'l-Nūn's ascent has a much more 'human' dimension as he seeks to separate himself from 'everything gross' and raise his soul to the 'heavens of the noble saints'.[32]

It is impossible to do justice to the beauty and fervency of this long praise-poem here, but from the standpoint of Dhū'l-Nūn's connection with the *Epistle of the Secret*, and ultimately ancient Egypt, enough needs to be mentioned now, starting right at the beginning with the imagery of his journey as a boat voyage. Stating his deep desire to set sail in the 'vessel of discernment', he longs to be blown by the 'winds of certainty' on the 'waters of salvation', his soul freely evolving in the heights, leaving far behind the 'dwelling of the unjust'.[33] And just as the *Epistle*'s alchemy is rooted in 'conception', 'birth' and 'rearing', and in an astrological view of 'timing' and the 'right moment', so is Dhū'l-Nūn's transformational path, which is evident when he expresses gratitude for his conception and the maternal care he has received, as well as the propitious timing of his birth during an era when people follow the 'good way':

> You have caused that I dwell in the loins [of my ancestors], and you have caused me to dwell in the maternal womb … by your abundant generosity you have caused me to be born, thanks to your mercy and goodness, at a time when people follow the good way… You have caused me to come into the world, formed and complete, you have ensured my protection when I was still a small child in the cradle, and you have permitted me to be nourished with generous milk. You have entrusted me to the maternal breast of women, in whose hearts you have placed tenderness and solicitude for me…[34]

His spirituality is founded on human love, which he has already experienced in his conception and birth, and he continues by celebrating his 'growth each year'—his nurture as his limbs become 'steady' and he becomes a 'complete human', watched over by the 'majesty' and 'magnificence' of his Lord. Protected by those expansive qualities, the hallmark also of a Jupiter-like being, his development here seems distinctly reminiscent of the *Epistle of the Secret*'s 'Jupiter' phase.

Dhū'l-Nūn knows he will be tested, that 'trials' will afflict him, as he begins his ascent to the 'royalty of the heavens', to the 'gardens of familiar relations where we pluck the fruits of our desire for you, where we drink from the vessels of your knowledge', and where 'the heart finds pasture'.[35] Guided by the 'science', he longs to reach this heavenly place of rest and refreshment. Indeed, at times his praise of his compassionate Lord sounds for all the world like praise of the ancient Egyptian sun god sailing with his companions, 'the justified ones', in their own 'royalty of the heavens' during the daytime ninth hour. 'The heart gains the interior of the breast', he exults, in the 'dwelling of Majesty'.[36] And in verse after verse, he pours forth the joy granted to him by dwelling in this heart region, living in the presence of an all-merciful, all-forgiving deity, the 'medicine of our infirmities', who 'treats us with kindness after we have done wrong', pardoning those who 'turn back'.[37] Uttering such words, he could be a reincarnation of the ancient Egyptian Sinuhe in the merciful presence of Senwosret I at his homecoming to the Egyptian court *(see pages 40–42)*.

So, too, Dhū'l-Nūn's subsequent praise of 'the Name' and the divine 'Face' distinctly echoes the ancient Egyptian 'work of the day'. For it is through the divine Name that all the abundant 'marvels of creation' are established in the 'inscrutable womb of the science'; through the light shining from the 'majestic beauty' of the divine Face all are sustained in a realm of rest where 'glory abounds and where the immortals dwell'.[38] As in ancient Egypt, the power of 'Names' sustains creation, though this knowledge, hidden from carnal eyes, is communicable only through the 'vision of hearts', praised through 'mysterious voices, and melodious, pure chants' in the matrix of 'the science'.[39]

Exactly like Theosebia in the *Epistle of the Secret*, Dhū'l-Nūn is a seeker of the water-bearing vessels of the wise, longing to be placed amongst those whose faces are turned towards the sanctuary (*mihrab*) of holiness, those

> Who have washed the vases of folly with the pure water of life nurturing the paradisal ways, vases that circulate in the assemblies where your Name is recited, moistened by the speech of those who invoke you. Place us, O my God, among those who rejoice in abundant nourishment … where the heart finds pasture and where the springs of the hidden realities gush forth.[40]

Enraptured, he moves in heavenly realms far removed from earth, for although his body is present below, his 'heart is absent', absorbed in the wonders of a creation sustained by the divine Name. Whenever he 'listens to the cry of animals, to the whisper of trees, to the murmur of water, to the harmonious songs of birds, to the delicious hospitality of shade, to the noise of the wind, to the rumble of thunder', he will always perceive in them the eternal light radiating throughout the world, the sublime revelation of divine 'Unity' in creation's generous abundance. His heart perceives signs of divine activity everywhere, and he longs to be united with his all-powerful, compassionate king through the same love a father shows to his child, to be clothed in the 'robe of honour of your love'.[41]

Among Sufis, Dhū'l-Nūn is regarded as the first to set out the 'states and stations' of the Sufi path to sainthood, and at first sight these 22 verses could easily pass as a purely Islamic heavenly ascension, yet there is also something utterly familiar about their 'three worlds' cosmology founded on 1) human birth and development in the physical realm; 2) ascent to the 'royalty of the heavens' (*Malakūt*); and 3) encounter with 'the Face' and 'Name' in the divine realm of 'essence' (*Jabarūt*), where the 'immortals reside', and where the true self is realized amidst the 'mysterious veils'.[42] In this flow of life, starting with his conception and birth, Dhū'l-Nūn grows, he ascends, he rises to a

peaceful state in blossoming fruitful gardens— a transformational journey through the heart's interior landscape, 'three worlds' perfectly in tune with the alchemy Hermes teaches Theosebia in the *Epistle of the Secret* and with the Sufi 'three worlds' more generally.[43]

Whether this is Ibn 'Arabī speaking through Dhū'l-Nūn or Dhū'l-Nūn speaking through Ibn 'Arabī matters little, for what this beautiful praise-poem reveals is the spiritual path that 'the greatest sheikh' associates with the Egyptian Sufi he reveres. Its similarities, moreover, with Hermes's revelation in the *Epistle* are simply too striking to ignore. Is it therefore coincidence that Ibn Umail specifically cites Dhū'l-Nūn as an alchemical authority when referring to the *Epistle*? Or that the *Epistle*'s alleged date of discovery at Akhmim coincides with Dhū'l-Nūn's lifetime? And that there was a mosque dedicated to him at Akhmim in the medieval era that was said to be 'impressed with the character of blessedness'?[44] For in the *Epistle* Hermes teaches Theosebia a remarkably similar way, albeit overtly through alchemy, leading to her heavenly illumination when the 'clouds are dispelled'. To be sure, Dhū'l-Nūn's alchemical activity was not documented historically in his lifetime, but Suhrawardī explicitly stated he was the bearer of Hermetic 'ancient knowledge', and Islamic authors connected him with alchemy. Are they simply mistaken, or is it that they knew very precisely his role in the transmission of this 'science'?

Yet it is Rūmī, in the second book of his *Masnavi*, in his long section devoted to Dhū'l-Nūn, who perhaps best illuminates how Sufis might have understood the *Epistle*'s alchemy, with its unnamed king and union with four ladies 'belonging to the fishes of the sea of wisdom'. Previously he had compared 'the illumined ones' to 'fishes in that sea' before asking: 'What pearls have you fetched from the bottom of the sea?'[45] And later he describes the biblical prophet Jonah's imprisonment in the body of a fish and escape, a 'resurrection from the tomb which is your heart'—a phrase that would surely have resonated with Akhmim's 'Man of the Fish'.[46] Indeed, he poetically begins his section on Dhū'l-Nun by referring to his well-known imprisonment,

269

due, Rūmī says, to his ecstatic states, which were anathema to the people, 'his fire was carrying off their beards'.[47] But then he continues:

> The great king rides alone, such a unique pearl
> in the hands of children!
> What pearl? The sea hidden in a drop, a sun
> concealed in a mote.
> A sun showed itself as a mote
> and little by little revealed its face.[48]

When the sun's face shines, the hidden manifests, for Rūmī knows that the imprisoning darkness in which Dhū'l-Nūn initially languished, the alchemical 'black' state when everything seems reduced to nothingness, ultimately turns to an illumination through light, and he ends this long section with a totally alchemical statement, as he must if he truly is to honour Akhmim's 'king': 'A friend is like gold, tribulation is like fire: pure gold is glad in the heart of fire.'[49]

His imagery more generally in the *Masnavi* is strewn with references to the dyeing vat and colours, to copper and gold, and to 'making the hidden manifest', so that reading this inspirational work in itself becomes an alchemical activity, transporting the seeker into the Hermetic art's transforming secrets. On their spiritual journey towards divine union, their quest to 'become mature' as Rūmī calls it, Islamic mystics found in alchemy a way of purifying their lives, knowing 'that bitter things become sweet', that 'copper becomes gold' and the 'dead are made alive' when burning hearts are touched by love's transforming power.[50] As Annemarie Schimmel observed, Rūmī's images 'go back to ancient rites of purification through fire, and lead, in turn to the vocabulary of alchemy.'[51] Which specific 'ancient rites of purification' she had in mind she left unsaid, yet the

same illumination 'science' is there centuries before at Abu Simbel. It is there in the southern crypt of Hathor's temple at Dendara, and there, too, in the *Book of Day* and the *Ritual of Hours*, a sequence of heartfelt praise through the 12 hours of the day as eloquent in their way as Dhū'l-Nūn's devotional praise-poem.

No longer is this 'ancient knowledge' built into the monumental stones of the Egyptian temples, though, but experienced in the interior of the human heart. Its ancient transformational wisdom is now expressed through an 'alchemy of happiness', with all its rich language of colours, metals and stones. Hence, it would be wrong to see this as some mistaken projection back onto ancient Egypt by seekers of 'ancient knowledge', since the

RIGHT fig.208 Queen Nefertari receives her scribal palette and water pot, 'and the mysteries which are in them', from Thoth, the Egyptian Hermes. (Wall-painting illustrating chapter 94 of the *Book of the Dead*. Tomb of Nefertari, Valley of the Queens, Western Thebes. 19th Dynasty.)

very same wisdom is inscribed in Hathor's temples.[52] And to live this journey in ancient Egypt is to bring feminine Wisdom (Maat) into the sun boat at dawn, to be born of Hathor in a transformational sea of copper and to soar with the sun god through the hours on the 'way of the green bird'.

'Green bird' wisdom is ingrained in Egyptian Sufism, too, since it is a deeply held belief that the souls of true believers live in green birds. 'Green birds of an unknown species' are reported to have flown above Dhū'l-Nūn's funeral cortège when he was being taken for burial.[53] Green and white birds miraculously hovered above Ibn al-Fārid's bier, too, as holy spirits prayed over it, honouring a Sufi whose devotion to Lady Wisdom had led him to

experience the mysteries of Unity.[54] Ibn Umail, too, mentions a green bird (a woodpecker) in his *Book of the Explanation of the Symbols*, reflecting a bird wisdom reaching deep into ancient Egypt.[55]

And there is something supremely fitting in the fact that the 'ancient knowledge' that so inspired them should be intertwined with Dhū'l-Nūn, a Sufi from Akhmim, whose father, Ibrāhīm, came from Nubia, where this ancient wisdom was so powerfully enshrined at Abu Simbel. For when this alchemy touches hearts, it speaks across the generations, offering beauty, healing, transformation and renewal to seekers on the way of the green bird, opening them to the mysteries of love's transforming power and Hathor's eternal treasure.

ABBREVIATIONS

AcOr *Acta Orientalia ediderunt societates orientales batava danica norvegica svecica.*

ÄA Ägyptologische Abhandlungen.

AH Ægyptiaca Helvetica.

AV Archäologische Veröffentlichungen, Deutsches Archäologisches Institut, Abt. Kairo.

BdÉ Bibliothéque d'Étude, Institut français d'Archéologie orientale du Caire.

BIFAO *Bulletin de l'Institut français d'Archéologie orientale du Caire.*

CALA Corpus Alchemicum Arabicum.

CdÉ *Chronique d'Égypte.*

CT A. de Buck, *The Egyptian Coffin Texts.* 7 vols. Chicago, 1935–1961.

DE *Discussions in Egyptology.*

FIFAO Fouilles de l'Institut français d'Archéologie orientale du Caire.

GOF Göttinger Orientforschungen: Veröffentlichungen des Sonderforschungsbereiches Orientalistik an der Georg-August-Universität Göttingen.

GM *Göttinger Miszellen: Beiträge zur ägyptologischen Diskussion.*

IFAO Institut français d'Archéologie orientale du Caire.

ISIS *Isis: International Review devoted to the History of Science and Civilization.*

JAOS *Journal of the American Oriental Society.*

JARCE *Journal of the American Research Center in Egypt.*

JEA *Journal of Egyptian Archaeology.*

JEOL *Jaarbericht van het Vooraziatisch-Egyptisch genootschap "Ex Oriente Lux".*

JHA *Journal for the History of Astronomy.*

JNES *Journal of Near Eastern Studies.*

JSAI *Jerusalem Studies in Arabic and Islam.*

JWCI *Journal of the Warburg and Courtauld Institute.*

MAGW *Mitteilungen der Anthropologischen Gesellschaft in Wien.*

MÄS Münchner Ägyptologische Studien.

MDAIK *Mitteilungen des Deutschen Archäologischen Instituts, Abteilung Kairo.*

OBO Orbis Biblicus et Orientalis.

OLA Orientalia Lovaniensia Analecta.

OMRO *Oudheidkundige Mededelingen uit het Rijksmuseum van Oudheden te Leiden.*

Or *Orientalia*, Nova series.

PT K. Sethe, *Die altägyptischen Pyramidentexte. Leipzig, 1908–22.*

RdÉ *Revue d'Égyptologie.*

RHR *Revue de l'histoire des religions.*

SAK *Studien zur altägyptischen Kultur.*

Urk *Urkunden des ägyptischen Altertums (ed. G. Steindorff):*

1. K. Sethe, *Urkunden des Alten Reichs.* Leipzig, 1903.

4. K. Sethe and W. Helck, *Urkunden der 18. Dynastie.* Leipzig and Berlin, 1906–58.

Wb A. Erman and H. Grapow, *Wörterbuch der ägyptischen Sprache.* 7 vols, Belegstellen 1–5. Berlin, 1926–63.

ZÄS *Zeitschrift für ägyptische Sprache und Altertumskunde.*

ZGAW *Zeitschrift für Geschichte der Arabisch-Islamischen Wissenschaften.*

NOTES

For full publication details, see Bibliography.

INTRODUCTION

1) Information on the temple is published by Desroches-Noblecourt and Kuentz, *Le petit temple d'Abou Simbel* (hereafter cited as *Abou Simbel*).
2) For the temple's date of construction, as indicated by the different forms of the king's nomen, see Spalinger, *JEA* 66 (1980), 95–8.
3) For this letter, see Kitchen, *Pharaoh Triumphant*, 80.
4) For this myth, see Roberts, *Hathor Rising*, 12 (with further references).
5) For translations of the *Story of Sinuhe, see below, chapter 3, n.7*. Cf. also the scenes of Queen Ankhesenamun on the small Golden Shrine, which show her embodying Hathor-Sekhmet's power and vitalizing Tutankhamun at the New Year, see Roberts, *Golden Shrine*, 11–14, 16–26, 38–46, with *figs.1, 8, 10–13, 25–6, 28, 30*.

CHAPTER 1

1) The modern-day titles were given to these 'books' by Piankoff in his original publication (*Le livre du jour et de la nuit*, 1942). The *Book of Night* has since been re-edited by Roulin (*Le livre de la nuit: Une composition égyptienne de l'au-delà*, 1996). For the re-edited *Book of Day*, see Müller-Roth, *Das Buch vom Tage*, 2008.
2) The relationship between the two books is controversial. For the books as a description of the complete 24-hour cycle, see Quack, *Die Welt des Orients* 28 (1997), 180; Betrò, *Or* 67 (1998), 510, 517; as separate compositions, Roulin, *op.cit.* 1, xvi, n.9; Müller-Roth, *op.cit.* 534–6. In my view, their juxtaposition suggests relationship, though possibly based on an alternating, rather than continuous, cycle through the 24 hours.
3) For the *Book of Day*'s geographical scheme, see Müller-Roth, *ibid.* 491–8.
4) For a list of the known versions (with bibliography), see *ibid.* 23–41.
5) Assmann first pointed out the *Book of Day*'s connection with the *Ritual of Hours*, see *Egyptian Solar Religion*, 26–30; also Müller-Roth, *op.cit.* 536–40. The *Ritual of Hours* is a modern title, though an abbreviated version, preserved on a papyrus from Tebtunis, has the title 'Instructions for knowing the names of the Hours of the Day' written in red ink above the text (Müller-Roth, *ibid.* 61). The ritual has no definitive form and variants in the Late-Period hymns (including which hymn belongs to which hour) indicate different versions of the ritual must have been in circulation. The hymns in the Edfu and Dendara temples also belong to a distinct group, and their relationship with the *Book of Day* is unclear, since, even though the hymns resemble the New Kingdom versions, the accompanying scenes simply show the different forms of Re in each hour boat, reflecting a Ptolemaic development (*ibid.* 46 for the Edfu version). Evidently there was a complex transmission over the centuries. For a current online publication of the ritual's known sources,

'Das Stundenritual Online', URL:http://www.uni.muenster.de/IAEK/forschen/aeg/proj/laufend/stundenritual.html. Assmann's German translation of all 12 hymns is the only one currently available (*Ägyptische Hymnen*, 97–112), and I have followed his sequence. For translations of some of the hymns, with an extensive commentary, see also Assmann, *Liturgische Lieder*, 113–64.
6) I have borrowed the term 'life world' from Abram, *Spell of the Sensuous*, 40–43, 65.
7) Assmann, *Ägyptische Hymnen*, 97–8 (my translation from the German).
8) *P. Berlin* 6750. See Smith, *Mortuary Texts*, 98.
9) For detailed evidence identifying the birthing goddess as Hathor rather than Nut, see Roberts, 'Invisible Hathor', 163–9, including Dendara's version of the *Ritual of Hours*, which allocates the two boats for the first hour to Hathor and Khepri (Cauville, *Dendara: Transcription*, 86). For the literature on the birthing goddess as Nut, see Roberts, *op.cit.* 163–4. Müller-Roth, for example, claimed Hathor had no significance in the *Book of Day* (*op.cit.* 89).
10) Assmann, *Ägyptische Hymnen*, 98.
11) For 'She who causes to rise' and the uraeus, see Roberts, 'Invisible Hathor', 167. For serpent power fuelling the sun god's ascent to the zenith, see Roberts, *Hathor Rising*, 66.
12) For this text, and translations of the ninth-hour's name, see Müller-Roth, *Das Buch vom Tage*, 100–101.
13) For this dawn synchronicity, see also Roberts, *My Heart My Mother*, 180–88.
14) For this youthful phase in the solar cycle, see *ibid.* 40–51.
15) From Assmann, *Ägyptische Hymnen*, 99. For clouds obscuring Horus's vision at Dendara and their removal, see Richter, *The Theology of Hathor of Dendera*, 302(84.5–6).
16) For the Horus–Seth conflict, see Roberts, *Hathor Rising*, 98–112 (with further references).
17) Quoted from te Velde, *Seth*, 75.
18) For this tenth-hour synchronicity, see Roberts, *My Heart My Mother*, 156–8. It is expressly stated in this hour: 'Those who adore Re on earth, and those who cense the gods in the *Dwat*, will be in the following of this god', *ibid.* 156.
19) For the third-hour hymn, see Assmann, *Ägyptische Hymnen*, 99–100.
20) For the Eye goddesses empowering the dawn sun king, see Roberts, *Hathor Rising*, 45–51.
21) For disturbances in vision associated with the Moon and Sun Eyes, see *ibid.* 112.
22) Seventh-hour hymn in Assmann, *Ägyptische Hymnen*, 105.
23) For the transformation from mother to daughter during solar ascent, see Roberts, *Hathor Rising*, 66–8.

CHAPTER 2

1) For the sun's south–north movement in the daytime cycle, see Assmann, *Liturgische Lieder*, 131–2.
2) For Khons as moon god, see Roberts, *Hathor Rising*, 78–80.
3) For this role of Sothis in the Field of Reeds, see *PT*§821b–22: 'You go forth with Orion in the east of the

273

sky, you descend with Orion in the west of the sky. Your third is Sothis, pure of thrones, and it is she who guides you on the beautiful ways which are in the sky in the Field of Reeds.' See also *PT*§341c, §935c–36b, for Sothis as the pharaoh's sister in the Field of Reeds.

4) *Papyrus Leiden 1 350*, 4, 21–2; Zandee, *Hymnen*, 87, with *pl.*4. See also Roberts, *My Heart My Mother*, 14.

5) For the symbolism of gold in ancient Egypt, see Daumas, *RHR* 149 (1956), 1–17.

6) Cf. Geb's gift of the Field of Reeds to the vindicated Horus ruler in the *Pyramid Texts*: 'The Field of Reeds, the mounds of Horus, the mounds of Seth, all belong to the king' says Geb (*PT*§943–4). For Horus's vindication in the Field of Reeds, see also chapter 110 of the *Book of the Dead*.

7) For this ninth-hour hymn, see Assmann, *Ägyptische Hymnen*, 107–108; Quirke, *Cult of Ra*, 55.

8) *Ibid.* 108; 55.

9) For the river axis in Ramesses IX's version, dividing the *Book of Day* into two halves ruled by Hathor-Sekhmet in her 'green' and 'red' aspects, see Roberts, 'Invisible Hathor', 167. Cf. also the deity called 'He who brings the Inundation', shown standing above the ninth-hour deities in Ramesses VI's version.

10) For the defeat of Apophis as a 'justification', see von Lieven, *Himmel über Esna*, 162.

11) Müller-Roth, *Das Buch vom Tage*, 380(409).

12) See *ibid.* 144 (lines 5–7), 151–2, 154, for the four *Bas* of Re, and the 'red *Ba*' associated with sunset. For the sun god's green/red colours associating him with the Eye goddesses, see Roberts, 'Invisible Hathor', 166–7.

13) Assmann, *Ägyptische Hymnen*, 109–10; Quirke, *op.cit.* 55–6.

14) For the nature of *sekhem*-power, see Roberts, *Hathor Rising*, 57–8.

15) *PT*§682e–f.

16) See Roberts, *My Heart My Mother*, 168–9.

17) Quoted in Daumas, *RHR* 149 (1956), 15.

18) For Hathor's role as 'Gold' during these nocturnal rites, see *Hathor Rising*, 26–9.

19) For the head being particularly associated with celestial life, *see chapter 11.*

20) Assmann, *Ägyptische Hymnen*, 110–11; Quirke, *Cult of Ra*, 56.

21) For 'Bull-of-his-Mother' and regeneration in the Theban harvest festival, see Roberts, *Hathor Rising*, 82–6.

22) For this name of Seth, see te Velde, *Seth*, 31–2.

23) For this text, see Müller-Roth, *Das Buch vom Tage*, 280–85.

24) For the Ramessid ritual for Amenhotep I, see Roberts, *My Heart My Mother*, 72–92. Since the writing of that chapter, a new edition of the ritual has become available (see N. Tacke, *Das Opferritual des ägyptischen Neuen Reiches*. Leuven, 2013), in which it is renamed 'The Offering Ritual of the Egyptian New Kingdom'.

25) For the pairing of these kings with Meretseger, see Müller-Roth, *op.cit.* 526–8.

26) English translation in Copenhaver, *Hermetica*, 49–54.

27) For a brief summary of the modern debate regarding the ancient Egyptian/non-Egyptian origins of the *Corpus Hermeticum*, see *ibid.* li–lix. Closely related to Book 13 is the Coptic Hermetic text known as *The Discourse on the Eighth and Ninth* from the Nag Hammadi Library, which situates the dialogue within a distinctly ancient Egyptian milieu, containing overt references to the Egyptian cults and magic (translation in Robinson [ed.], *The Nag Hammadi Library*, 321–7). On the similarities with Book 13, see Mahé, *Hermès*, 21, 41–4. Cf. also Bourgeault's discussion of 'rebirth through the image' in non-canonical early gospels, the unifying experience of the 'two become one' marking the supreme inner transformation of the 'completed human being' (*Mary Magdalene*, 54, 78, 106–107, 129–31, 172–3).

28) For the 12th-hour hymn, see Assmann, *Ägyptische Hymnen*, 111–12; Quirke, *op.cit.* 56–7. Cf. the prominence of serpent goddesses in the West in Ramesses IX's abbreviated version of the *Book of Day*. Whereas in the 'green' eastern half Hathor is shown giving birth to the sun, in the 'red' western half there are five anonymous figures, including a lioness-headed goddess. Beneath them, four cobras raise their arms in praise, all facing the king enclosed in Nut's arms (see *fig.12* in this book; also Roberts, 'Invisible Hathor', 167). For the sun god's return with his Eye, see Roberts, *My Heart My Mother*, 84–5, 180–88.

CHAPTER 3

1) It should be pointed out that my interpretation here differs from the one in the temple's publication, which Desroches-Noblecourt based on two scenes in the transverse chamber showing Ramesses offering to the birth goddess, Taweret, and Nefertari being crowned by Hathor and Isis. She emphasized the importance of the *Goddess in the Distance* myth, and the queen's identification with Isis-Sothis as 'the great mother' or 'great flood' (Mehet Weret), who ensured the king's renewal and the return of the inundation at the New Year (*Abou Simbel* I, 111–18). However, she offered no analysis of the other reliefs in the temple as an integrated, coherent sequence following the pattern of the *Book of Day*.

2) For the naos sistrum and 'daughter' aspect of solar goddesses, see Roberts, *Hathor Rising*, 54–64.

3) See Moret, *Rituel*, 140–44.

4) For Hathor's *menit*-necklace transmitting 'attraction' to the dawn sun king, see Roberts, *op.cit.* 46–50.

5) For similarities between the Egyptian ascent and the Tantric tradition, see *ibid.* 63–4, 66–7.

6) Cf. an unfinished statue of Akhenaten tenderly kissing his daughter seated on his lap; *ibid.* 61, *pl.*70.

7) For English translations of the *Story of Sinuhe*, see Lichtheim, *Ancient Egyptian Literature* I, 222–35; Parkinson, *Tale of Sinuhe*, 21–53.

8) I first set out this relationship between the Sinuhe chant and the hall's four scenes in my unpublished doctoral thesis ('Cult Objects of Hathor' 1, 156–8), though without connecting them to the *Book of Day*.

9) For these Middle Kingdom developments, see Roberts, *Hathor Rising*, 73–5.

10) For Sinuhe's characterization of Senwosret I in terms of Hathor's loving-destructive power, see *ibid.* 48.

11) For Ptah's creation of Ramesses as a divine image, see Roberts, *My Heart My Mother*, 25 (with further references).

12) *Ibid.* 22, 26–7.

13) See Kees, *ZÄS* 65 (1930), 73 (lines 9–13); also Rundle Clark, *Myth and Symbol*, 136–7.

14) Kees, *op.cit*. 74 (line 36).

15) The name of the city is written *Iwn*, which can refer to both Heliopolis and Dendara, though the latter seems preferable here in this north to south sequence.

16) It is unclear when Hathor's fourfold aspects first emerged, but four Hathors, including Hathor of Dendara, are shown suckling Ramesses II at Abydos (see Roberts, *My Heart My Mother*, 22, *pl.16*). Cf. also the 18th-Dynasty Hathor columns with four faces at Serabit el-Khadim, and a faience bowl decorated with four Hathor columns (Pinch, *Votive Offerings*, 158). For her four aspects in the Graeco-Roman era; see Derchain, *Hathor Quadrifrons*.

17) For this relationship between Hathor and Maat, see Roberts, *Hathor Rising*, 32–6.

18) Cf. also the 'energy field' created by interacting spells inscribed on the west gable of the sarcophagus chamber and the east wall of the antechamber in the Old Kingdom pyramid of Unas; Naydler, *Shamanic Wisdom*, 293–5.

19) For the interrelationship between scenes on either side of the same wall in Egyptian temples, see *ibid*. 225, 404, n.102, with reference to Schwaller de Lubicz's discussion of 'transposition'.

CHAPTER 4

1) For a parallel to the three Horus gods, cf. the three falcons on standards in a cryptographic inscription of Ramesses III at Medinet Habu, which are used to write the word 'gods'; see Darnell, 'Ancient Egyptian Cryptography' (forthcoming). The same writing occurs at Dendara, see Richter, *The Theology of Hathor of Dendera*, 428 (Doc.23).

2) For the 'cow and marsh' motif, see Pinch, *Votive Offerings*, 175–9.

3) For this power of the *Ka*, see Roberts, *Hathor Rising*, 93–4.

4) *Abou Simbel* I, 111–18.

5) Quoted in Bell, *JNES* 44 (1985), 254. The text is on an architrave in the court of Amenhotep III.

6) *Ibid*. 272, 273, *fig.6*. The 'Magistrates' Chamber' was traditionally where the divine tribunal convened to resolve the dispute between Horus and Seth, deciding in favour of Horus, which culminated in his coronation (*ibid*. 272). For suckling by goddesses marking the king's transition from human to divine status and his birth into kingly rule, see Leclant, 'Sur un contrepoids', 251–84, esp. 263–7. Cf. Philip Arrhidaeus's *rite of passage* depicted at Karnak, in which the king is crowned by Amun-Re and the goddess Amaunet suckles him on her lap (see Roberts, *op.cit*. 62–3, with *pl.71*).

7) The quote is from Troy, *Patterns of Queenship*, 149.

8) For the cobra's association with the ninth-hour chamber in the Crypt South 1 at Dendara, *see chapter 7*.

9) Cf. a scene of Neferabet 'propitiating' the uraeus associated with flowing water in his tomb at Deir el-Medina. He kneels before the falcon god Re-Harakhti, simultaneously touching the crowned uraeus perched before him and the uraeus coiled around the god's sun disk, whilst, to the left, an arm, emerging from tree branches, holds a vase to pour cool water over him (see Roberts, *op.cit*. 22, *pl.27*).

10) For Khnum's role in assisting women giving birth, cf. a hymn inscribed in Esna temple: 'He makes women give birth when the womb is ready' (Lichtheim, *Ancient Egyptian Literature* 3, 112). Cf. also his role as companion of the four divine midwives who deliver the first three kings of the Fifth Dynasty, as told in the *Westcar Papyrus* (*ibid*. I, 220–21).

11) For Nefertari being identified with Isis-Sothis, *see chapter 3, n.1*.

12) For Isis as the partner of Bull-of-his-Mother, see Roberts, *op.cit*. 92–3. For the complementary pairing of Hathor and Isis in the solar cycle, see *ibid*. 94–6.

13) For the mother–daughter duality in New Kingdom queenship, see Troy, *op.cit*. 112, 125; and for queens being identified with Hathor and Isis, see *ibid*. 68–70.

14) For 'divinization' being associated with 'becoming young again' in a scene of Amenhotep III offering incense and flowers to Amun in the Luxor temple, which leads to his coronation, see Bell, *op.cit*. 281–3, with *fig.9*. This word-play between youth (*rnp*), the year and the return of the inundation reappears at Dendara, being associated with Hathor's role to bestow life and renewal, particularly in her aspect as divine cow; see Richter, *The Theology of Hathor of Dendara*, 176–9.

15) Hathor's role in sustaining the continous 'greening' cycle of life is beautifully expressed in the 'strewing *tjehenet*' rite enacted in Graeco-Roman temples at harvest time in the month of Epiphi, when the annual food-bearing cycle had ended and a new 'greening' one was anticipated. It honoured her as the returning Eye goddess from the South, and involved strewing a green powder (probably derived from malachite), together with gold, wheat and barley. See Goyon, 'Répandre l'or', 85–100, esp. 91, 96.

16) For Horus's justification and peaceful reconciliation with Seth, see te Velde, *Seth*, 63–73, esp. 70–71.

17) From the New Kingdom *Crossword Hymn to Mut*; see Stewart, *JEA* 57 (1971), 103 (line 60, vertical text). For Mut's association with the serpent goddesses and New Kingdom queenship, see Roberts, *Hathor Rising*, 76–8.

18) Cf. Bell's observation about Amenhotep III's reversed names when he is shown with his *Ka* in the Luxor temple. There his *Ka*-name is ascribed to his Horus name and his Horus name to his *Ka*-name, a switch 'intended to convey to us the unity of the figures visualized separately' (*op.cit*. 278, with n.135).

19) For the importance of the 'name' in ancient Egypt, see Roberts, *My Heart My Mother*, 85–6, 157–8, 229, n.16.

20) El-Sawi, *MDAIK* 43 (1986), 226, *fig.1*; illustrated also in Roberts, *op.cit*. 82, *pl.66*.

21) For the curved ram's horn symbolizing divinization, see Bell, *op.cit*. 269, with n.84. For burning incense as a rite identifying the reigning monarch with his divine ancestors, see *ibid*. 281–5.

22) For the symbolism of Ahmose-Nefertari's black complexion, see Manniche, *AcOr* 40 (1979), 11–19, who associates it with her posthumous role as patroness of the Theban necropolis and mother of the dynastic line (*ibid*. 17–18).

23) For the king being identified with Kamutef as renewer of the royal line, see Roberts, *Hathor Rising*, 82–96 (with further references).

24) Hornung, *Buch von den Pforten* 1, 36–7; *ibid*. 2, 68. This unity is clearly stated elsewhere in the *Book of Gates*: 'The one who gives them offerings is one who has a place in their cavern' (*ibid*. 1, 23–4; *ibid*. 2, 50), quoted also by

Wente, *JNES* 41 (1982), 173. Or again: 'The one who gives them offerings is in Igeret in the West' (Hornung, 32–3; 59; Wente, 174).

25) For the king's divinity being expressed through his identification with Amun-Re-Kamutef, see Bell, *op.cit.* 258–9.

26) *See chapter 7* for the Ptolemaic king's *Ka*-union with the Sixth-Dynasty ruler Pepi I in the Crypt South 1 at Dendara.

27) For the ancestral kings residing in Memphis, see Roberts, *My Heart My Mother*, 67–8, 82–3.

28) For Memphite creation through the Word, see *ibid.* 16–17, 85–6, 160.

29) Cf. the depiction of Osiris wearing the White Crown as ruler of the *Dwat* in the *Book of Night*'s eighth hour (*ibid.* 149–51, figs. *118, 120*).

30) Quoted from Spell 331, CT 4, 172–6.

31) See Roberts, *My Heart My Mother*, 85.

32) Cf. also the paired Hathor-headed tree branches framing the scene of a woman and child, as depicted on a Middle Kingdom birth-brick from Abydos, see Wegner, 'A Decorated Birth-Brick from South Abydos', 458–63, with figs.*4, 7, 15*. The woman has evidently just given birth and, as Wegner noted, by showing her between these emblems she becomes a manifestation of Hathor giving birth to the sun god in the eastern horizon (*ibid.* 479–80).

CHAPTER 5

1) For the sun's annual north–south journey, see Kurth, *RdÉ* 34 (1982–3), 71–5; GM 83 (1984), 39–41; Leitz, *Studien*, 8–9.

2) For the shrine's orientation to the southern solstice, see Shaltout and Belmonte, *JHA* 36 (2005), 293; Belmonte Avilés, *Pirámides*, 152–3, with fig.*4.31*. Ziegler, however (in Desroches-Noblecourt, *Ramsès le Grand*, 159), associated the baboon and scarab in the naos with the sun's descent in the West and replacement by the moon at night. For the scenes of Re-Harakhti and Thoth on the side walls of the naos, see *ibid.* 156–7.

3) For the orientation of Upper Egyptian temples towards the southern solstice, see Shaltout and Belmonte, *op.cit.* 283–6; Belmonte Avilés, *op.cit.* 146–53. For solstitial sunrise illustrations, see *ibid.* colour fig.*13*, opp. page 274 (Deir el-Bahri), colour fig.*14* (colossi of Memnon). For Egyptian beliefs and ritual associated with the beginning of *Akhet*, see Roberts, *Golden Shrine*, 30–1, 64–5.

4) See Salazar, *Astronomia Inka*, 54–8. I am grateful to Helen Maimaris and Sam Gladstone for bringing this Inca astronomy to my attention.

5) For the harvest festival being associated with Kamutef and Isis, see Roberts, *Hathor Rising*, 84–6, with pls.*89–95*.

6) Willeitner associated these two dates in the agricultural calendar with Abu Simbel; see Schmidt and Willeitner, *Nefertari*, 79. See also Shaltout and Belmonte, *op.cit.* 293, and Belmonte Avilés, *op.cit.* 151–2.

7) For this festival as an ancient Ptah festival subsequently transferred to Amun, see *Lexikon der Ägyptologie* 2, col.177. For Ptah as 'raiser of heaven', see Darnell, *Enigmatic Netherworld Books*, 360, n.373 (with further references). For the festival's relationship with the *Book of Night*'s seventh night hour, see Roberts, *My Heart My Mother*, 170. According to the festival calendar in

Graeco-Roman temples, it was still being celebrated during this period, and at Esna almost the same rituals were celebrated then as at Edfu on New Year's Day (I *Akhet* day 1); see Leitz, *Astronomie*, 21, n.52.

8) For the festival's association with Osiris's burial at Busiris, see Alliot, *Culte d'Horus à Edfou*, 217 (line 5), 229. It is mentioned twice in the Ptolemaic *Book of Traversing Eternity*; see Herbin, *Parcourir l'éternité*, 55 (III: 19–20), 66 (VI: 28, 29), 162, 233, 361.

9) Von Bomhard (*The Egyptian Calendar*, 24) suggested juxtaposed representations of Orion and Sothis in their respective boats conveyed the New Year juncture in time, symbolized by Orion turning his face away from Sothis while twisting his body back towards her, as if to proclaim the end of the preceding year as she looked towards him to inaugurate the new one.

10) Neugebauer and Parker, *Egyptian Astronomical Texts* 3, 17–18, pl.5; von Bomhard, *op.cit.* 84–5.

11) For these scenes, see Ziegler in Desroches-Noblecourt, *Ramsès le Grand*, 156–7.

12) For this astronomical ceiling at Deir el-Haggar, *see also below, chapter 12, n.25*.

13) It is beyond the scope of this book to discuss the complexities surrounding the Egyptian calendar, including whether it was the solstitial solar cycle or the cycle of Sothis that led to a year of 365 days. But during the Archaic Period and the Old Kingdom, the heliacal rise of Sothis coincided with the summer solstice (von Bomhard, *op.cit.* 26, n.3).

14) Shaltout and Belmonte (*JHA* 36 [2005], 293) thought the festival of 'Raising the Sky', celebrated on the first day of the seventh month, would have coincided with the southern solstice in Ramesses II's reign; also Belmonte Avilés, *op.cit.* 152–3.

15) There is not, however, sufficient evidence to link the October sunrise with Ramesses II's first Sed Festival, as some authors have suggested (e.g. Krupp, 'Astronomers', 208).

16) Quoted from *Politicus*, 271b–c. For the whole account of this resurrection, see 270b–71c (trans. Skemp, 147–9). For similarities with Egyptian notions of the reversal of time in the *Amduat*, see Uždavinys, *Philosophy*, 266.

17) For the conception, procreation and rearing of children being associated with the 'age of Zeus', see *Politicus*, 273–4d, trans. Skemp, *op.cit.* 152–4.

18) For the 'solstice' connotations of Plato's account, and his deliberate extension of the astronomical meaning, see *ibid.* 147, n.2.

19) Strabo, *Geography*, 17.1.29 (trans. Jones, 82–5). For Strabo's account of Plato's visit to Egypt and his indebtedness to the ancient Egyptian mysteries, see Naydler, 'Plato, Shamanism and Ancient Egypt', 76–92.

20) Naville, *Aegyptische Todtenbuch* 1, pl.*133* (lines 21–3).

21) According to the Edfu calendar of Hathor festivals, Ptah performed the festival of 'Raising the Sky' at the side of Harsaphes of Herakleopolis on the first day of the seventh month (1. Phamenoth), when the burial of Osiris at Busiris was also celebrated; see Alliot, *op.cit.* 217 (line 5), 229.

22) For the solar boat's standstill being associated with the deceased's justification and a critical moment for the cosmos, see Darnell, *Enigmatic Netherworld Books*, 287–8, n.59 (with further references, though not to the

sixth and seventh months). Darnell also refers to a scene on the sarcophagus of Padiese from Salamieh in the Cairo Museum, which shows the deceased standing between the goddesses of the East and West with his arms raised in the gesture of 'justification' and a sky sign placed above the tips of his fingers (*ibid*. 409).

23) Schott, *Urkunden*, 138–9 (lines 19–22). See also Leitz, *op.cit*. 87. Cf. a faience heart amulet in the British Museum portraying a face and a *Benu*-bird on one side and inscribed on the other with chapter 30B of the *Book of the Dead*, in which the deceased implores the 'heart of the mother' not to be hostile before the tribunal, illustrated in Taylor (ed.), *Journey though the Afterlife*, 229 (118).

24) See Budge, *Book of the Dead*, 314–17. Illustrated in Faulkner, *Book of the Dead*, 130 (Ptolemaic papyrus of Haremhab, BM 10257).

25) Budge, *op.cit*. 315 (4–5).

26) See Roulin, *Livre de la nuit* I, 120; Roberts, *My Heart My Mother*, 116. For Atum as the ageing sun god, see *ibid*. 55, 56, 64.

27) For the *Book of Night*'s journey from the third to the seventh hours, see *ibid*. 116–48. For the seventh-hour birth of a child, see *ibid*. 143–4, pls.114, 115.

28) The term 'right' wall here refers to the point of view of an observer facing the sarcophagus chamber's rear wall. The modern title *Book of the Earth* stems from Hornung, though confusingly it is sometimes called the *Book of the Creation of the Solar Disk*. The most complete version is in Ramesses VI's tomb (first published by Piankoff, *La création du disque soleil*, 1953, and since re-edited by Roberson, *The Ancient Egyptian Books of the Earth*, 2012). When it was originally composed is unclear, but a very fragmentary version is in the Cenotaph of Seti I at Abydos, located on a wall beneath the *Book of Night* and the *Book of Nut* on the ceiling. It appears in New Kingdom royal tombs, and various extracts continued to appear through into the Graeco-Roman period, including in Late Period private tombs and on sarcophagi (see Roberson, *ibid*. 9–11, 461–2).

29) It should be noted that Roberson interprets Ramesses VI's version in terms of the sun's diurnal journey, arguing it starts on the chamber's left wall at sunset and ends on the right wall with sunrise (*ibid*. 55–9). Stricker also commented extensively on the vignettes and texts, describing them as an 'embryological' treatise, and included many references to Hermetic and Classical philosophical sources (*Geboorte van Horus, passim*). Neither scholar mentions annual 'reversal' between the sixth and seventh months, associated with the 'filling' of the Moon Eye and Osiris's regeneration (right wall), nor the Sun Eye's role at this reversal, as depicted on the left wall (*see chapter 15*).

30) For the *Aker* group in the bottom register, see also Roberson, *op.cit*. 145–54 (*fig.5.5*).

31) For these lions in Seti I's temple at Abydos being associated with 'Yesterday' and 'Tomorrow', Osiris and Horus, and the transfer of power between father and son, see Roberts, *My Heart My Mother*, 54–6.

32) For this vertical relationship, see Roberson, *op.cit*. 39–40, 139.

33) For this name of Nun, see *ibid*. 149, 374 (Text 66).

34) See *ibid*. 142–5, for the accompanying inscriptions.

35) For the group towing the boat, see *ibid*. 139 (*fig.5.3*).

36) *Ibid*. 247–53, with *fig.5.55*.

37) For the shrew-mouse as the blind aspect of Horus of Letopolis, see *ibid*. 279, with n.955 (without reference, however, to its meaning here in the annual cycle). For a scene of seven deities, each with the head of a shrew-mouse, shown adoring the ram-headed sun god in Ramesses IX's tomb, see *ibid*. 278–9 (*fig.5.73*). For Horus as an eyeless deity in the *Book of Caverns*, see Darnell, *op.cit*. 172–3.

38) See Roberson, *op.cit*. 250–51 (Texts 12, 14).

39) For Horus of Letopolis in the seventh night hour, see Roberts, *My Heart My Mother*, 144–6.

40) Roberson, *op.cit*. 206[8]. Cf. also the enigmatic writing of the epithet 'the one who raises the sky and descends inverted into the *Dwat*', associated with 'the Ram of rams' in *Papyrus Salt 825* and symbolizing access to the chthonic and celestial realms. The hieroglyphic group is comprised of an upside-down figure enclosing a *Dwat* sign and a standing figure raising a sky-sign; see Darnell, *op.cit*. 410, with n.176, 427–8.

41) For this scene, see also Roberson, *op.cit*. 179–88, with *fig.5.19*. Barguet (*RdÉ* 30 [1978], 53) convincingly suggested the figure was standing within a clepsydra depicting the night hours.

42) See Roberts, *My Heart My Mother*, 144–8, with *pls.114–16*.

43) Roberson, *op.cit*. 185–6, 504, pls.34a–b. According to the *Amduat*, the seventh hour is when Horus correctly orients the stars and the plummeting inverted souls in the netherworld (see Darnell, *op.cit*. 298).

44) See Roberts, *Hathor Rising*, 111 (with further references).

45) For the Moon Eye and 'seed' in the conflict of Horus and Seth, see te Velde, *Seth*, 32–59.

46) *Ibid*. 52–3.

47) Cf. also a scene of Osiris in the tomb of Ramesses IX which is placed next to a scene of the king offering Maat to Ptah (Guilmant, *Tombeau de Ramsès IX*, pl.63; Roberts, *My Heart My Mother*, 133, pl.107). It shows him with an erect phallus, lying on an earth mound in Ptah-Sokar's earth realm and evidently associated with annual reversal and regeneration, as conveyed by the figures in the three registers directly to the left. In the top register are eight disks, each one enclosing an inverted figure; in the middle register enemies are being vanquished by arrows issuing from a boat containing the two Eyes and a solar scarab; the bottom register includes four figures leaning backwards, each with a solar disk rolling across the chest region, and accompanied by a tiny child and scarab. For the scene's association with the sixth night hour (though not its relationship with annual reversal), see Roberts, *ibid*. 132–5.

48) For this scene, see also Roberson, *op.cit*. 269–71, with *fig.5.68*. For the mummy as Osiris, see *ibid*. 271, with n.907.

49) This version with seven disks is depicted in the tombs of Merneptah, Tawosret and Ramesses III; *ibid*. 271.

50) *See above, n.8*, for Osiris's burial being associated with the festival of 'Raising the Sky' at the beginning of the seventh month. *See also chapter 10* for this association in the Khoiak Festival at Dendara.

51) This 'reversal' is also expressed in tiny details in Seti I's monuments at Abydos. For example, in the *Book of Night*, the eighth-hour text describes the sun god passing close to the gate of Naref (associated with Herakleopolis) and his 'way turning aside'. Here the 'legs walking' determinative of the verb 'turn aside' (*stnm*) faces in the opposite direction to the rest of the hieroglyphs, as if to reinforce the sun boat's reversed movement during the eighth hour, which is also when Horus brings life to the enthroned Osiris (see Roulin, *Livre de la nuit* 1, 236–7, with note d; *ibid*. 2, 99). Cf. also the six reliefs in the central inner Osiris shrine in the main temple, which are placed three by three on the side walls. Initially they are read from right to left, until the middle scene on the left wall, showing the *Inmutef* priest burning incense for Seti's transfiguration, with Isis shaking her sistrum behind him. Depicted in the sistrum's superstructure are the *Aker*-lions flanking the young sun child. The sequence then reverses, with the end scene on each wall being read from left to right. (See Roberts, *My Heart My Mother*, 58–69, 73–88, 90–91, though when writing that chapter, whilst noting the reversal in the sequence, I had not understood the relationship with the sun's annual southern reversal.)

52) Cf. also the 'world-encircler' snake in the *Amduat*'s 12th hour, through which the sun god, the deities and the blessed dead all travel, entering by its tail as old and frail and leaving from its mouth rejuvenated as small children, due to the reversal of time occurring in its body (see Hornung, *Valley of the Kings*, 90–91). Much later, this regeneration from old man to child, associated with a time period of 180 days, reappears in the *Book of Pictures*, an Arabic alchemical treatise from Egypt; and also in sermon 58 of the Latin alchemical treatise the *Turba philosophorum*. The passage says that if the soul of a 100-year-old man is enclosed for 180 days in a house surrounded by dew, he will be transformed into a youth, 'so that the father becomes the son' (see Abt *et al.*, *Explanation of the Symbols* 2, 195–7, with n.468; Roberts, *My Heart My Mother*, 204, with further references). Abt recognized the 180 days here goes back to ancient Egyptian beliefs and the sun god's journey through the night hours, without referring, however, to the *Book of the Earth* and the sun's annual southern 'reversal'.

53) For the scene of Huh and Hauhet, see *ibid*. 162–3, with *pl.129*. For the ninth night hour being associated with the ancestors, see *ibid*. 150–53, and being synchronized with the daytime first hour, *ibid*. 180–85.

54) *Politicus*, trans. Skemp, 92. For Plato and Egypt, *see above*, n.19.

55) *Ibid*. 270c, 147.

56) Schwaller de Lubicz (*Sacred Science*, 126–37) observed how Ramesses II's version of the Battle of Kadesh was synchronized with the reversal of time in the sun god's night journey, referring particularly to the king's two superimposed figures facing in different directions on the west wing of the pylon at Luxor. Of Abu Simbel, he wrote: 'It is precisely the characteristic of reversal, moreover, that underlies the deeper meaning of this rock-hewn temple' (*ibid*. 132), though without reference to reversal at the southern solstice. See also Naydler, *Temple of the Cosmos*, 112–20, for his interpretation.

57) For Akhenaten's religious revolution, see Roberts, *Hathor Rising*, 130–68 (with further references).

58) Cf. the year 3 inscription in the Luxor temple, in which Ramesses explains how he discovered, by reading certain books in the House of Life, 'the secrets of heaven and all the mysteries of earth', which he could then apply to his building work in the temple's first court (Kitchen, *Ramesside Inscriptions* 2, 346, lines 5–7). As Darnell observed: 'Learning the nature of the macrocosm, Ramesses knows how to reproduce it in the temple' ('Ancient Egyptian Cryptography', [forthcoming]). I would like to thank Professor Darnell for sending me a pre-publication copy of his paper.

59) For the literature on the myth's relationship with the sun's annual passage and the solstices, see Richter, *The Theology of Hathor of Dendera*, 3, with nn.6,7.

60) Utterance 405, PT §705a–c. Cf. the 'legs walking backwards' determinative of *innt*, indicating this 'bringing of the years' occurs through an inverse movement.

CHAPTER 6

1) For the temple's construction and decoration between 54 and 21/20 see Quaegebeur, *GM* 120 (1991), 49–72.

2) For Hathor's face, the sistrum and *sekhem*-power, see Roberts, *Hathor Rising*, 57.

3) For the crypt's name as 'secret palace', see Daumas, *ASAE* 51 (1951), 386, n.1, and Waitkus, *Texte*, 94, 97, n.4, 251.

4) For representations of a queen (often anonymous) in the other crypts and elsewhere in the Dendara temple, see Quaegebeur, *op.cit*. 49–72.

5) Cf. a Demotic inscription at Philae, dated to 373, in which a scribe of the divine book of Isis states he arranged for a statue of Cleopatra to be covered in gold; see Quaegebeur, 'Reines ptolémaïques', 256.

6) For this Late Period trend, see Troy, *Patterns of Queenship*, 70. For the birth scenes at Dendara, see Cauville and Ibrahim Ali, *Dendara*, 273–77.

7) For crypts as the *Dwat*, see, for example, Waitkus (*op.cit*. 265, 273–4); Richter, *The Theology of Hathor of Dendera*, 190, 199. Waitkus compared the statues kept in Dendara's crypts to corpses buried in the netherworld.

8) 'Zu den Darstellungen', 1–23, though regrettably Kurth's interpretation, which was concerned primarily with the scenes of Pepi I in the Dendara temple, has received scant attention. It did not include, however, a complete analysis of the Crypt South 1's decoration.

9) See Daumas, *op.cit*. 383–4, 393; Cauville and Ibrahim Ali, *op.cit*. 145–8.

10) For Hathor's rooftop illumination during the Khoiak Festival, see Cauville, *Dendara: Commentaire*, 223–4.

11) See Daumas, *op.cit*. 384–93; and *ibid*. 382–3 for the timing with the eighth hour.

12) For this staircase inscription, see Junker, *ZÄS* 43 (1906), 107.

13) For the Dual Shrines and Hathor, see Roberts, *op.cit*. 41–7, 75, 77.

14) For these texts around the *Per-Neser*'s entrance, see Cauville, *Dendara* 3: *Traduction*, 278–91.

CHAPTER 7

1) Information on the crypt is published in Chassinat, *Le temple de Dendara* 5(1). IFAO. Cairo, 1952, reprinted 2004, 115–60 (inscriptions); *ibid*. 5(2), *pls.413–50* (reliefs).

For a German translation with commentary, see Waitkus, *Texte*, 94–165, without reference, however, to the 12-hour daytime solar cycle.

2) For these four avatars represented on the *Wabet*'s east and west walls, see Meeks, *RdE* 15 (1963), 35–47.

3) For New Kingdom antecedents of Hathor as fourfold goddess, *see chapter 3, n.16.*

4) For this text from the southern frieze inscription, see Waitkus, *op.cit.* 125. For the 'child in the nest' and the epagomenal days at the year's end in the *Pyramid Texts*, *see also chapter 14.*

5) For the chamber's link with the New Year period and its festivals, see Kurth, 'Zu den Darstellungen', 10–11 (chamber C in his plan).

6) Chassinat, *op.cit.* 5(1) 139, 5–6; Waitkus, *op.cit.* 131.

7) *Ibid.* 130.

8) For the Memphite *Ka*-avatar being associated with the son's elevation to the throne and representing 'birth' in the *Wabet*, see Meeks, *op.cit.* 37(1).

9) From an inscription at the entrance to Harsomtus's shrine in the main temple; see Cauville, *Fêtes*, 16.

10) Only one oval shape is included in the complementary scene across on the north wall, where Horus is shown as a wingless bird *(fig.137)*, perhaps because he is here still in an incomplete state.

11) E.g. in the crypt's end chamber (west side), Hathor's son is said to wear 'an amulet of the female *Djed* at his throat' (Waitkus, *Texte*, 152). For Hathor as *Djed*-pillar, see also Frankfort, *Kingship*, 178.

12) For this description, see Cauville and Ibrahim Ali, *Dendara*, 96–7, 121. Waitkus (*op.cit.* 137, n.55) calls the vessel a kind of '*Blase*', meaning 'bubble' or 'alembic'.

13) *See above, n.6.* The same preposition occurs in the New Year inscription associated with Isis at the crypt's entrance; see Daumas, *ASAE* 51 (1951), 391, n.2.

14) See *Wb.* 1.147. For the Dual Shrines, see Kees, *ZÄS* 57 (1922), 120–36, and Frankfort, *op. cit.* 95–6, 371, n.33. Kees referred to Dendara's scenes of Harsomtus (*op.cit.* 133), interpreting them as 'raising the snake stones'.

15) See Roberts, *Hathor Rising*, 41–7, 124, and *My Heart My Mother*, 188.

16) Cf. an inscription in the *Per-Wer* accompanying the offering of the Double Crown to the king: 'The Lords of the *Iterty* come in order to see you, and your awe-inspiring terror travels around the Two Lands', see Richter, *The Theology of Hathor of Dendara*, 365 (Doc.65). For further references, see also *ibid.* 284 (Doc.23), 292 (Doc.26), 384 (Doc.78), 386 (Doc.79). Referring to Gardiner's translation of *Iterty*, Richter associates the name solely with the two shrine rows on either side of the Sed Festival court (e.g. *ibid.* 385 n.606), but see Frankfort (*op.cit.* 371, n.33) for the limitations of Gardiner's translation.

17) Cf. a 19th-Dynasty scene in Seti I's funerary temple at Qurna, showing a female figure bearing a *Ka*-symbol on her head, within which is written the temple's name. Embracing Seti, she tells him: 'Behold, I am around you. I am your temple forever.' See Roberts, *My Heart My Mother*, 233, n.21 (with further references).

18) For the founding of a temple being compared to human birth, see Roth and Roehrig, *JEA* 88 (2002), 135–6.

19) Borghouts, *Magical Texts*, 39–40 (Spell 62). Wegner ('A Decorated Birth-Brick from South Abydos', 480) noted the eastern horizon context of this spell, referring to the well-known association between the temple pylon and the *akhet*.

20) Cf. chapter 25 of the *Book of the Dead*, in which becoming a 'builder' in the eastern sky is associated with 'naming' in the *Per-Wer* and *Per-Neser*; see Roberts, *My Heart My Mother*, 156–8, 196.

21) For this half-hearted approach of the Egyptian priesthood, see Winter, 'Der Herrscherkult', 158.

22) For the *menit*-necklace and 'attraction', see above, chapter 3, n.4.

23) For the Memphite *Ka*-avatar embodying 'life' on the *Wabet*'s east wall, see Meeks, *RdÉ* 15 (1963), 37(2). The avatar is said to give 'the breath of life to the one he loves'.

24) Cf. Thoth bringing the *Shebet* (or *Wensheb*) to Hathor, another of her cult objects at Dendara, as described in the southern frieze inscription (see Chassinat, *Dendara* 5[1], 134, 14–15; Waitkus, *Texte*, 114). Elsewhere it is depicted as a baboon crouching beside a *hen*-receptacle, though it is no longer thought to represent a clepsydra. Here Thoth brings it to symbolize the healed Moon Eye 'complete with its parts'. For the healing of the Moon Eye being associated with Re rising in the East; *see chapter 1.* For Thoth healing the Moon and Sun Eyes, see Roberts, *Hathor Rising*, 110–12.

25) Griffiths (ed.), *De Iside et Osiride*, 219; Roberts, *op.cit.* 57.

26) For the four sistra symbolizing Hathor's power to sustain the sun boat's 'movement', see Derchain, *Hathor Quadrifrons*, 46.

27) For *phr* and *m3'* occurring together in the context of the solar boat, see Darnell, *Enigmatic Netherworld Books*, 265.

28) Jacq, *Egyptian Magic*, 4, quoted also in Naydler, *Temple of the Cosmos*, 125.

29) For 'encircling the heart' in Egyptian magic, see Ritner, *Magical Practice*, 66–7.

30) Lines from the southern frieze inscription; see Waitkus, *op.cit.* 114.

31) Illustrated in Chassinat and Daumas, *Dendara* 6, pl.461.

32) For the sistrum ruling the main temple's central axis, see Derchain, *op.cit.* 16–18. For its association with *sekhem*-power, see Roberts, *Hathor Rising*, 57. Richter (*op.cit.* 36, with n.123) noted that the light shaft in the ceiling of the *Per-Wer* allows the sun to shine through into the sanctuary precisely at noon on the summer solstice, again a feature enhancing this area's association with the sun at its greatest height.

33) Waitkus, *op.cit.* 104. For the 'places of drunkenness', see *ibid.* 102. For the fear engendered when encountering a deity, see Roberts, *Hathor Rising*, 57–9.

34) Waitkus, *op.cit.* 107; Chassinat, *Dendara* 5(1), 128, 6–7. For the uraeus defining solar sovereignty see Roberts, *Hathor Rising*, 41–2, 66.

35) Waitkus, *op.cit.* 104; Chassinat, *op.cit.* 123, 5.

36) Waitkus, *op.cit.* 109; Chassinat, *op.cit.* 130, 1–2. Hathor's association with Thoth's writings has a long history; see *My Heart My Mother*, 124, 138. For these writings as the '*Bas* of Re' or 'emanations of Re' in the Graeco-Roman temple cults, see Fowden, *Egyptian Hermes*, 58–9.

37) For the rite of 'seeing the deity' in the *Per-Wer* at Dendara and the king's knowledge of Thoth's writings, see Richter, *op.cit.* 220–25, 227–33.

38) *Ibid.* 373 (Doc.72).

39) For New Kingdom associations of the *Per-Wer* with Hathor and the solar circuit, see Roberts, *Hathor Rising*, 40–47; *Golden Shrine*, 11–26, 38–46.

40) Waitkus, *Texte*, 104; Chassinat, *Dendara* 5(1), 123, 10–11.

41) For the name of this chamber, see Waitkus, *op.cit.* 146, n.1.

42) Text from the northern frieze inscription; see Waitkus, *ibid.* 140; Chassinat, *op.cit.*, 148, 18–19.

43) For the precise measurements of the crypt's five chambers, see Kurth, 'Zu den Darstellungen', 6. The two chambers flanking the 'House of the Sistrum' are 3.2 metres long, in contrast to the two end chambers (4.6 metres long) and the central chamber (4.3 metres long), thus architecturally reinforcing the complementarity between chambers that is also worked out thematically.

44) For Isis on the *Isheru* lake, see Waitkus, *op.cit.* 141; Chassinat, *op.cit.* 147, 1–2. For a similar scene of Isis in the *Per-Wer*, see Richter, *op.cit.* 150–52, with *fig.5.31*. Cf. also the inscription at the crypt's entrance stating that the inundation god, Hapy, encircles the processional way when the statue of Isis is carried to the roof on the fourth epagomenal day (Waitkus, *ibid.* 96; Daumas, *ASAE* 51 [1951], 391). According to Daumas (*ibid.* 390, n.2), this statue was brought from the crypt's 'Pure House'. If so, then the chamber's 'watery' theme would have carried through into the procession itself.

45) Correspondingly, the Memphite *Ka*-avatar's 'water' aspect is associated with 'fruitfulness' in the *Wabet*. He is 'lord of vital things, rich in food, who inundates this land with beneficial things' (see Meeks, *RdÉ* 15 [1963], 37[3]).

46) Waitkus, *op.cit.* 142; Chassinat, *Dendara* 5(1), 147, 8–9. For other references to Hathor and the moon at Dendara, see Waitkus, *op.cit.* 147, n.26.

47) Waitkus, *op.cit.* 140; Chassinat, *op.cit.* 5(1), 148, 19–149, 1 (northern frieze inscription).

48) Hathor's enthronement between Re and Ptah seems not to have been represented before the Ptolemaic period. For texts relating to this enthronement, see Cauville, *Fêtes*, 80–97.

49) The songs for Hathor are inscribed in the Room of offerings, accompanying a scene of the king offering the sacred *menou*-vessel. For this offering 'making the inner outer', see Daumas, *RdÉ* 22 (1970), 76; Sternberg el-Hotabi and Kammerzell, *Hymnus*, 103–104.

50) Waitkus, *Texte*, 144; Chassinat, *Dendara* 5(1), 150, 12–13. Cf. also the praise of Hathor as 'the female king of Upper and Lower Egypt, the female sun, who reveals the interior' in the *Per-Wer*, see Richter, *op.cit.* 370 (Doc.70).

51) This chant is preserved in the southern frieze inscription; see Waitkus, *op.cit.* 140; Chassinat, *op.cit.* 5(1), 145, 13–146, 2. For earlier 'Awake in peace' chants to the serpent goddesses, see Erman, *Hymnen, passim*.

52) For the complementary relationship between texts and images in neighbouring rooms in a temple, *see also chapter 3, n.19.*

53) Waitkus, *op.cit.* 152; Chassinat, *op.cit.* 157, 10–11. For the sun god as 'judge in the West', see Roulin, *Livre de la nuit* 1, 26–7, 30–31. For the Memphite *Ka*-avatar being associated with 'burial', see Meeks, *op.cit.* 37[4].

54) Waitkus, *op.cit.* 150; Chassinat, *Dendara* 5(2), *pl.438*.

55) For Ihy's nocturnal regeneration, see Roberts, *Hathor Rising*, 29–32. Kurth recognized the chamber's 'western'

theme ('Zu den Darstellungen', 15–16) without mentioning Ihy's long association with regeneration. Cf. Amenhotep III manifesting as an Ihy musician in nocturnal regeneration rites performed during his first Sed Festival; see Roberts, *op.cit.* 26–9.

56) *See above, n.43.*

57) Cf. the king's union with four goddesses in Utterance 205 of the *Pyramid Texts (see chapter 21, with n.11).* Kurth suggested Pepi's inclusion here reflected his donation of a statue to the temple, which was still preserved in the crypt (*op.cit.* 5, 20–22, with further references).

58) See Waitkus, *op.cit.* 156; Chassinat, *Dendara* 5(1), 158, 7–9, and 5(2), *pl.447*.

59) For this abundance, see Waitkus, *op.cit.* 151; Chassinat, 5(1), 153, 5–9. For the offering of Maat as the 'seed of the bull', meaning the dissemination and propagation of Maat in the world, specifically to guarantee 'fruitfulness', see Kurth, '"Same des Stieres"', 280–81. The same association occurs in the western Osiris rooms on the roof terrace, *see chapter 10, with n.27.*

60) For this notion of an 'enveloping field of presence' in which space and time are not distinct, see Abram, *Spell of the Sensuous*, 201–204.

61) Cf. the secret mystery Isis transmits to her son in an alchemical Greek treatise called the *Letter of Isis to her Son Horus*. This mystery has been revealed to her by an angel called Amnael, who tells her she should only share it with her 'dear and legitimate' son 'so that he [the angel] becomes you and you become him' (see Mertens, *RHR* 205[1] [1988], 7[f], 18–23).

62) Waitkus, *Texte*, 157; Chassinat, *Dendara* 5(1), 159, 11–12. For the heart as the source of regeneration, see Roberts, *My Heart My Mother*, 77–9, 126, 133, 234, n.14 (with further references).

63) For this suckling scene, see Waitkus, *op.cit.* 154; Chassinat, *Dendara* 5 (2), *pls.443–4*.

64) *P.Graec. Mag. IV*, 719–23. Translated in Betz (ed.), *Greek Magical Papyri*, 52. Cf. a passage in the *Pyramid Texts*: 'O Osiris king, you go that you may return, you sleep that you may wake, you die that you may live' (*PT* §1975a–b).

65) For the correlation of Hathor's annual festivals with particular chambers, see Kurth, *op.cit.* 10–12.

CHAPTER 8

1) Information on the Osiris rooftop chambers is published in Cauville, *Dendara 10: Les chapelles osiriennes*, and *Le Temple de Dendara: Les chapelles osiriennes.*

2) Cauville, *Dendara* 10(1), 26–49, and (2), *pls.3–5*, 25–30; *Transcription*, 14–28.

3) *Dendara* 10(1), 46 (6–7); *Transcription*, 26.

4) *Ibid.* 28(6)–29(6); *Transcription*, 15–16. The remarkable underwater excavations carried out by the European Institute for Underwater Archaeology at Thonis-Herakleion have discovered many ritual objects from the Khoiak Festival performed there during the Ptolemaic era. For a pink granite garden tank, see Goddio and Masson-Berghoff, *Sunken Cities*, 170–71. For a Late Period Osiris *vegetans* figure in a falcon-headed coffin from Middle Egypt, *ibid.* 168–9, and for various ritual objects, *ibid.* 174–5.

5) *Asclepius* 38; see Copenhaver, *Hermetica*, 90.

6) Cauville, *Dendara*, 10(1), 31 (2–14); *Transcription*, 17.

7) Because the Egyptians did not add a day every fourth year, their 365-day civil calendar, consisting of 12 months of 30 days and five epagomenal days, gradually fell out of step with the seasons and the heliacal rising of Sothis. Hence, it was only after a period of 1,461 years that 1. Thoth in the 'wandering' civil calendar again coincided with the astronomical rising of Sothis. According to the Roman author Censorinus, the heliacal rising of Sothis fell on 1. Thoth in 139 and this crucial information has enabled the calculation of key dates in Egyptian chronology. According to Geminus of Rhodes, writing in the first century BCE, a whole month separated the important festival of Isis, celebrated from 17 to 20 Hathyr (the third month of the Egyptian year), from the winter solstice (see Youtie, 'The Heidelberg Festival Papyrus', 194). It follows then that the Khoiak Festival in the fourth month must have coincided with the winter solstice. See also the table for the Khoiak Festival and the winter solstice in Neugebauer, *Ancient Mathematical Astronomy* 2, 580.

8) See Bleeker, *Egyptian Festivals*, 70.

9) It should be pointed out that my discussion here differs considerably from Cauville's interpretation in her publication of the Osiris shrines, and elsewhere. She argues that they were built between 54 BCE and the death of Ptolemy XII (51 BCE), and that all the decoration was complete by the time of their inauguration on 28 December 47 BCE. However, these dates depend primarily on the correlation of imagery in the 'round' zodiac (depicted in the eastern central chamber on the roof terrace) with actual solar and lunar eclipses in the middle of the first century. This astronomical approach to the zodiac is by no means universally accepted, however, since it takes no account of the planets being shown according to their astrological 'exaltations', *see chapter 12, n.1*.

10) For the tableau's relationship with the solar cycle of the hours, see Roberts, *My Heart My Mother*, 111–12, with *pl.86*. See also Roberson, *The Awakening of Osiris*, esp. 131.

CHAPTER 9

1) Cauville, *Dendara* 10(2), *pls.7, 17*; *Transcription*, 2. Cauville (*Commentaire*, 7) noted these protective deities were not associated with the hours elsewhere.

2) *Ibid*. *Transcription*, 2–3.

3) Cf. also the identification of Osiris's boat with the solar day and night boats in the eastern court; see *ibid*. *Commentaire*, 17.

4) *Ibid*. 10(1), 67(6–7); *Transcription*, 38 (eastern frieze inscription).

5) Daumas, *Dendara* 9(2), *pls.829–30, 847–8*. See also Cauville and Ibrahim Ali, *Dendara*, 221–4 (including the temple construction scenes shown on the exterior east wall).

6) See Roberts, *My Heart My Mother*, 196–7 (with further references).

7) Cf. also the release of messenger birds in the New Year ceremony confirming the pharaoh's power; see Roberts, *Golden Shrine*, 51–2.

8) Cauville, *Dendara* 10(1), 58(8–10); *Transcription*, 32.

9) The text is partially damaged, and only the first, fifth and sixth-hour names are intact. For these boats associated with the *Ritual of Hours*, *see above chapter 1, n.5*.

10) For these six daytime hours and Re-Harakhti's speech, see Cauville, *Dendara* 10(1), 162; *Transcription*, 84. Cf. the scene of Nefertari at Abu Simbel shaking her naos sistrum before Anukis *(fig.28)* immediately before the king's zenithal offering of Maat to Amun-Re.

11) (West lintel scene) *Ibid*. 10(1), 163, and (2), *pls.58, 84*; *Transcription*, 85; (praise of Hathor) *ibid*. (1), 164; *Transcription*, 85.

12) Cauville, *Dendara* 10(2), *pls.39–42, 65–8*: *Transcription*, 42–5 (Hathor's procession), 46–51 (Ptah's procession).

13) Seth's destruction is repeatedly stated; see *ibid*. 52–6. For his demonization in the Late Period, see te Velde, *Seth*, 140–51. He was, however, still worshipped in Western Desert temples (*ibid*. 115–16, 140, n.1).

14) For Osiris's descent as a falcon, see Cauville, *Dendara* 10(1), 151–2; *Transcription*, 79–80 (eastern frieze inscription).

15) *Ibid*. 10(1), 249(5); *Transcription*, 133 (eastern frieze inscription). For Horus as the *Ba* of Osiris, see Roberts, *My Heart My Mother*, 78, 79, 128, *pl.62*.

16) For the vigil's sixth night hour, see Cauville, *Dendara* 10(1), 132–3; *Transcription*, 72. For the location of the vigil's hours around the windows, see *ibid*. (*Commentaire*, 72), though Cauville's interpretation of the Khoiak Festival takes no account of this alternating cycle of day and night.

17) For this unification in the *Emerald Tablet*, see Roberts, *op.cit*. 219–22.

18) For the heart commanding the body in the *Memphite Theology*, see *ibid*. 14–18.

19) For these doorway inscriptions, see Cauville, *Dendara* 10(1), 181–2; *Transcription*, 94–5.

20) For Nut 'giving birth to the gods', see *ibid*. 10(1), 185(8), 212(12); *Transcription*, 96, 112.

21) Cf. the west wall scene of four goddesses suckling the Horus child, identified as two Hathors, Heket and Isis (*ibid*. 10[1], 207–208; *Transcription*, 109).

22) For this ninth-hour text, see *ibid*. 10(1), 170(4); *Transcription*, 88.

23) *Ibid*. 105 (200). For Thebes as the traditional birthplace of Osiris, see *ibid*. *Commentaire*, 236–8. The text is also quoted in Cauville and Ibrahim Ali, *op.cit*. 175.

24) *Faust* (Part 2, Act I, Scene 5). Significantly, when descending into this realm, Faust must touch the flaming Tripod there with his key. Raphael (*Goethe and the Philosophers' Stone*, 142–4) associates the Tripod with Apollo, and a reference to the sun is certainly appropriate, resonating further with the triadic nature of this maternal realm, both in alchemy and ancient Egypt.

25) Cauville, *Dendara* 10(1), 260; *Transcription*, 139. The text is quoted also in Cauville and Ibrahim Ali, *op.cit*. 176.

26) Copenhaver, *Hermetica*, 4[17].

27) See Roberts, *Hathor Rising*, 163–4.

28) See Abt *et al.*, *Explanation of the Symbols* 2, 225–7.

29) *Ibid*. 120.

30) The astronomical ceiling is illustrated in Cauville, *Dendara* 10(2), *pl.115*. For Orion in his mother's womb, see *ibid*. 10(1), 261(3); *Transcription*, 140.

31) For this recipe, see Daumas, *Dendara* 9(1), 124–6. For Amun's god-like presence in the palace revealed by his fragrance, see Roberts, *Hathor Rising*, 123–4. The divinity of the Osirian king's body is already affirmed in the

Pyramid Texts, cf. Utterance 723: 'O king, raise yourself … for this body of yours belongs to a god … may your flesh be born to life.'

32) Cauville (*Commentaire*, 110, with n.245) noted the symbolism of the room's location above the 'laboratory'.

33) For this application of colours, see Cauville, *Dendara* 10(1), 45(10–14); *Transcription*, 26.

34) For this passage, see Festugière, *Hermétisme*, 241–2, 244. See also Roberts, *My Heart My Mother*, 68, for the text's relationship with Osirian renewal. The date of its composition is uncertain. Festugière (*op.cit.* 213) tentatively dated it to the fourth or fifth century CE, and Letrouit ('Chronologie', 83–5) to the seventh century or later, arguing it showed the influence of the alchemist Stephanus of Alexandria.

35) Festugière, *op.cit.* 244.

CHAPTER 10

1) Cauville, *Dendara* 10(1), 268(11)–9(3); *Transcription*, 143–4.

2) *Ibid.* 289, 15–290; *Transcription*, 155.

3) For the *kefen-* bread and ploughing texts in the Khoiak Festival inscription, see *ibid.* 19, 20.

4) Cf. a poignant scene in the *Ramesseum Dramatic Papyrus* when bread and beer are brought immediately after Horus has embraced the 'tired' Osiris. As the beer is brought, the weeping Horus turns to Geb, saying, 'They have put this father of mine into the earth.' Then, as a loaf of bread, identified with Osiris, is brought, he says to Geb, 'They have made it necessary to bewail him.' Also present is Isis, 'Mistress of the house', associated with the beer; see Frankfort, *Kingship*, 136. Cf. also the nocturnal dance ritual performed at Amenhotep III's first Sed Festival for the king and Hathor, which includes a song describing the pounding and transformation of grain; Roberts, *Hathor Rising*, 28.

5) Cauville, *op.cit.* 10(1), 290(8); *Transcription*, 155.

6) Griffiths (ed.), *De Iside et Osiride*, 171 (chapter 33, 364c).

7) For the etymological origins of the word 'alchemy', *see chapter 19, n.4.*

8) *Koré Kosmou*, §32; see Nock and Festugière (eds.), *Fragments Stobée*, 10. As Festugière noted (*Hermétisme*, 247), this 'gift' is to be understood as alchemy. The treatise is preserved in the *Anthology* of Stobaueus from Macedonia, who perhaps lived in the fifth century. For the relationship of its title to Isis-Hathor as 'pupil of the eye', see Roberts, *Golden Shrine*, 149, n.18 (with further references).

9) Note, however, that Cauville takes no account of this fifth/sixth month symbolism and 'reversal' at the end of the sixth month (*Commentaire*, 148–9; *Guide archéologique*, 78–83).

10) Cauville, *Dendara* 10(1), 314–15; *Transcription*, 169.

11) For this name of Jupiter, see Neugebauer and Parker, *Egyptian Astronomical Texts* 3, 177–8. Cauville (*Commentaire*, 148, 213) interprets Jupiter's presence here in terms of factual astronomical data, i.e. the planet's rising in the axis of Dendara temple on 28 December 47 BCE, when Orion disappeared from the sky (her view quoted also in Cauville and Ibrahim Ali, *Dendara*, 212). *See, however, chapter 12* for Osiris's identification with Jupiter's astrological 'fall' in Capricorn here in the western court.

12) Thoth's role to reunite Osiris's *Ba* and body is stated in a niche on the central chamber's west side (Cauville, *Transcription*, 199).

13) For the Egyptian *Benu*-bird as forerunner of the phoenix, see *Lexikon der Ägyptologie* 4, cols. 1030–39, and Quirke, *Cult of Ra*, 27–30.

14) Identification of the somersaulting figure is controversial. Cauville (*Commentaire*, 177, 278) interprets it as the fœtal Osiris enclosed in Nut's womb (see also Cauville and Ibrahim Ali, *op.cit.* 208) but, if so, it really belongs in the eastern end chamber. Von Lieven similarly identified the figure as Osiris (*Grundriss des Laufes der Sterne*, 137–9). For the figure as Geb, see Daumas, *ASAE* 51 (1951), 373, n.1; Kaper, *JEA* 81 (1995), 179–82. Geb is normally shown beneath Nut in Egyptian iconography and, in my view, his acrobatic pose here symbolizes the reversal of time he rules, which Plato associates with the 'age of Kronos' and resurrection (*see chapter 5*). For Kronos being identified with Geb, cf. a Greek Isis aretalogy in which Isis declares she is the eldest daughter of Kronos, the wife and sister of Osiris, and the mother of Horus; see Roberts, *Golden Shrine*, 99. The thematic correspondences with Osiris's resurrection at Dendara are not considered in Kaper's interpretation of the Deir el-Haggar astronomical ceiling. For a brief description of its scenes and Kaper's interpretation, *see below chapter 12, nn.24, 25.*

15) Cauville, *Dendara* 10(1), 378–84; *Transcription*, 204–207. For the *Litany of Re*, see Hornung, *Buch der Anbetung des Re*; *Valley of the Kings*, 87–9, 121–2.

16) Cauville, *op.cit.* 336, 4–5; *Transcription*, 180 (speech of the Heliopolitan nome). For the unification of Osiris and Re associated with the right and left Eyes in the Esna temple, see von Lieven, *Himmel über Esna*, 84–7.

17) Cauville, *op.cit.* 376; *Transcription*, 203.

18) *PT* §208c–10c.

19) Cauville, *op.cit.* 376, 14–15; *Transcription*, 203.

20) *Ibid.* 311, 9–10; 168.

21) Cf. also the correlation of the *Book of Night's* seventh hour with chapter 74 of the *Book of the Dead*, which is an invocation to Sokar celebrating the *Ba's* separation from the earthbound body and power to move freely. This is also the nocturnal hour when a new child is born and the martial Horus protects Osiris from his enemies; see Roberts, *My Heart My Mother*, 144–8.

22) For the association with rites celebrated between 24 and 26 Khoiak, see Cauville, *Commentaire*, 177, Cauville and Ibrahim Ali, *op.cit.* 201–208.

23) For this nome procession, see Cauville, *Dendara* 10(2), *pls.184–7, 213–18; Transcription*, 174–82.

24) Cf. also the vertical axis, encompassing the 'zenith' and Osirian 'deep water', being celebrated in the New Year ritual confirming the pharaoh's power (Roberts, *Golden Shrine*, 55, 61–2, 64–5).

25) Cauville, *Dendara* 10(1), 82, 4–5 = *Transcription*, 45. The paste's 'alchemical' resonances were noted by Cauville (*Commentaire*, 34).

26) *PT* §2145c. For the scenes of the *Opening of the Mouth Ritual*, see *Dendara* 10(2), *pls.188–91, 219–22.*

27) For these lintel scenes, see *Dendara* 10(2), *pls.240, 265; ibid.* 10(1), 392–5; *Transcription*, 212–13. Cf. also the identification of the Maat offering with the 'seed of the bull' in the Crypt South 1's western chamber (*see chapter 7, n.59*).

28) For Ptah creating the world through his 'Word', see Roberts, *My Heart My Mother*, 16–17.

29) Cauville, *Dendara* 10(1), 390(2–3); *Transcription*, 210.

30) *Ibid*. 395, 13–396, 1; 213. This is the opening line of a long invocation to Osiris to 'stand erect'.

31) Quoted *ibid*. 397, 9–10, 398, 2–3; 214, 215.

32) For this Isis incantation, see *ibid*. 428–30; 232–3.

33) *Ibid*. 428–30; 232–3

34) *Ibid*. 396, 7; 214.

35) This sequence of Osiris's burial is described in the frieze inscription, see *ibid*. 426–7; 232.

36) Cf. also the deliberate word-play at Dendara between the term 'nine-stranded' linen (*psḏ*), which derives from *psḏ* meaning 'nine', and the word for 'light' (*psḏ*); see Richter, *The Theology of Hathor of Dendera*, 28–9. The nine rays shining down on Hathor's image on the ceiling of the *Wabet* likewise convey this sense of light, clothing and adornment (see *fig.74* in this book). *See also below chapter 13, n.29.*

37) See Gardiner, *Egyptian Grammar*, Sign list Aa 30.

38) A remarkably similar resurrection is described in an Arabic alchemical treatise entitled the *Silvery Water and the Starry Earth* by the early tenth-century alchemist Ibn Umail. He refers to the 'Ethiopian earth' in which gold is sown and from which gold is extracted. He also associates this black earth with 'the ashes which come from ashes' and the 'lunar Jupiter', as well as 'the great king who needs to be given back his crown' and returned to gold in the eternal, heavenly waters with 'his sister', purified from the darkness of the tomb; see Abt *et al.*, *Explanation of the Symbols* 2, 145. As Abt observed, 'Alchemy continued the crucially important symbolism that was developed in Ancient Egypt, a culture that was devoted to the constant renewal of their highest principle embodied in the Horus-Pharaoh' (*ibid*. 267).

CHAPTER 11

1) *Timaeus*, 90c–d (trans. Cornford), 354.

2) Cauville, *Dendara* 10(1), 397(5); *Transcription*, 214. For the restoration of the head as a rite of revirilization, see Berlandini, *OMRO* 73 (1993), 29–41, and Naydler, *Shamanic Wisdom*, 244–6. For headless beings in the netherworld and heads being associated with stars, see Darnell, *Enigmatic Netherworld Books*, 111–17.

3) For Sokar as the 'golden remedy', see Ritner, *Magical Practice*, 55, with n.257. For the possible derivation of the Greek word *pharmakon*, meaning 'remedy' or 'charm', from the Egyptian word *pekheret*, *see chapter 17, n.14.*

4) For this passage about rising from Hades, see Festugière, *Hermétisme*, 241, 243.

5) *PT* §285c–86a, 286c. Naydler (*op.cit.* 244) associates the king's descent into the earth and the sun god's rebirth here with the ploughing ceremony in the Khoiak Festival.

6) Inscriptions in the eastern end chamber accompanying a scene of the king offering the day boat to Hathor, with Osiris shown in the adjacent scene; *Transcription*, 128; *Dendara* 10(1), 242–3, and (2), *pl.111*. For the corresponding evening boat, see *ibid*. 132; (1) 248, 7–8, and (2), *pl.114*. For Hathor-Sekhmet as 'Lady of the red cloth', see Roberts, *Golden Shrine*, 135, n.22.

7) See Cauville, *Commentaire*, 223–4. For Hathor protecting the ceremonies during the night of 25 to 26 Khoiak,

see Cauville and Ibrahim Ali, *Dendara*, 207. This flow of Hathorian 'sympathy' and 'antipathy' influencing plant life is paralleled in Theophrastus's botanical Greek writings based on the doctrine of the 'sympathies' and 'antipathies' in nature, and also in the work of Bolus of Mendes, who applied it to the world of plants for purposes of magic, healing and ritual. Very little is known about Bolus's life, though it is now no longer thought he is the same person as an early alchemist called Democritus *(see below, chapter 16, n.13).*

8) Griffiths (ed.), *De Iside et Osiride*, 219 (chapter 63, 376c–d), and Roberts, *Hathor Rising*, 57. Griffiths (*op.cit.* 525) doubted whether Plutarch's Greek etymological explanation was valid in Egyptian, yet Dendara's priests evidently connected the sistrum with movement; *see also chapter 7, n.26.*

9) Derchain, *Hathor quadrifrons*, 10 (no.18).

10) Book 12 (15–16); see Copenhaver, *Hermetica*, 46–7.

11) The vessel's imagery was discussed by Desroches-Noblecourt, *Monuments … de la Fondation Eugène Piot* 47 (1953), 1–34.

12) For the juxtaposition of marsh 'hunting' scenes and Hathorian imagery on Tutankhamun's small Golden Shrine from the same period, see Roberts, *Golden Shrine*, 16–26.

13) Desroches-Noblecourt suggested the *Tilapia* symbolized the gestation phase in the mummification rites leading to rebirth. For numerous references to fish, Nun and the placental waters, see Aufrère, *L'univers minéral* 2, 467–71. See also Pinch, *Votive Offerings*, 187–8, for fish symbolism and Hathor, and *ibid*. 312–15, for 'marsh bowls' associated with Hathor, the waters of Nun and regeneration.

CHAPTER 12

1) The most influential studies for this 'astronomical' interpretation are Aubourg, *BIFAO* 95 (1995), 1–10, and Aubourg and Cauville, 'En ce matin', 767–72. See also Cauville, *Dendara* (*Commentaire*), 79–80; Cauville and Ibrahim Ali, *Dendara*, 164–73. This astronomical analysis continues to be cited, e.g. Kaper, *JEOL* 41 (2008–2009), 34, with n.17. For a critique of it, see von Lieven, *Himmel über Esna*, 157, n.458, and Leitz, *SAK* 34 (2006), 285–318, esp. 286–7. As von Lieven also noted (*op.cit.* 176–7), the term 'zodiac' is not really appropriate for Dendara's round zodiac, since it incorporates far more symbolism than simply the Babylonian zodiacal signs, and she suggested the Egyptian term for round 'sky images' might have been 'golden sky' ('goldener Himmel').

2) For a good overview of Egypt as the home of astrology, see Barton, *Astrology*, 25–9. For astrological consultations in Egyptian Graeco-Roman temples, see Evans, *JHA* 35(1) (2004), 24–37.

3) *Library of History* 1, 81 (trans. Murphy, *Antiquities of Egypt*, 102). Murphy rejects the notion that Mesopotamian astronomy or astrology was an offshoot of Egyptian wisdom (*ibid*. 102, n.157), but see below, n.7.

4) For these Theban horoscopes, see Neugebauer, *JAOS* 63 (1943), 115–27. The complete list of 'places' is preserved in Ostracon 3.

5) See Neugebauer and Parker, *Egyptian Astronomical Texts* 3, 203.

6) *Asclepius* 30; see Copenhaver, *op.cit*. 85.

7) For the Greek *hypsomata* being derived from the planetary 'places of secret', as known from Babylonian omen, and late astrological, texts, see Rochberg-Halton, *JAOS* 108 (1988), 53–7. See, however, Conman (*DE* 64 [2006], 7–20), who argued the Babylonian 'places of secret' were influenced by Egyptian astral concepts.

8) See *P. Michigan* 149, col.xvi, 23–35. This papyrus, dating from the second century, was published by Robbins in Winter (ed.), *Papyri in the University of Michigan Collection*, 62–117. For this unique terminology, see *ibid*. 106 (23–35), also Powell, *Hermetic Astrology*, 33, nn.9, 15. For the planets represented in their 'exaltations' in the Esna temple, see von Lieven, *op.cit*. 156–7.

9) *Wb*. 4.54.

10) Cf. also the Zoroastrian springtime rites celebrated at No-Roz, in which Rapithwin, the 'Spirit of Noon' and 'Lord of Ideal Time', re-emerges from his five-month winter sojourn underground to bring new life from the seeds he has been warming. In the Zoroastrian cycle of the day, the watch ascribed to him is from noon until three o'clock (thus corresponding to the Egyptian sixth to ninth hours); see Mistree, *Zoroastrianism*, 84, 115.

11) For this identification with the 'Headless One' in Classical sources, see Leitz, *SAK* 34 (2006), 300, who rejected Aubourg's suggestion (*BIFAO* 95 [1995], 5) of the constellation Equuleus.

12) See Gardiner, *Egyptian Grammar*, Sign list F26.

13) For Hapy being associated with Aquarius, see Desroches-Noblecourt, *Archéologia* 292 (1993), 34.

14) See Frankfort *et al*., *Cenotaph* 2, pls.75–80 (*Book of Night*), pl.81 (*Book of Nut*), and Hornung, *Zwei ramessidische Königsgräber*, 89–100, with pls.68–75. See also the later representations in the tomb of Mutirdis (Assmann, *Grab der Mutirdis*, 85–9, with pl.39).

15) Identification of the decan stars is problematic, not least because opinions differ as to where the 'decanal belt' was located. According to Neugebauer and Parker (*op.cit*. 1, 97–100), it was slightly south of the ecliptic, but this is not universally accepted; see von Bomhard, *Egyptian Calendar*, 54.

16) For the purification of the decans and their rebirth as fish in the *Book of Nut*, see Neugebauer and Parker, *op.cit*. 1, 75–6, with pl.53; also *ibid*. 68, for other descriptions of a star's life in the lake of tears. For discussion of this cycle, see Roberts, *My Heart My Mother*, 96–104, and in addition to the references cited there, see von Lieven, *Grundriss des Laufes der Sterne*, 89–91 (§116–24), 169–70.

17) For this Demotic commentary and translation, see von Lieven, *ibid*. 61. However, she overlooked the significance of the 'sheep' stars in the text, preferring to interpret *sriw* as a reference to stars in general.

18) That both Seti's brief text and the later Demotic commentary are concerned with the passage of the year through the months, rather than simply Sothis's heliacal rising, is indicated by the absence of 'day 1' in the citation of the first month of *Akhet*. Von Lieven (*ibid*. 61 n.278) noted this unusual feature without further explanation.

19) *Asclepius*, 24. See Copenhaver, *op.cit*. 81[24].

20) On the tomb's dating to the early second century CE, see Whitehouse, 'Roman in Life, Egyptian in Death', 262, with n.7.

21) *See above*, chapter 5, n.22, for 'justification' associated with annual 'reversal'.

22) For the inscription, see Osing *et al*., *Denkmäler*, 92. Cf also an inscription close to a zodiac in Petubastis's neighbouring tomb, expressing the wish for his *Ba* to have unimpeded power of movement between heaven and earth. See *ibid*. 80–81 (Text A), with *pl*.70.

23) In the publication, Pingree (*ibid*. 100) associates the falcon with putative Mithraic beliefs about the soul's escape from the 'evil of the material creation', symbolized by the double-headed 'ouroboros'. For a summary of Harris's critique of Pingree's view, see Whitehouse, *op.cit*. 266–7.

24) The ceiling lay in pieces on the sanctuary's floor when discovered, though, due to Kaper's painstaking work, most of the scenes have been reconstructed on paper (see *JEA* 81 [1995], 176, *fig.1*). Kaper interpreted the symbolic imagery differently from here, associating it with Theban ideas of the opposition of sun and moon linked with Amun and Khons (*ibid*. 190–93). His argument, however, that 12 hour goddesses must have been intended, and that the reduced number towing the night boat was due to lack of space (*ibid*. 184–5), is unlikely, as six hour goddesses are also depicted in Petosiris's nearby tomb. Interestingly, the astronomical ceiling shows the goddesses processing towards a baboon and scarab, the very same creatures housed in Ramesses's II's solar sanctuary at Abu Simbel, which was oriented towards the sun's southern solstice. For Kaper's identification of the acrobatic figure as Geb, *see chapter 10, n.14*.

25) For the *Book of the Earth* on the right wall of Ramesses VI's tomb, *see chapter 5*. Its registers are arranged as follows: 1) Osiris encircled by stars (top section); 2) the raising of the sky and procreation of a child associated with the healed Moon Eye and annual reversal (middle section); 3) the sun god's rejuvenating night journey in the sun boat (bottom section). The correspondences with the Deir el-Haggar astronomical ceiling are: 1) in the top register Osiris-Orion and Isis-Sothis sail in their boats. To the right of them is Geb's somersaulting figure and then a *Benu*-bird and a decanal star group; 2) in the register beneath, 16 deities process towards a lunar Eye within a disk, here coming to celebrate the filled Eye; 3) beneath this lunar register is the nocturnal course of the sun, including Atum and Khepri sailing in their respective boats, accompanied by six hour goddesses. In addition, the month deities are depicted in the bottom register, though these are not included in Ramesses VI's tomb.

26) Manilius, *Astronomica* 2: 218–20 (Goold edn, 98–9). For a diagram of the 'northern' and 'southern' signs, see Beck, *Ancient Astrology*, 51, *fig.5.1*.

27) For the figures as Saturn and Venus, see Pingree's interpretation, *op.cit*. 100 *(above, n.23)*. For Plutarch's reference to Kronos and Aphrodite, see Griffiths (ed.), *De Iside et Osiride*, chapter 69.

28) For Horus on the crocodile representing the 'right moment' or 'fullness of time', and being identified with Kairos, the son of the Greek god Kronos, see Barb, *MAGW* 82 (1952), 20–21. My thanks are due to John Harris for this reference.

29) On Ipet as the mother of Osiris, see Bulté, *RdÉ* 54 (2003), 1–20.

30) For the three scenes of Osiris in room 2, see Osing *et al.*, *Denkmäler*, *pls.28b, 29a*. For Osiris's three forms in the *Book of Day*'s ninth hour, *see chapter 2*.

31) *PT§1987a–c*.

32) For late New Kingdom evidence typifying an 'astrological outlook', see Kákosy, *Oikumene* 3 (1982), 187–91.

33) For this development in Hatshepsut's reign, see Roberts, *Hathor Rising*, 118–29.

34) Evans (*op.cit.* 42, n.63) suggested Petosiris's tomb might have been used for astrological consultations, though this use is unproven.

35) Von Lieven (*Himmel über Esna*, 189–90) noted the spread of Egyptian decanal knowledge in astrology, though her statement that its continuing 'vitality' is not reflected in 'any other element of the Egyptian religion' is wide of the mark.

36) Cauville and Ibrahim Ali, *Dendara*, 226–7.

CHAPTER 13

1) EA 19. See Moran, *Amarna Letters*, 44[59–70].

2) See Roberts, *Hathor Rising*, 26–7.

3) Chassinat, *Dendara* 4, 73, 9–10. The hieroglyph for 'copper' here is the 'sledge bearing ore', which must be read *bi3*. In adhering to *bi3* as 'iron', Graefe (*Untersuchungen*, 145) unconvincingly suggested the addition of *m3ꜥ* in Hathor's epithet might reflect iron's rarity. For the expression 'true copper' used by the early alchemist Maria the Jewess, *see chapter 18, with n.45*. For Hathor as a 'great wonder' (*bi3it ꜥ3t*) at Dendara, see Graefe, *op.cit.* 146.

4) The two most influential studies presenting evidence against *bi3* as 'copper' are Harris, *Lexicographical Studies*, 50–62, and Graefe, *op.cit.* 26–9, who argued it originally meant 'meteoric ore' and 'metal', especially meteoric iron. This has resulted in vague and confusing translations of *bi3*, e.g. Aufrère wavers between 'marvellous mineral' (*L'univers minéral* 1, 103) and 'ferrous metal' (*ibid*. 2, 431), and treats it as a generic term based on the idea of 'miracle' (*ibid.* 1, 106). In fact, he omits *bi3* from his section on copper altogether (*ibid.* 2, 449–50).

5) Harris, *op.cit.* 50.

6) For the names of malachite, see *ibid.* 102–104, 132.

7) For the Coptic names of copper, *ibid.* 51, 61.

8) See El-Raziq *et al.*, *Inscriptions d'Ayn Soukhna*, 40–41 (Inscription 4b). The remains of furnaces at the site provide valuable evidence for the stages of copper production (see *ibid.* 5). The authors also note the rarity of references to copper in Egyptian mining inscriptions (*ibid.* 101).

9) Gardiner *et al.*, *Inscriptions of Sinai* 1, *pl.10* (23); *ibid.* 2, 66.

10) See Chassinat, *Mystère d'Osiris* 2, 464, 466–7. For *bi3* meaning 'copper', see also Lalouette, *BIFAO* 79 (1979), 333–53, and Nibbi, *JARCE* 14 (1977), 59–65.

11) See Herbert, *Red Gold*, 10, with n.26.

12) For a survey of the use of iron in ancient Egypt, see Ogden, 'Metals', 166–8. Rothenberg (ed.), (*Sinai*, 164, 166) observed how the ancient Egyptians left the iron ores untouched in the southern Sinai region, since only copper ore interested them.

13) For 'iron' as the 'bone of Typhon', see Griffiths (ed.), *De Iside et Osiride*, 217 (chapter 62). For copper's

luminosity being compared to the shining sun, see Lalouette, *op.cit.* 339.

14) Even though Aufrère maintained *bi3* meant 'ferrous metal', he nevertheless translated it as 'copper' when it occurs with gold in a decan list at Dendara (*L'univers minéral* 1, 180 [xxvii]). Harris (*Lexicographical Studies*, 59) faced the same predicament regarding a yellow-coloured metal vessel named *bi3*, which Thutmose III offered to Amun. He noted the yellow colour was 'strange' and 'less certainly iron'. Similarly, when *bi3* follows 'gold' and 'silver'in Hathor's epithets at Dendara and Edfu, Graefe's addition of a question-mark after 'iron' indicates he realized the implausibility of his translation. In fact, he queried iron's relevance here (*Untersuchungen*, 144–5).

15) Cf. the famous Middle Kingdom stela of Horwerra at Serabit el-Khadim, which refers to the difficulties of finding turquoise's best colour in the blazing summer heat; see Gardiner *et al.*, *op.cit.* 1, *pl.25a* (no.90), and 2, 97–8; and Aufrère, *op.cit.* 2, 492–5. For *Bi3* as the name of the Sinai peninsula, see Graefe (*op.cit.* 35).

16) For Hathor's temple at Timna, see Rothenberg, *Timna*, 125–201; *The Egyptian Mining Temple at Timna*. For Hathor at copper-mining sites, see also Pinch, *Votive Offerings*, 59–70. For the sacred nature of copper-working at Kition on the island of Cyprus during the Late Bronze Age, see Karageorghis (*Kition*, 74–5, 169–71), who noted the parallels with the worship of Hathor at Timna (*ibid.* 75).

17) Rothenberg *et al.*, *Sinai*, 137.

18) Gardiner *et al.*, *op.cit.* 1, *pl.10* (no.23), and 2, 66.

19) For these two root meanings, see Graefe, *op.cit.* 1–5.

20) For 'red' and 'green' as Hathor's colours, see Roberts, *Hathor Rising*, 10–13.

21) Chassinat, *Dendara* 4, 152(10), quoted in Aufrère, *op.cit.* 1, 134.

22) For the association of 'mine' and 'womb', see *ibid.* 67–8, 73. For the 'birthing' of metals, see also Daumas, 'L'alchimie', 115–16.

23) Cf. the juxtaposition of a procession of personified mining regions with a procession of Ramesses II's royal children led by the queen depicted in the first court of the Luxor temple; see Darnell, 'Ancient Egyptian Cryptography' (forthcoming). The sistrum-shaking queen stands directly behind a food-bearing fecundity figure (illustrated in Roberts, 'Cult Objects of Hathor' 2, *pl.90*), thus reinforcing the king's role to guarantee fecund life, both in the palace and in the mineral-bearing regions. It also suggests his identification with Min-Amun, the fertile god presiding over these regions, whose festival is depicted on the pylon at the temple's entrance.

24) The complete text is preserved in the 'corridor' version of the *Book of Day* in Ramesses VI's tomb, placed beneath the gods praising Hathor as the first-hour birthing goddess; see Müller-Roth, *Das Buch vom Tage*, *pls. III–IV* (Text C). For an abbreviated version in the sarcophagus chamber, see *ibid. pl. XVII* (Text C).

25) *Medinet Habu* 6, *pls.420–21*.

26) For the complete text, see Müller-Roth, *op.cit.* 142–4 (with further references).

27) Müller-Roth's translation of *m bi3* as 'aus dem Ehernen' is ambiguous, since Ehernen in German has connotations of 'iron', 'bronze' and 'brass'. In the accompanying note

(*ibid*. 149, note g), he refers both to Graefe's publication (which rejects *bi3* as a name for copper; *see above, n.4*), and to Lalouette's study (which accepts it; *see above, n.10*), thus leaving the meaning indeterminate here. This association of metallurgy with Re's role as 'Lord of the Palace' is paralleled in the juxtaposition of scenes in the Luxor temple associated with Ramesses II *(see above, n.23)*.

28) British Museum (no.1163); see Budge, *Hieroglyphic Texts*, pl.19; Graefe, *Untersuchungen*, 47 (Document 70). Budge remarked on the text's remarkable phraseology, and Graefe offered no translation for *bi3* here, noting that what is meant by Hathor's joy and Thoth's vitality 'unfortunately remains obscure' (*ibid*. 47–8). The cartouches on the stela are those of the 17th-Dynasty ruler Sobkemsaf I.

29) For this 'veiling of *bi3*' as 'clothing' or 'enveloping' the sky in light, see Graefe (*op.cit*. 48–50). He cites various examples of 'clothing the sky', though none mentions *bi3* again. He also refers to much later scenes at Dendara and Edfu (*ibid*. 51–3) showing the king offering the *menkhet*-cloth to Hathor and other deities, which is associated with clothing the divine body in light and illuminating the temple. For the relationship of clothing, light and adornment, *see also chapter 10, n.36*.

30) Utterance 469, *PT§906–908*.

31) Spell 731, *CT* 6, 363p–q; Graefe, *op.cit*. 47 (Document 69).

32) Spell 335, *CT* 4, 292b–c, 295a–b; see also Lalouette, *BIFAO* 79 (1979), 352.

33) For *nbi* being associated with metalworking, see *Wb*. 2.241.

34) See Herbert, *Red Gold*, 54, for this 'copper eye' description.

35) For these lines, see Lalouette, *op.cit*. 352. Quirke's translation, 'metal', however, obscures copper's importance: 'The glittering [faience] is crafted, Ptah is at his metal. O Re, laugh' (*Going Out in Daylight*, 154).

36) Kitchen, *Ramesside Inscriptions* 2, 266(5–10). The word for 'iron' here is *bi3 n pt*, possibly a manufactured iron (see Harris, *Lexicographical Studies*, 59).

37) For this terminology, see Aufrère, *L'univers minéral* 1, 65.

38) Chassinat, *Dendara 5*, pl.425. Although *ḥmty km* is often translated as 'black bronze', its meaning is rather 'black copper' (see Giumlia-Mair and Quirke, *RdÉ* 48 [1997], 95–108), the word for 'bronze', *ḥsmn*, seemingly never occurring with *km*, 'black'. It refers to an intentionally darkened copper-based alloy, which is treated chemically to give it a black patina, thus heightening the impact of various coloured metal inlays. For 'black bronze' in alchemy, see Martelli, *Four Books*, 67.

39) For Ihy's regeneration in the *Coffin Texts*, see Roberts, *Hathor Rising*, 29–32.

40) 'Seed of Life', 373 *passim*. Here in these traditional American cultures, unlike in ancient Egypt, copper is associated with the moon and lunar deities.

CHAPTER 14

1) The statues are made from almost pure copper with only small amounts of iron and arsenic present; see Eckmann and Shafik, "*Leben dem Horus Pepi*", 73. The reason for two statues has been variously explained and there has been debate over whether the smaller statue represents Pepi's successor, Merenre (*ibid*. 37). The authors conclude both statues represent Pepi (*ibid*. 46–7). For the possibility of a winged falcon originally protecting the smaller statue's head, see *ibid*. 49. For a statue of King Khafre with a falcon protecting his head, see Roberts, *Golden Shrine*, fig.38.

2) *Urk*.1, 294. For the translation 'coloured like gold' rather than 'overlaid with gold', see Lalouette, *BIFAO* 79 (1979), 336. Eckmann and Shafik (*op.cit*. 35) noted there must have been a long tradition of copper-working before the Sixth Dynasty, even though no earlier statues are so far known. A copper statue of the Second-Dynasty ruler Khasekhemwy is mentioned on the Palermo Stone, and copper boats apparently belonged to Neferirkare's sun temple at Abusir. See also Lalouette, *op.cit*. 336–7.

3) *PT§1964d–1965c*. The text is damaged and it is difficult to know which deity is actually speaking, but I have followed Graefe's emended version, parts of which he restored from its adaptation in Spells 682 and 989 of the *Coffin Texts*; see *Untersuchungen*, 60–61 (Document 97).

4) *PT§1966a, d, 1967*. Because *bi3* is often not translated as 'copper' here, the text's meaning is completely obscured. Graefe (*op.cit*. 61) simply leaves it untranslated. Faulkner (*Pyramid Texts*, 284) translates it as 'iron'; Graindorge-Héreil (*Le dieu Sokar*, 317) as 'marvel' or 'ferrous metal' (here influenced by Aufrère's interpretation of *bi3*; *see chapter 13, n.4*). Cf. however, Lalouette's translation as 'copper' (*op.cit*. 345), describing copper in this text as the 'nécessaire armature divine'. Similarly, Mercer (*Pyramid Texts* 1, 293) translates it as 'copper'.

5) *PT§1967*. Cf. the special knot an initiate ties for Hathor in the *Coffin Texts*: 'I raise the *menit*, I tie the knots for Hathor, so that the sky shines with your beauty' (Spell 753; *CT* 6, 382i–j).

6) *PT§1969b–c, 1971*. If Pepi I's smaller copper statue did originally have a falcon attached behind his head *(see above, n.1)*, it would have perfectly mirrored this text in which Horus is empowered to fly. Such a flight was perhaps ritually enacted at the king's Sed Festival, which is mentioned in the inscription originally attached to the larger statue's base.

7) *PT§1944a–b, 1945d*. Cf. also Spell 991 of the *Coffin Texts* for 'Becoming Sobek', the crocodile god: 'I am that seed which goes forth from the encircling clothing. I am one whose teeth break forth, who cuts the copper' (*CT* 7, 201e). Graefe (*op.cit*. 62) suggests this refers to 'breaking the egg'. Cf. a description of the smith working at his furnace in the *Satire of Trades*, which says he has 'fingers like claws of a crocodile' (Lichtheim, *Ancient Egyptian Literature* 1, 186), so possibly the same nexus of metallurgical ideas underlies Sokar's role in Spell 991.

8) Waitkus, *Texte*, 127. For 'concealed wings', rather than 'equipped wings', see *ibid*. 136, n.53.

9) *Ibid*. 130.

10) *Ibid*. 126–7, 129. Waitkus translates *bi3* here simply as 'metal' without further comment. Cf. however, Cauville's translation of 'bronze' when similar imagery is depicted in the main temple, which at least retains a link with copper (*Dendara 3: Traduction*, 233). There Harsomtus is named as 'the serpent issuing from the Eye of Re' and is thus explicitly associated with Hathor as Eye of Re.

11) Waitkus, *op.cit*. 125.

12) For Hathor's unifying power, see Roberts, *Hathor Rising*, 48–50; *Golden Shrine*, 20, 70.

13) For this scene, see Roberts, *My Heart My Mother*, 22, with *pl.*16.

14) Cf. the winged and wingless two birds depicted on the stone tablet that the tenth-century Islamic alchemist Ibn Umail saw during his visit to an Egyptian temple *(see chapter 16)*. An inscription beneath them states: 'The female is the spirit, extracted from the male, carrying it, flying away with it'. Just as Hathor's copper gives the wingless falcon flight at Dendara, here the female does the same for this alchemical wingless bird.

15) For the techniques used to assemble metal statues, see Ogden, 'Metals', 157–60.

16) For Mesopotamian casting techniques during this period, see Eckmann and Schafik, *"Leben dem Horus Pepi"*, 35–6.

17) For a contemporary metalworker's account of creating Hindu bronze cult images by lost-wax casting, which he experiences as a sacred birth from the womb, see Dalrymple, *Nine Lives*, 196–7. Dalrymple's meeting with this devout smith, called Srikanda, gives an insight into how the ancient Egyptians might also have experienced this metallurgical art, or 'act of devotion', as Srikanda calls it.

18) *CT* 6, 309l–q, bringing copper *(bi3)* from the *Henu*-boat for tying the limbs and bringing 'the egg' *(swht)*; *ibid.* 312g–q (solarization themes).

19) For this myth, see Roberts, *Hathor Rising*, 10–12. For the shift to the 'earth-related' solar circuit from the Middle Kingdom onwards, see *ibid.* 73–8.

20) See Roberts, *My Heart My Mother*, 24–6.

21) The date of this falcon sculpture is controversial, but Eckmann and Schafik place it in the Old Kingdom and suggest the crown may have been added at a later date *(op.cit.* 65).

22) Cf. also Aufrère's observations about *bi3* and electrum in the *Opening of the Mouth Ritual* (*L'univers minéral* 2, 358). He notes how, during the first opening of the mouth and eyes, an implement made of *bi3* is used, which he describes as a 'corruptible' metal, in contrast to the incorruptible 'finger of electrum' during the second opening. Unfortunately, as throughout his study, he interprets *bi3* as a ferrous metal, and thus overlooks the important juxtaposition of copper and gold here in the ritual. Falchetti also notes how copper has 'characteristics of "mortality", which relate to humanness, in opposition to gold with its "immortal" qualities, usually linked to deities' ('Seed of Life', 348).

23) For the complementary relationship between the Moon and Sun Eyes, see Roberts, *Hathor Rising*, 112.

24) For this scene's association with chapter 162 of the *Book of the Dead*, see Osing *et al.*, *Denkmäler*, 83.

25) For this Edfu text, see Germond, *Sekhmet*, 281, and Roberts, *Golden Shrine*, 149, n.24.

CHAPTER 15

1) For this group, see Roberson, *Books of the Earth*, 268–9, *fig.*5.67. He calls them 'the two arms of Nun', understanding the feminine ending -t as superfluous (*ibid.* 268, n.894), thus obscuring the feminine context. Note that Roberson's interpretation of this left wall, as with the right wall *(see chapter 5, n.29)*, differs considerably from the one offered here.

2) *Ibid.* 174–6, *fig.*5.16; 332–3 (Text 24). The text is placed between the Hathor vignette and the scarab emerging from the disk.

3) For this scene, see Moussa and Altenmüller, *Grab des Nianchchnum*, 135. For metalworking scenes, and Hathor's role in the night journey, see Roberts, *My Heart My Mother*, 140–42.

4) Herbert, *Red Gold*, 40. Herbert referred to Monsignor de Hemptinne's unique early 20th-century account of copper-smelting in the Katanga region of central Africa, in which he quoted a Yeke chant for copper-smelting: 'On the summit of Kalabi rises a high furnace, a high furnace with a large womb, the heritage of our father Lupodila, a high furnace where copper trickles and billows. O my Mother, O my Mother!' (*ibid.* 39). She also included de Hemptinne's rare pictures of Katanga copper-working using malachite (*ibid.* 35–8, *figs.*1–7). For the furnace as womb, see also Eliade, *Forge and the Crucible*, 57–60; Lindsay, *Origins*, 290–94, 300.

5) This embryological association casts light on a strange episode in a Coptic magical text for the healing of stomach pains (*P. Berlin P8313*, recto, col.2, 19–20). It describes an encounter on the Heliopolitan mountain between Isis and her son (here called the 'third demon Agrippas') in which Isis, wearing an 'iron crown', is 'heating a copper oven beneath'. Referring to the well-known role of Isis in alchemy, Richter (*JEA* 93 [2007], 259–63) ingeniously emended the text to interpret it as an allusion to Isis as alchemist working beside her copper distillation oven, a late antique 'update' of her role as 'Great of Magic' (*ibid.* 260–61). What he omitted, however, was the well-known episode when Horus brutally beheads his mother during his conflict with Seth, and, in one version, Isis is healed when Thoth gives her a cow's head, thereby transforming her into Hathor (see Roberts, *Hathor Rising*, 106). Probably this esoteric Coptic magician knew this tradition, but, rather than a cow's head, he provided a metallurgical allusion to Hathor's womb furnace in order to convey this healing of Isis. For the use of copper in Egyptian medical remedies, see Harris, *Lexicographical Studies*, 56.

6) Gardiner, *Egyptian Grammar*, Sign list N41. Nibbi (*JARCE* 14 [1977], 61, with *fig.*1) suggested the vessel represented a furnace for smelting copper when the sign for 'water' was indicated within it. See also Graefe, *Untersuchungen*, 84–5, for discussion of this sign.

7) Though minimizing the meaning of *bi3* as 'copper' *(see chapter 13)*, Graefe did note its association with the heavenly firmament and water (*ibid.* 46). Thus, in the *Pyramid Texts*, not only is the determinative the 'water-filled vessel', it is also said to be crossed by boat, as Allen noted ('Cosmology of the Pyramid Texts', 9), though his translation of *bi3* as 'basin' obscures the copper meaning, as does his identification of *bi3* with 'meteoric iron'.

8) Herbert, *op.cit.* 285.

9) *Ibid.* 287, 368, n.45.

10) *Ibid.* 281.

11) See also Roberson, *Books of the Earth*, 277–9, with *fig.*5.72.

12) *Ibid.* 216–17, with *fig.*5.32.

13) *Ibid.* 343 (Text 31); see also 286–8, with *fig.*5.76.

14) Hornung, *Amduat* 1, 37 (lines 3–6), 38 (lines 4–5); *ibid*. 2, 55.

15) Zosimus's three visions, comprising five dreams, are recorded in his three lessons *On Excellence*, see Mertens, *Zosime de Panopolis*, 34–47. The two dreams here are in his first lesson. For 'excellence' or 'perfection' as the goal of alchemy, see *ibid*. 213–14. The priest's name is sometimes translated as Aion, but Mertens rejects this, relating it rather to the *ios* of alchemical transmutation (*ibid*. 36, n.14). For Jung's extensive commentary on the visions from a psychological perspective, see 'Visions of Zosimos', 57–108. Von Lieven (*Himmel über Esna*, 131–2) compared the 15 steps leading up to the altar in the visions with ancient Egyptian depictions of the moon steps, suggesting Osiris's death and dismemberment, associated with the moon's phases, is the prototype for Zosimus's experience. Abt (*Book of Pictures* 2, 40) noted the connection with the Osirian mummification process.

16) For a cauldron as a vessel of punishment in ancient Egypt, see George, *Vorstellungen*, 42–4.

17) Roberson noted the oval was designated as *nnw.t*, a word originally meaning the scarab's dung ball (*op.cit*. 173, with n.306). It can also refer to the sun disk, the burial mound of Osiris and even the underworld itself—all locations connected with gestation and solar rebirth (*ibid*. 252–3).

18) For Horus's birth as a revelation of Osiris's *Ba*-power, see Roberts, *My Heart My Mother*, 78, with *pl*.62.

19) For this scene, see Roberson, *op.cit*. 279–84, with *fig*.5.74.

20) Mertens, *op.cit*. 46. This dream vision is in the third lesson *on Excellence*.

21) For Osiris's regeneration and the heart, see Roberts, *op.cit*. 78–9, 132–4, with *pl*.107.

22) From *PT*§2244–5.

23) For this scene, see *Roberson, Books of the Earth*, 198–210, *fig*.5.25; for the text, see *ibid*. 318 (Text 3).

24) Cf. Plato's visionary passage describing the winged soul's ascent to the celestial divine realm in the *Phaedrus*: 246–7 (trans. Rowe, 60–63). For ancient Egyptian influence on Plato's account, see Naydler, *Shamanic Wisdom*, 51–3, and 'Plato, Shamanism and Ancient Egypt', 88–90.

25) For this reading, see Hornung (*Valley of the Kings*, 93–4); Roberson, *op.cit*. 201. Unfortunately, Hornung omits the upper sun, hence the 'two suns' meaning is lost.

26) Stricker recognized the 'heavenly' and 'earthly' aspects of Ramesses III's 'two suns' (*Geboorte* 4, 370–77, esp. 371). He also connected the 'earthly' sun's rebirth specifically with the female and the womb's fertilization, describing it as the foundation of the world and the place of the god's manifestation (*ibid*. 372).

27) Roberson, *op.cit*. 253–6, with *fig*.5.56a.

28) *Ibid*. 321–2 (Text 8).

29) Hornung, *op.cit*. 76, 95, *colour pl*.57. Darnell (*Enigmatic Netherworld Books*, 274) refers to a Late-Period scene on the sarcophagus of Tadipakem showing a scarab and a ram-headed human figure enclosed in a sun disk, from which a serpent and crocodile emerge. Here it is explicitly stated the crocodile and serpent protect the gate of the West in the *Dwat*.

30) For the concept of 'two *Bas*' being associated with the unification of opposites in the *Amduat* and the *Book of Gates*, see te Velde, *Seth*, 70. He noted that 'the dead on their journey through the other world are confronted

with the mystery of totality, in which the contrasts are subsumed.'

31) Cf. also a similar scene of a goddess, her body painted gold, in the *Book of Caverns* in Ramesses VI's tomb (Hornung, *op.cit*. 83, *colour pl*.48). Here 'ascent' and 'descent' are conveyed by the four crocodiles facing upwards behind the goddess and the descending figures spiralling downwards in front of her, including a young child. Cf. also Plato's statement in the *Phaedrus* that the immortal winged souls partaking in the divine 'stand upon the outer part of the heavens, and positioned like this they are carried round by its revolution' (*see above, n.24*).

32) For the revelation of the 'name' or 'essence' at the close of the *Book of Night*, see Roberts, *My Heart My Mother*, 158.

33) For this tableau's association with creation through the Word, see Stricker (*op.cit*. 1, 61). Roberson (*op.cit*. 134) somewhat arbitrarily interprets it as the *Book of the Earth*'s opening tableau, initiating the sun's descent in the West which then continues in the other scenes on the left wall, with the narrative sequence then culminating in Osiris's resurrection at sunrise in the East across on the right wall.

34) Quoted from Book 12 [14–16]; Copenhaver, *Hermetica*, 46–7 [14–16].

35) *In Timaeus* 3, 321, translation after Siorvanes, *Proclus*, 232.

36) *Timaeus* 43A, 3–6.

37) *In Timaeus* 3, 322. For this section of Proclus's commentary see Festugière (trans.), *Proclus: Commentaire sur le "Timée"* 5, 202–204.

38) According to Siorvanes (*op.cit*. 232), Proclus is the source of this famous alchemical maxim. However, as Uždavinys noted (*Philosophy*, 262), this is unlikely, since he 'and other Neoplatonists are themselves recipients and interpreters of various ancient cosmological traditions', and he then referred briefly to Osiris's death and rebirth. Lindsay (*Origins*, 60–61) emphasized Proclus's familiarity with alchemy, noting how he used various technical terms occurring also in the Leiden and Stockholm alchemical papyri. For Proclus's worship of the sun and interest in ancient Egyptian religion, see Fowden, *Egyptian Hermes*, 127, 185.

39) 'All things are interlaced and all unlaced. All things are mixed and all things unmixed. All things are mingled and all unmingled. All things are moistened and all things dried. All things flower and shed their blossoms in the phial-altar … it is a natural method, breathing in and breathing out … when in a word, everything is in accordance with separation and union, and if the method is respected, nature is transmuted.' For this passage, see Zosimus's first lesson *On Excellence* (Mertens, *Zosime de Panopolis*, 38–9[4] [my translation from the French]).

40) From the 'Fifth Book about the Magnesia', see Abt (ed.), *Book of Pictures* 2, 339–40, 342.

41) *Ibid*.

CHAPTER 16

1) For this text, see Scharff, *ZÄS* 62 (1927), 104–105; Darnell, *Enigmatic Netherworld Books*, 477–8. Scharff suggested the stela came from Athribis, on the west bank across from Akhmim (*op.cit*. 88), as the goddess Repit (Triphis) is

depicted in the group of deities (*ibid*. 91–2). She was worshipped in her own temple at Athribis *(see chapter 20).*

2) It is summed up by Haarmann ('Islam and Ancient Egypt', 191): 'Any continuity from ancient to Islamic Egypt was irretrievably and doubly cut off, first by the adoption of Christianity in Egypt in the fourth century and then, three centuries later, by the Islamic conquest.'

3) See Abt, *The Great Vision of Muḥammad Ibn Umail*; Abt *et al.*, *Explanation of the Symbols* 2, 15–64; Berlekamp, 'Painting as Persuasion', 35–59.

4) For the Saqqara region as the location of the visit, see Stricker, *AcOr* 19 (1943), 101–37.

5) For the pharaoh's 'quest for the stone' in the 'New Year' ritual and alchemy, see Roberts, *Golden Shrine*, 51, 54, 61, 68, 79–80.

6) For these two phases of the work, see Abt *et al.*, *Explanation of the Symbols* 2, 58; Berlekamp, *op.cit*. 53.

7) See Abt *et al.*, *op.cit*. 39–40; Berlekamp, *op.cit*. 51.

8) This is stated in the inscription above the left side of the tablet: 'Water, air and fire. Therefore they drew it as three, to indicate thereby that it is one, within which is three…'; see Berlekamp, *ibid*. 52.

9) For 'lead-copper' in the work's first phase, see Abt *et al.*, *op.cit*. 33, 80–85, with *fig.27*. Abt includes many comparisons with ancient Egypt in his commentary, though not with copper's importance in pharaonic times.

10) Quoted *ibid*. 22, n.22.

11) For a summary of this pervasive view, see Roberts, *My Heart My Mother*, 218–19. For a general overview of the evidence, see also Hornung, *Esoterische Ägypten*, 40–48.

12) The Greek texts are edited (with a French translation) in Halleux, *Les alchimistes grecs 1: Papyrus de Leyde, Papyrus de Stockholm, fragments de recettes*.

13) The two Byzantine treatises are *Physika kai mystika* ('Natural and Secret Questions') and *Peri asēmou poiēseōs* ('On the Making of Silver'); see Martelli, *Four Books*, 1–2. Martelli argued against the identification of Democritus with the Egyptian Bolus of Mendes, and hence against the attribution of the *Four Books* to him (*ibid*. 36–44).

14) The quotation is from Mertens, 'Alchemy, Hermeticism and Gnosticism', 168. Cf. also Lindsay's statement (*Origins*, 124) that, whilst alchemy originated from craft recipes, there was 'the attempt to make an esoteric doctrine or philosophy out of the data of craft-recipes and the like.' Such statements are repeated continually in present-day literature on alchemy.

15) For alchemy as a 'pseudo-science', see Festugière, *Révélation* 1, 217, 238. This view of alchemists as prototypal chemists underpins Principe's book, which, apart from a brief reference to its artisans (*Secrets of Alchemy*, 9–10), noticeably excludes ancient Egypt, and also Ibn Umail, one of the chief transmitters of Egyptian alchemy.

16) Festugière, *op.cit*. 238. There is a long scholarly tradition of promoting the ancient Egyptians as 'non-philosophical' and 'practical', which was rightly criticized by Naydler, *Shamanic Wisdom*, 40–44 (with further references).

17) The Greek text was originally published in Berthelot and Ruelle, *Alchimistes grecs* 2, 239–46, and 3, 231–7; and re-edited by Festugière (*op.cit*. 363–8). See also *ibid*. 275–82, for a French translation. The treatise's Greek title is problematic, since the word 'count' in Greek can also be

interpreted as 'account', 'quittance, 'receipt', 'distance' and 'abstinence' (see Mertens, *Zosime de Panopolis*, lxv), hence the variations in its title in present-day publications.

18) For this passage see Festugière, *op.cit*. 275[1]. In his commentary on a now-lost work by Zosimus, Olympiodorus gives a version of this passage, which Letrouit ('Chronologie', 19–20) re-edited. In it he calls the two Egyptian arts 'the art of colouring' and 'the art of natural minerals'. He also refers to priestly knowledge: 'For the so-called divine art, in other words the doctrinal art, to which all those adhere who seek out all the manual procedures and the noble arts—I mean the four arts, those that seem to produce some effect—has been entrusted only to the priests.' For the two versions, see Martelli, *op.cit*. 45.

19) Festugière, *op.cit*. 277[2].

20) See Derchain, *CdÉ* 65(130) (1990), 233–4; English translation in Martelli, *Four Books*, 65–6.

21) For the mining of metals as a royal prerogative, see Aufrère, *L'univers minéral* 1, 316–21; Martelli, *op.cit*. 64–7. Aufrère (*op.cit*. 317–18), referring to Zosimus's account, described the Egyptian king as 'the great alchemist'. For the secrecy surrounding Egyptian metalworking, see also von Lieven, *SAK* 36 (2007), 147–55.

22) For early Jewish interest in metalworking, see Patai, *Jewish Alchemists*, 41–6. For foreigners in the Egyptian cults, see Vittmann, 'Beobachtungen', 1231–50. For Chaiapis (Haihep), see *ibid*. 1244–6; Roberts, *Golden Shrine*, 78, *fig.53*, 139, n.11. For Phoenicians at Memphis in Hellenistic times, see also Thompson, *Memphis*, 88–93. Vittmann (*op.cit*. 1241–4) also discussed a Yemenite trader called Zayd'il, who seems to have served as a priest in a Memphite temple during the early Ptolemaic period.

23) For the names of alchemists in Greek treatises as transcriptions of genuine Egyptian names, see Daumas, 'L'alchimie', 112–13.

24) When Dendara temple ceased functioning is unknown, but according to the Greek Heidelberg Festival Papyrus (*P. Heid. 1818*), dating to the second century CE, festivals for Isis, Bes and Seth-Typhon were still being celebrated in the region, though whether connected to the temple is unclear; see Frankfurter, *Religion in Roman Egypt*, 55–6, 67–8. For the dating of the original *Four Books* to the first century CE, possibly in the reign of Nero, see Martelli, *op.cit*. 29–31.

25) I have borrowed the term 'astrologers in reverse' from Bachelard, *Psychoanalysis of Fire*, 40, who refers to Novalis's idea that the miner is an 'astrologer in reverse'.

26) Jung, *Mysterium*, 342, with n.327 (para 483).

27) Quoted in Berthelot and Ruelle, *Alchimistes grecs*, 3, 33; Festugière, *Révélation* I, 259[6].

28) Quoted in Abt *et al.*, *Explanation of the Symbols* 2, 146.

29) *Ibid*. 255.

30) Emblem 6. See de Jong (ed.), *Atalanta Fugiens*, 81–7, with further references to alchemy and agriculture, and 382 (*fig.6*).

31) Cauville suggested a mud-brick building in the precincts of Hathor's temple at Dendara, which Daumas had previously identified as a sanatorium, might have been a dye works (*BSFE* 161 [2004], 28–40; Cauville and Ibrahim Ali, *Dendara*, 264–66). If so, in the light of the belief in dyes as 'remedies' in antiquity, a 'magical pillar' located there, inscribed with a healing magical text on one face

and a text describing the restitution of Osiris's body on another, is particularly interesting (*op.cit.* 39). For the name of the dyer in ancient Egypt, see Germer, *Textilfärberei*, 134–5.

32) For these texts describing the substitution of metals, see Derchain, *CdE* 65(130) (1990), 235–6. Derchain also noted their alchemical associations (*ibid.* 222–3, 232). The room's ritual function is unclear, though Waitkus (*Texte*, 267–8) suggested an 'Opening of the Mouth' rite might have been performed there for the cult statues being taken to the roof. For a similar view, see also Cauville and Ibrahim Ali, *op.cit.* 213.

33) For discussion of these mummies, including Winlock's initial response, see Riggs, *JEA* 86 (2000), 121–44, with *pls.*16–19.

34) *Ibid.* 140–41.

35) *Ibid.* 141. Sezgin (*ZGAW* 16 [2004/5], 192) observed pagan religion must have continued for a long time in Upper Egypt, citing the baptismal rites performed on 'idol-worshippers' by an early seventh-century bishop from Armant.

36) For alchemy's secret as 'the mysteries of the queen', see Jung, *Mysterium*, 379, with n.429 (para 536), 381 (para 542).

CHAPTER 17

1) E.g. the Greek treatise *On the Philosophers' Stone* quotes Maria as saying: 'Philosophers have carried out the operations of the stone in four phases: blackening, whitening, yellowing and tincturing in violet' (see Berthelot, *Alchimistes grecs* 3, 194[2]). See also Hopkins, *Alchemy*, 92–9, for this colour sequence, and 115–17 for the shift to red in alchemy.

2) *Natural History*, Book XXXV: 32:50 (trans. Rackham, *Natural History* 9, 298–9).

3) For a proposed reconstruction of the alchemical *kerotakis* and its use, see Mertens, *Zosime de Panopolis*, cxxxv, *fig.9*.

4) Cf. the treatise *Of the Nature of Man*, attributed to Hippocrates (or his son-in-law, Polybus), which associates the four humours with blood (red), bile (yellow), bile (black) and phlegm, and also with the four seasons, and the universal qualities of cold, moist, warm and dry, a connection extended also to the four ages of the human being (see Klibansky *et al.*, *Saturn and Melancholy*, 8–10). Cf. also a fragmentary poem by a Hellenistic astrologer, Antiochus of Athens, which correlates the four seasons and signs of the zodiac (see Dronke, 'Tradition and Innovation', 63–4).

5) Democritus's colour theory does not survive in the original, but Theophrastus gives an account in his *De Sensibus*, 73–82. There is some uncertainty as to whether the fourth colour, called *chloron*, should be interpreted as 'yellow' or 'green'. For *chloron* as 'yellow', see Gage (*Colour and Culture*, 272, n.7). For the overlap between 'yellow' and 'green', see also Martelli (*Four Books*, 247[23]). Hershbell (*Ambix* 34 [1987], 9, 14) suggested Democritus's interest in colour theory could have contributed to his later connection with alchemy.

6) Dronke (*op.cit.* 64). Mertens (*op.cit.* cxxx–cxxxi) rightly urges caution about linking the alchemical *kerotakis* directly with Greek painting.

7) See Schenkel (*ZÄS* 88 [1963], 140–46), who argued the Egyptian language had only four basic colour terms: black (*km*), white (*ḥḏ*), 'red' (*ḏsr*) and 'green/blue' (*w3ḏ*).

8) The Byzantine chronicler George Syncellus and Synesius both provide accounts of Democritus's initiation; see Martelli, *op.cit.* 69. In Phoenician tradition, however, Ostanes appears as a transmitter of Thoth's wisdom (see Roberts, *Golden Shrine*, 150, n.3, with further references). Quack (*JNES* 65 [2006], 282) suggests Persians living in Egypt may have composed the texts written in the name of Zoroaster or Ostanes.

9) This *iosis* phase is notoriously difficult to define. For example, Martelli (*op.cit.* 244[13]) associates it with a way of treating metals to produce 'rust' on the surface, thus linking it with the word *ios*, meaning 'rust' or 'venom' in non-chemical works. According to Hermes, the ancients understood the stone's purple colour to mean 'the rust of copper' (Festugière, *Révélation* 1, 242[4]). For the several meanings attached to *iosis*, see Patai, *Jewish Alchemists*, 555, n.17. For *ios* as 'rust', see *ibid.* 554, n.11.

10) *Colour and Culture*, 25. Gage observed how words for 'purple' in early Greek texts seem to have connotations of 'movement' and 'change'. For the symbolism of purple in antiquity more generally, see *ibid.* 25–6.

11) Quoted in Abt *et al.*, *Explanation of the Symbols* 2, 154.

12) For this episode, see Martelli (*op.cit.* 82–5).

13) For the Egyptian tradition of finding secret knowledge, see Roberts, *My Heart My Mother*, 219–20. For the episode's possible Egyptian background, see Martelli (*op.cit.* 213[21a]), who cautiously concluded the original *Four Books* probably included this initiation in some form (*ibid.* 213–14[21b]). Quack (*op.cit.* 280, with n.97) noted a similar theme in an unpublished Egyptian papyrus in which a festival-goer recounts how he discovered a book of Imhotep, son of Ptah, in the temple of Heliopolis when a block of stone fell from the wall.

14) For these meanings of *pharmakon*, see Roberts, *Golden Shrine*, 63, 113, 148, n.12 (with further references). For a possible derivation of the Greek word from the Egyptian word *pekheret* meaning 'remedy', see *ibid.* 136, n.7. For the relationship between substances used by alchemists and those used by medical practitioners, see Martelli, *op.cit.* 242[9]. In *Book 5 of Democritus to Leucippus*, alchemical recipes for treating copper are compared with pharmaceutical procedures (Berthelot, *Alchimistes grecs* 3, 57–60). The date of this treatise is unclear, though its author is not thought to be the Democritus who wrote the *Four Books* (see Martelli, *op.cit.* 17).

15) For this treatise, *see above*, chapter 9, n.34.

16) For the 'divine water reviving flowers in Ibn Umail's work, see Abt *et al.*, *op.cit.* 94. On 'dyes' as 'flowers', see *ibid.* 106, 145, 149, 175; and for colours symbolizing the return of life after a state of death, *ibid.* 119.

17) Translation in Lichtheim, *Ancient Egyptian Literature* 3, 98. See also Roberts, *My Heart My Mother*, 194, for these waters associated with the ancestors.

18) For the *Gospel of Philip* and alchemy, see Roberts, *Golden Shrine*, 112–18 (with further references).

19) Ferguson (trans.), *Aurora*, 29; von Franz, *Aurora*, 94–5.

20) Copies of the treatise are preserved in much later manuscripts, the oldest being *Ms. Marcianus Graecus 299* (M), dating from the tenth or 11th century, where it is

entitled *Olympiodorus, philosopher of Alexandria, Commentary on the Book 'On the Operation' of Zosimus, and on the Sayings of Hermes and the Philosophers*. For a list of the known manuscripts, see Viano, 'Olympiodore l'alchimiste', 137. The Greek text was published in Berthelot, *Alchimistes grecs* 2, 69–104, and 3, 75–113 (French translation), with the title *On the Sacred Art*.

21) The debate centres on whether he is to be identified with the sixth-century Alexandrian philosopher Olympiodorus, or whether he lived earlier. Viano summarized the different views (*op.cit.* 100–102), and concluded there was not sufficient evidence to settle the question definitively.

22) Berthelot, *Alchimistes grecs* 3, 86–7[17–18].

23) *Ibid*. 93[30].

24) *Ibid*. 87[18].

25) *Ibid*. 94–5[31].

26) *Ibid*. 93–4, n.5. This version is preserved in *Ms. Laurentianus graecus* 86, 16 (known as L), now in the National Library in Paris, which was copied in 1492 by Antoine Dranganas (see Mertens, *Zosime de Panopolis*, xxxviii–xl). Neither the place where it was copied nor the origins of its copyist are known.

27) For the *Book of Day*'s fourfold structure, see Müller-Roth, *Das Buch vom Tage*, 462–3.

28) *De mysteriis* VII: 3. See Iamblichus, *De mysteriis* (trans. Clarke *et al.*), 294–7. Cf. also Hippolytus's statement that a Gnostic sect, the Peratae, worshipped Isis as ruler of the daytime 12 hours (Doresse, *Secret Books*, 273–4).

29) For Nekhbet and the White Crown, see Roberts, *Hathor Rising*, 75–7.

30) Assmann, *Sonnenhymnen*, 74–5 (Text 53, 6). For the symbolism of silver more generally, see Aufrère, *L'univers minéral* 2, 409–28.

31) Waitkus, *Texte*, 114 (northern frieze inscription).

32) Cairo Museum JE 46356; see Aufrère, *op.cit.* 2, 469.

33) For the terminology of gold and silver, see Harris, *Lexicographical Studies*, 32–50; Schorsch, *JEA* 87 (2001), 56.

34) As Schorsch noted, 'Gold for the ancient Egyptians clearly had some "quality" of redness' (*ibid*. 68), and she cited instances of 18th-Dynasty jewellery where copper had evidently been used to redden gold (*ibid*. 68–9).

35) For the mother goddess's changing colours in the *Pyramid Texts*, see Motte, 'La vache multicolore', 147–8. Thus, when Nut gives birth at dawn, so the sky becomes red like wine (*PT*§1082); when Re impregnates her, she becomes filled with luminous seed (*PT*§990a); and during her gestation she becomes 'green' (*PT*§569 a–b).

36) For this colour symbolism in hymns to the uraeus, see Troy, 'Mut Enthroned', 308–309.

37) Berthelot, *Alchimistes grecs* 3, 95[31].

38) Abt *et al.*, *Explanation of the Symbols* 1, 155 (lines 11–12). For references to 'one day' in European alchemy, see Jung, *Mysterium*, 335, with n.288 (para 472).

39) Berthelot, *op.cit.* 109–10[51]. For Manilius's correlation of the zodiac signs with the human body, see *Astronomica* 2, 453–65; *ibid*. 4, 701–709 (trans. Goold, 118–19, 278–9). *See above, n.14*, for alchemy and medicine.

CHAPTER 18

1) Eliade, *Forge and the Crucible*, 148–9.

2) For these 'washing' and 'uniting' processes, see Berthelot, *Alchimistes grecs* 3. 75–9[1–8]. For the ancient Egyptian

technique of washing alluvial gold, see Ogden, 'Metals', 161–2. For chrysocolla used in the 'soldering gold' process, *see chapter 19, nn.16, 17*.

3) For this section on copper-colouring processes, see Berthelot, *op.cit.* 81–5[12–15].

4) *Ibid*. 97–8[35].

5) For the Sulphur–Mercury theory in Arabic alchemy, see *Encyclopaedia of Islam* 5, 113.

6) *Ibid*. 114.

7) For this statuette and Cyprian sacred copper-working, see Karageorghis, *Kition*, 74–6, 171, with *pl.56*.

8) For various objects found on Cyprus depicting Hathor, see Sophocleous, *Atlas*, 124–37 (*pls.31–4*). For Aphrodite in Egypt, see *Lexikon der Ägyptologie* 1, col.337.

9) Quoted in Jung, *Mysterium*, 305, n.205.

10) Berthelot, *op.cit.* 196[11].

11) For these descriptions of malachite, see Finlay, *Colour*, 295.

12) Berthelot, *Alchimistes grecs* 3, 59[4]. For Ibn Umail's reference to the 'stone' as 'the chameleon', see Abt *et al.*, *Explanation of the Symbols* 2, 113, 118–19.

13) Berthelot, *op.cit.* 168[1], quoted also in Patai, *Jewish Alchemists*, 83. Stephanus of Alexandria says the same; see Taylor, *Ambix* 1 (1939), 129–31. Agathodaimon is frequently cited as an authority in Graeco-Egyptian alchemy (Lindsay, *Origins*, 301–22).

14) *Papyrus Harris* 1, 78, 1–5. See Daumas, 'L'alchimie', 116–17. Cf. also the copper vessels called *bi3* coloured yellow, with handles of silver, listed in a New Kingdom inventory of objects (*Urk.* 4. 638). Harris (*Lexicographical Studies*, 59) thought the yellow colour strange for an 'iron' vessel (*see chapter 13, n.14*), but here it probably indicates the prized gold-coloured 'copper'.

15) For copper and red gold in ancient Egypt, *see chapter 17, n.34*. For the tinting of copper in African metallurgy, see Herbert, *Red Gold*, 78. Traditional African smiths particularly prize the red, not gold, colour of copper, and use various techniques to obtain this redness (*ibid.* 78). Herbert (*ibid.* 164–5) noted that the light colour of Sudanese copper, probably due to the presence of gold in it, may have detracted from its value.

16) For *pneuma* as the colouring agent, see Lippmann, *Entstehung* 1, 39. Papathanassiou (*Chrysopœia* 7 [2000–2003], 25–6) noted the different epithets given to *pneuma* in Stephanus's work. It is 'vegetal', 'vital', 'psychic', 'ardent' and 'tinctorial', and associated also with the healing, tincturing *pharmakon*, or 'poison'. For life-giving *pneuma* animating bodies in the alchemically inspired *Gospel of Philip*, see Roberts, *Golden Shrine*, 115.

17) Berthelot, *op.cit.* 129[5], quoted in Jung, *Psychology and Alchemy*, 295 (para 405).

18) Quoted in Martelli, *Four Books*, 245–6[17]. This passage is discussed by Papathanassiou (*op.cit.* 23–4).

19) Maria's quotation is preserved in *The Four Bodies are the Nourishment of the Tinctures* attributed to Zosimus (see Berthelot, *op.cit.* 169[2]; Festugière, *Hermétisme*, 239). It belongs in a group of texts known as the *Chapters to Eusebia* (see Mertens, *Zosime de Panopolis*, lviii[x], who classifies it as a work of a later epitomist selecting the most interesting passages from Zosimus's writings [*ibid*. lx]). For copper being associated with human beings in Islamic alchemy, see Abt (ed.), *Book of Pictures* 2, 293, 302.

20) Berthelot, *op.cit.* 124. The alchemist Pelagius also directly quotes Zosimus about copper as a fruitful tree; see *ibid.* 250[12].

21) On copper's 'nourishing' qualities in African metallurgy, see Herbert, *op.cit.* 51. For dyeing as a nutritive process, see Roberts, *Golden Shrine*, 113. Lindsay (*op.cit.* 238–9) observed how the work of the kitchen underlay much metallurgical and dyeing terminology, and also influenced the development of alchemical apparatus.

22) See Mertens, *op.cit.* 23[2]. For various references to Memphis associated with astrology and alchemy in antiquity, see *ibid.* 188–9.

23) Papathanassiou, *op.cit.* 24, with n.84.

24) *Ibid.* 24, with n.85.

25) Book 12 [11], [16]; Copenhaver, *Hermetica*, 45[11], 47[16]. Cf. Plato's view of the soul's 'aliveness' being defined by self-movement (*Laws*, 895e–6a).

26) Quoted in Abt *et al.*, *Explanation of the Symbols* 2, 46.

27) *Ibid.* 1, 7 (line 1); 2, 77, and, for 'lead-copper' in Islamic alchemy, 78–85.

28) For this episode in the *Book of Krates*, see Berthelot, *Chimie au moyen âge* 3, 61–4; Kahn, *Hermès Trismégiste*, 118–22; Roberts, *Golden Shrine*, 121–2.

29) Rūmī, *Masnavi* 2. 3255–6; Nicholson (trans.), *The Mathnawi*, 390. Cf. also the similar statement by the tenth-century Andalusian Sufi, Ibn Masarrah, in his *Epistle on Contemplation*, that, whilst the female soul sustains the motion of life, it does not possess Intellect, see Stroumsa and Sviri, *JSAI* 36 (2009), 222[27–8]. A. Williams is preparing a new edition and translation of the *Masnavi*, including Book 2, for publication with I.B. Tauris.

30) Schimmel, *As Through a Veil*, 64. Ibn Umail refers to copper as 'mother of the gold'; see Abt *et al.*, *Explanation of the Symbols*, 2, 79. Cf. also Democritus's instruction to Zosimus to 'turn copper into gold' in the Arabic *Book of the Keys to the Art* (*ibid.* 84, with n.173).

31) It is summed up in these lines: 'And how can alchemy be seen if copper's/low-grade, inferior nature is not known?/ Deficiencies are mirrors of perfection;/the vilest things are mirrors of His glory.' (*Masnavi* 1. 3222–3; Williams [trans.], *Spiritual Verses*, 296.) For the lower soul being associated with the feminine and base instincts in classical Sufism, see Schimmel, *op.cit.* 152.

32) Herbert, *Red Gold*, 282–3. These copper spirals are also associated with 'water' and the 'Word'.

33) *Ibid.* 283. For copper and water in a Kuba genesis myth, *see chapter 15, with n.9.*

34) See Falchetti, 'Seed of Life', 351–3.

35) For composite bodies being associated with change and movement, see Book 12 [14]; Copenhaver, *Hermetica*, 46[14].

36) For the transformations of the *tetrasomia* as the 'four in one', see Lippmann, *Entstehung*, 1, 38, 344; Abt *et al.*, *Explanation of the Symbols* 2, 90–93.

37) For copper–tin and copper–lead alloys in ancient Egyptian metallurgy, see Ogden, 'Metals', 153–5. Traces of arsenic also occur (*ibid.* 152–3).

38) Sermon 41. See Ruska, *Turba*, 148, 228. Cf. also the significance of four in the indigenous copper tradition of Vancouver Island, where four beads are attached to each end of the death cord because 'four is a full number, a true number' (Cameron, *Daughters of Copper Woman*, 181).

39) Berthelot, *Alchimistes grecs* 3, 104–105[44]. Olympiodorus mostly steers away from negative Gnosticism, but his description here of the 'unfortunate one, fallen and chained in the body of the four elements' seems tinged with Gnostic ideas of the body as a prison. For the fourfold nature of humankind's earthly, or material, part in the *Asclepius*, see Copenhaver, *op.cit.* 70[7], 221, n.7. For copper marking transitions in the human lifecycle from child to ancestor in African societies, see Herbert, *op.cit.* 266–76, 279.

40) In the non-canonical *Gospel of Mary*, Jesus teaches a similar reversion of material forms back to their origins: 'All of nature with its forms and creatures exist together and are interwoven with each other. They will be resolved back, however, to their own proper origin, for the compositions of matter return to the original roots of their nature.' See Robinson (ed.), *Nag Hammadi Library*, 471–4; Bourgeault, *Mary Magdalene*, 46, 48. Bourgeault compares this passage with the teachings of Empedocles and Parmenides (*ibid.* 247, n.6). Cf. however, the ageing sun god's reversion to his material origins at the beginning of the ancient Egyptian *Myth of the Destruction of Humanity*, becoming like a metallic cult image, 'his bones silver, his flesh gold, his hair true lapis lazuli', as he prepares to enter the waters of Nun for renewal (Lichtheim, *Ancient Egyptian Literature* 2, 198). What is later expressed philosophically is here expressed mythically.

41) Quoted in Berthelot, *Alchimistes grecs* 3, 105[44].

42) Lippmann (*op.cit.* 344) noted how early alchemists associated this completion with nine hours or with a year, time periods in accordance, therefore, with ancient Egyptian notions of 'second birth' in the daytime cycle (not mentioned by Lippmann). The poet and philosopher Samuel Taylor Coleridge encapsulates this alchemical process in a letter to his Bristol publisher dated 7 March 1815: 'The common end of all narrative, nay, of all, Poems is to convert a series into a Whole: to make those events, which in real or imagined History move on in a strait Line, assume to our Understandings a circular motion—the snake with it's [*sic*] Tail in it's [*sic*] Mouth.' See Griggs (ed.), *Collected Letters* 4, 545. I am grateful to Jack Herbert for this reference.

43) For the four separate elements transforming into the unified 'three', see Jung, 'Psychology of the Transference', 207–208.

44) Quoted in Mertens, *Zosime de Panopolis*, 40[6]; also in Jung, *Psychology and Alchemy*, 386 (para 472).

45) Maria's words are quoted by Zosimus in the *Book of Sophe*. See Berthelot, *op.cit.* 206; Patai, *Jewish Alchemists*, 89(2).

46) Quoted in Berthelot, *op.cit.* 105[44]. For gold as the 'red blood' of silver in Greek alchemy, see Jung, *Mysterium*, 15, n.52 (para 11).

47) On copper's association with blood, the female and transformation in the Desana copper-working communities of the northwest Amazon region, see Falchetti, 'Seed of Life', 350. For copper and blood in traditional African metallurgy, see Herber, *Red Gold*, 279, 281.

48) *Psychology and Alchemy*, 386 (para 472). Jung also notes how the alchemical symbol of the work, 'the squaring of the circle', involves breaking down 'the original chaotic

unity into the four elements' and then recombining them into a 'higher unity'. This unity 'is represented by a circle and the four elements by a square' (*ibid*. 124 [para 165]).

49) Quoted in Abt *et al.*, *Explanation of the Symbols* 2, 227. Abt here traces this egg symbolism back to ancient Egypt.

50) Rūmī, *Selected Poems*, 279.

51) For this whirling dance, see Schimmel, *As Through a Veil*, 132–3.

52) It has even been debated whether the Leiden and Stockholm papyri should be regarded as alchemical papyri at all or simply the 'technical' recipes of artisans. See Halleux, *Alchimistes grecs* 1, 24–30.

53) Eliade, *Forge and the Crucible*, 182–3. Eliade's study was based on iron and ferrous ores, and completely excluded copper, because he considered 'one cannot speak of an Age of Copper, for only minute quantities of the metal were produced' (*ibid*. 25). Inevitably, his account of alchemy had to omit ancient Egypt, since, as he recognized, iron technology was only introduced there at a late date (*ibid*. 22–3).

CHAPTER 19

1) The *Book of Sophe* is preserved in two manuscripts: *Ms. Parisinus graecus 2327*, now in the National Library, Paris, and, according to its colophon, copied in 1478 on the island of Crete by Theodore Pelekanos from Corfu; and *Ms. Laurentianus graecus 86*, 16, now in Florence, copied in 1492 by Antoine Draganas, though where is not known. For details of the manuscripts, see Mertens, *Zosime de Panopolis*, xxxi–xliii. For Sophe as the Greek form of Khufu's name, see *ibid*. lxvii. Various works attributed to Khufu circulated in the third century CE (*ibid*. lxvii–lxviii).

2) Panopolis belonged to the Thebaid division of Egypt, which probably explains why Zosimus is called 'the Theban' in the treatise's title (rather than 'from Panopolis'). Only the title of the first part specifically names him, which has led to uncertainty about whether he or Democritus might be the author of the second part. Mertens wavers between attributing the *Book of Sophe* to Zosimus ('Alchemy, Hermetism and Gnosticism', 167) and a more cautious approach (*Zosime de Panopolis*, lxviii–lix, with n.201). The second part's reference to both Egyptian and Jewish wisdom is perfectly consistent, however, with his interest in esoteric Judaism elsewhere, cf. *On the Letter Omega*, where he states that the primordial man is called Adam by the Jews and others, and 'called by us Thoth' (see Mertens, *ibid*. 5[9]).

3) The treatises 'framed' by the *Book of Sophe* are the *Final Count*, the *Letter of Isis to her Son Horus* and *Book 5 of Democritus to Leucippus* (which deals specifically with the 'arts of the Egyptians', especially preparing copper for tincturing purposes). For the list of the treatises, see Mertens, *ibid*. xxxvi, xl. Interestingly, these texts are not included in two other Byzantine manuscripts preserving Greek alchemical texts, and it seems, from the fact that they are grouped together, that the copyists knew they all stemmed from the same Egyptian 'copper' tradition.

4) Zosimus uses the word *chemeia* here to characterize alchemy, though the etymological origins of the word 'alchemy' itself, as it is known today, are controversial. It appears in medieval Latin texts in various forms—*alkimia, alquimia, alchimia, alchemia*—being composed of the Arabic definite article 'al' and an uncertain root word. Some have traced the latter back to the ancient name for Egypt as the 'Black Land' (*Kemet, KHME* or *XHMI* in Coptic), others to the Greek word *cheo*, meaning to 'melt' or 'smelt' metals, in which case alchemy's origins would derive primarily from metallurgical practices; see Lindsay, *Origins*, 68–89; Halleux, *Textes alchimiques*, 45–7; Principe, *Secrets of Alchemy*, 23–4.

5) For this text, see Festugière, *Révélation* 1, 261, with n.2.

6) *Ibid*. 261–2. See also Lindsay, *op.cit*. 182–3.

7) For this Greek belief, see Roberts, *My Heart My Mother*, 229, n.2 (with further references).

8) Berthelot, *op.cit*. 3, 196[13]. See also Patai, *Jewish Alchemists*, 66.

9) For this passage, see Taylor, *Ambix* 2 (1938), 41; Papathanassiou, *Chrysopœia* 7 (2000–2003), 18; Martelli, *Four Books*, 249[33]. Whether Stephanus the alchemist was the same person as the Neoplatonic philosopher Stephanus, who was active in both Alexandria and Constantinople, is debated. Papathanassiou (*op.cit*. 11, 12, n.5) concluded he was, contrary to Letrouit ('Chronologie', 59–60).

10) Cf. also Zosimus's quotation from Maria the Jewess in the Arabic *Book of Pictures*: 'If copper is burnt and then put back, and if this is done several times, it becomes better than it was before' (Abt [ed.], *Book of Pictures* 2, 481). See also *ibid*. 497, for copper becoming 'better than gold without a shadow'.

11) See Telle (ed.), *Rosarium* 1, 20, and 2, 23–4[20] (my translation from the German). Cf. also the 'Green Lion' text in the same treatise: 'Our copper is namely not the usual copper; the usual copper is perishable and corrodes everything with which it is placed. The copper of the philosophers, however, perfects and colours that white one (the stone), with which it unites. Therefore Plato says: "All gold is copper, but not all copper is gold." Therefore our copper has a body, soul and spirit, and these three are one, because they are all out of one, from the one, and are united as one, which is its root. The copper of the philosophers is therefore their elixir, from spirit, body and soul, completed and perfected' (see *ibid*. 1, 176–7, and 2, 148). Clearly, the *Rosarium*'s alchemy is not simply based on the Sulphur–Mercury theory, as often stated (e.g. Telle, *ibid*. 2, 177; Principe, *op.cit*. 75), but the assumption that it is based on a 'pure quicksilver theory', which supplanted the old Sulphur–Mercury theory (see Gamper and Hofmeier, *Alchemische Vereinigung*, 79–80), completely misses copper's importance in this text. *See also chapter 23, n.32*. For the association of mercury and copper in the Arabic *Epistle of the Secret, see below, chapter 21, n.8*.

12) For alchemy being defined as 'goldmaking', see, for example, Principe, *op.cit*. 13.

13) See Berthelot, *op.cit*. 3, 103[42]. Cf. a bronze statuette of Osiris incised with alchemical symbolism now in the Chambéry Museum in France; see Ratié, *Revue du Louvre* (1980), 219–21. Ratié suggested a fourth-century date, and concluded it was probably not a fake.

14) Quoted in Berthelot, *op.cit*. 3. 103–104[43]. I am grateful to the late Dr Ulla Jeyes for help with the Greek text. For these oracles, see also Roberts, *Golden Shrine*, 80–82.

15) Kingsley refuted the scholarly dismissal of Olympiodorus as a charlatan (*Ancient Philosophy*, 61, with further

references), noting his 'self-conscious principle of composition' when discussing the Presocratics in his commentary on Zosimus's treatise (*ibid.* 64).

16) Pliny describes a glutinous gold-solder (chrysocolla), especially used by goldsmiths, which was made from Cyprian copper verdigris and the urine of a boy who had not yet reached puberty, with the addition of soda (*Natural History*, Book XXXIII: 29:93, trans. Rackham, 71–2). See also Martelli, *op.cit.* 224[47].

17) For chrysocolla used as a solder in ancient Egypt, see Roberts, *Golden Shrine*, 82, 140, n.23. Stapleton *et al.* (*Ambix* 3 [1949], 85) suggested chrysocolla's use in goldmaking, as mentioned in Greek alchemical texts, might have had an ancient Egyptian origin, going back to methods used in the temple workshops.

18) Quoted in Jung, *Mysterium*, 377 (para 533).

19) For the lower sun's 'suffering' facial expression, see Sherwood, *Spring* 74 (2006), 246–7. Obrist associates the three faces on *Aurora*'s coat of arms in the Prague manuscript with the moon (*Débuts*, 216, with *ill.*38). But cf. the solar eagle with three heads shown in the *Book of the Holy Trinity*, described as 'God, light eternal, black sun with three necks', which Obrist associates with Saturn as the black sun (*ibid.* 149, with n.156). See also Birkhan, *Lehrdichtung* 1, 217–18, for Saturn's three faces.

20) Trismosin, *Splendor Solis*, 95.

21) *Ibid.* 21.

22) Obrist, *op.cit.* 180, n.295, as quoted from *Nuremberg Ms.* 80061, fol.26va.

23) For this Venus text in the Nuremberg manuscript, see Ganzenmüller, *Buch der heiligen Dreifaltigkeit*, 242, n.43 (my translation from the German). Ganzenmüller briefly noted the connection with copper (*ibid.* 255, n.91).

24) Dronke, 'Tradition and Innovation', 84.

25) Hildegard's concept of 'greenness' (*viriditas*) is often said to be specifically hers. See, however, Ploss *et al.* (*Alchimie*, 89–90), who noted that, whilst she never used the word 'alchemy', she referred to the '*opus*' throughout her writings. They also cited her emphasis on the human being as the '*opus plenum*' responsible for nature, and her references to Mercurius and other philosophers. (*See also Part 5* for Hildegard's influence in *Aurora consurgens*.) For 'greenness' in Ibn Umail's work, see Abt *et al.*, *Explanation of the Symbols* 2, 247. Ibn Umail associates it with the 'soap of wisdom' and the green substance growing 'out of copper' (verdigris). He also compares the 'divine water' with the 'green woodpecker' (*ibid.* 253). Arnald of Villanova refers to the 'greenness' in gold that makes it 'true gold' not 'common gold' (see Jung, *op.cit.* 432 [para 624]).

26) *Scivias* (trans. Hart and Bishop), 123 (first part, vision 4, section 25).

27) For Plutarch's Egyptian 'third love', see *Moralia* 9 (sections 764–5D). For Plutarch's association of Aphrodite with the moon, and the 'passive' partner of the 'active' male sun, see Roberts, *Golden Shrine*, 82–3, 140, n.26.

28) Plutarch, *op.cit.* section 762A (lovers returning from Hades).

29) *Ibid.* section 765B–C.

30) *Asclepius* [21]; Copenhaver, *Hermetica* 79[21]. This passage is placed quite deliberately between Hermes's description of the fecund, androgynous, creator god

(*Asclepius* [20]) and a passage about the twofold nature of human beings, both divine and mortal, able to span the realms (*Asclepius* [22]). Cf. also Dorn's description of the alchemical marriage as the hidden 'third thing' uniting heaven and earth, 'partaking of both their extremes … one thing out of three', thus restoring the 'one world' (*unus mundus*) that existed at the beginning of creation; see Jung, *Mysterium*, 462.

31) For the *Emerald Tablet*, see Roberts, *My Heart My Mother*, 219–22 (with further references); Principe, *Secrets of Alchemy*, 30–32.

32) Quoted in Taylor, *Ambix* 2 (1938), 43.

33) Hillman, *A Blue Fire*, 302.

34) Festugière, *Révélation* I, 279[6].

35) For green, yellow and red as the alchemical colours of copper and Aphrodite, see Berthelot, *op.cit.* 1, 83[6]. Festugière (*op.cit.* 279, n.1) compared these colours with Democritus's statement about the 'multicoloured chameleon', without noting that Democritus associated the chameleon with copper (*see chapter 18, n.12*).

36) See Berthelot, *Chimie au moyen âge* 2, 226[19].

37) See Taylor, *Ambix* 1 (1937), 129. For the soul understood as a 'cloud' rising up from a purified, distilled substance, see Martelli, *Four Books*, 245[15].

38) Taylor, *Ambix* 2 (1938), 45.

39) See Festugière, *op.cit.* 280[8]; English translation in Fowden, *Egyptian Hermes*, 122–3.

40) For Zosimus's alchemy as an inner work of perfecting the soul, see Festugière (*op.cit.* 282). Principe completely debunked the idea of alchemy as 'self-transformation' (*op.cit.* 94–106), calling it a modern association 'simply not supported by the historical record' and 'now rejected by historians of science as valid descriptions of alchemy' (*ibid.* 104–105). He briefly cited the *Final Count* (here called the 'Final Account') to illustrate Gnostic influence on Zosimus's alchemy (*ibid.* 20), without, however, mentioning the Egyptian 'arts' or copper.

41) Mertens noted that alchemical ideas of *pneuma* and the transmutation of metals underlie Theosebia's spiritualization here ('Alchemy, Hermetism and Gnosticism', 168). Fowden (*op.cit.* 123) mistakenly separated Theosebia's soul work from 'technical' procedures, interpreting the latter as simply a preliminary stage in the soul's purification. However, as Kingsley rightly pointed out, what Zosimus is saying here is that 'the procedures of conventional alchemy are to be reinterpreted as, themselves, symbolizing the purification and perfection of the soul' (*Ancient Philosophy*, 390, n.56).

42) Festugière, *op.cit.* 281[8].

43) For this Gnostic aspect of Zosimus's alchemy, see Mertens, 'Alchemy, Hermetism and Gnosticism', 170–5. Papathanassiou (*Chrysopœia* 7 [2000–2003], 15, with n.27) detected a similar detestation of the body in Stephanus's work.

44) *Asclepius* [6]; Copenhaver, *Hermetica*, 69[6].

45) Book 1[18]; *ibid.* 4[18].

46) Cf. the tirade, in the treatise *Rosinus ad Sarratantam episcopum*, against Venus, who not only corrupts metals, but also, unlike the clean male seed, weakens the child in her 'corrupt' womb, making it 'leprous and unclean' (quoted in Jung, *Mysterium*, 304–305).

47) For the problems associated with the Byzantine transmission of Zosimus's alchemy and how little is

left of his vast work, see Mertens, 'Graeco-Egyptian Alchemy', 209–15.

48) For colouring images with black copper, see the Syriac Sixth Book attributed to Zosimus, *On the Working of Copper* (Berthelot, *op.cit.* 2, 223[2]); for his interest in talismans, see the 12th Book, *On Electrum* (*ibid.* 264–6[5]).

49) *Ibid.* 226.

50) On 'love' (i.e. copper) as the first principle and binding agent for all substances in the work of Raymond Lull, see Pereira, *Alchemical Corpus*, 68. In another (pseudo)-Lull manuscript, mercury is the required single matter (*ibid.* 73). See also Principe (*Secrets of Alchemy*, 64–9) for alchemists working with mercury as the first principle.

CHAPTER 20

1) For this description of Theosebia, see Fowden, *Egyptian Hermes*, 125.

2) *Ibid.* 125–6, 167.

3) Quoted in Festugière, *Révélation* 1, 279–80, with n.6.

4) See Berthelot, *Alchimistes grecs* 3, 186–8[7–8]. In the Arabic *Book of Pictures*, Zosimus refers negatively to Nilus as 'the priest'; see Abt (ed.), *Book of Pictures* I, 21, with nn.35, 36.

5) Bachelard, *Psychoanalysis of Fire*, 53. For women in medieval alchemy, cf. the partnership between Nicolas Flamel and his wife, Pernelle. Birkhan noted Gratheus's treatise was addressed to both men and women, hence indicative of an inclusive alchemy (*Lehrdichtung* 2, 85 [line 1410], 101 [line 1588]). A couple are also represented working together in the *Mutus Liber* (*ibid.* 1, 191).

6) The text has been edited, with an extensive commentary and German translation, by Vereno, *Studien zum ältesten alchemistischen Schrifttum: Auf der Grundlage zweier erstmals edierter arabischer Hermetica* (cited hereafter as *Studien*). For Theosebia's family members, see *ibid.* 136[1].

7) *Ibid.* 138[18]. The names of Greek alchemists in Arabic manuscripts are sometimes difficult to recognize, and here the letter writer's name is *Amtūṯāsiya*. For her identification with Theosebia, see *ibid.* 246, with n.274. For various renderings of Theosebia's name in Arabic texts, see Mertens, *Zosime de Panopolis*, lxxx, n.246.

8) In the Chester Beatty version of the *Epistle*, Hermes is called 'Hermes of Būdašīrdī', though variants of the latter have 'Būdašīr'. Vereno associates this with Busiris, an important cult centre of Osiris in the Delta (*op.cit.* 241–6). Sezgin, in in her review of Vereno's publication, suggested it might be an Arabic form of 'Petosiris', referring either to the illustrious high priest of Thoth, Petosiris, (fourth century BCE), whose tomb at Tuna el-Gebel was a well-known place of pilgrimage, or to the renowned astrologer Petosiris (*ZGAW* 12 [1998], 364–5).

9) Vereno's edition is based mainly on a manuscript in the Chester Beatty collection in Dublin (*Ms. 3231*), dated in the colophon to the year 907/1501–2 (*op.cit.* 158[178]), which he collated with *Ms. Bursa, Umumi* 813, dated 959/1551–52. He suggested the original text was composed sometime between 850 and 960 (*ibid.* 36).

10) For details of its discovery, see *ibid.* 136[2–7]. The text is said to have been written in the *Musnad* script. According to Ibn an-Nadim, the tenth-century Islamic bibliographer in Baghdad, this was the script of the Himyarites in Yemen and also the script in which the 'ancient sciences' were

written in the Egyptian temples (see *ibid.* 260, with n.342; also El Daly, *Egyptology*, 64–5).

11) Vereno, *op.cit.* 140[31]. In the Arabic *Book of Pictures*, Zosimus tells how he dreamed he brought the head of a murdered man to Theosebia, which she then buried in her tomb (see Abt, *op.cit.* 2, 216–17).

12) Sezgin (*op.cit.* 363–4) pointed out the author's familiarity with actual burial practices at Akhmim during the Roman period.

13) See Vereno, *op.cit.* 250–51, 279–80.

14) *Ibid.* 282–4.

15) *Ibid.* 337–9. Sezgin, however, doubted the *Epistle*'s link with Gnosticism (*op.cit.* 353–4), observing its quest was for 'knowledge', not a Gnostic hope for a redeemer or return to God.

16) Vereno, *Studien*, 11.

17) For this view see Lory, *Alchimie*, 9.

18) Since 2003 extensive archaeological work has been carried out at Athribis by a joint German–Egyptian mission and the work is still ongoing. For the publication of the temple, see Leitz *et al.*, *Athribis* 2.

19) For the sistrum-shaking scene, see *ibid.* 2/1, 192 (C3, 40), 195. For the text of the fleeing Eye goddess and Re as a blind god, see *ibid.* xxv, with n.71, and 350 (C5, 40–42).

20) See *ibid.* 2/2, 482; text C3, 106. For the tradition of textile production at Akhmim, see *ibid.* 2/1, xxxii. At some point in late antiquity, after the closure of Repit's temple, workshops for the dyeing and weaving of textiles were installed there, though when exactly is unclear.

21) *Ibid.* 2/2, 436 (C2, 51, and C2, 57 = day hours), and 476 (D1, 26 and D1, 30 = night hours).

22) See *ibid.* 2/2, 523 (Text D2, 1), also 2/1, xxxiv. For the importance of the divine child's birth in the temple, see *ibid.* 2/1, xxix. For suckling scenes, see *ibid.* 2/3, 102 (D1, 2), and 105 (D1, 7).

23) For Min's epithet 'he who sees gold', see *ibid.* 2/1, xxii–xxiii; and see *ibid.* xii, for a string of epithets characterizing the god's virile nature.

24) *Ibid.* 2/1, 259 (Text C3, 102). Only the lower half of this scene remains. For Min as a self-regenerating fertile god associated with Amun-Bull-of-his-Mother and Isis in New Kingdom Thebes, see Roberts, *Hathor Rising*, 82–6.

25) *P. Fouad 80*. See Smith, *Following Osiris*, 429, with n.57. Greek documentary texts from the same era mention temples, priests and priestesses at Panopolis, including a temple of Amun; see *ibid.* 429, with n.53.

26) For Ibn Umail's reference to the *Epistle*, see Vereno, *Studien*, 186; *also below chapter 24, with n.19.*

27) The problems surrounding the transmission are evident in Birkhan's publication of Gratheus's 14th-century untitled treatise (*Lehrdichtung* 1, 153–89). He recognized the influence of Arabic alchemy (*ibid.* 164) and suspected Hermetic influence (*ibid.* 75), but was unable to find any sources. In fact, the *Epistle of the Secret*, which Vereno published in the same year as Birkhan's edition, offers direct parallels to Gratheus's text. The attention now being given to Hermetic Arabic treatises in the *Corpus Alchemicum Arabicum* series, edited by Abt and Madelung, should mean the transmission begins to stand on firmer ground. Cf. for example, Abt's discussion of a dialogue between Zosimus and Theosebia in the Arabic *Book of Pictures*, which resurfaces in the Latin treatise

295

Rosinus ad Euthiciam and influenced the *Rosarium philosophorum*; see Abt (ed.), *Book of Pictures* I, 59–63. This illustrated Arabic manuscript is now in the library of the Istanbul Archaeology Museum (*Ms.1574*), and its colophon states it was copied in 668/1270, evidently in Egypt, since both the Islamic and the Coptic month names are included. According to Abt, it is an Arabic translation of an authentic Greek work, written by Zosimus himself, or by Theosebia, or a disciple (*ibid*. 2, 73–138). By contrast, in his review of Abt's publication, Hallum argues it was composed after 900 as a later Islamic compilation of texts written by Zosimus (*Ambix* 56 [2009], 76–88).

28) *See chapter 24, n.2,* for the transmission associated with Marianus/Morienus.

29) For the influence of 'Frau Minne' (Lady Love) in medieval European spirituality, see Newman, *God and the Goddesses*, 10–12, 138–89, 291 (with further references).

30) Quoted in Abt *et al.*, *Explanation of the Symbols* 1, 95 (line 8). Cf. also the description of alchemy as 'a special transmission outside of dogma' in Chinese alchemy; Cleary (trans.), *Secret of the Golden Flower*, 10[6], 138.

31) Berthelot, *Alchimistes grecs* 3, 105[45, 46].

32) For these principles, see Vereno, *op.cit*. 140[32–45].

33) Quoted in Berthelot, *op.cit*. 58[1]. It is not thought Democritus was the author of *Book 5, see chapter 17, n.14*

34) For Ibn Umail's statement about reddened copper as the goal of the work, see Abt *et al.*, *op.cit*. 1, 29 (lines 15–16).

35) From the British Library's 15th-century manuscript of *Donum Dei* (*Ms. Sloane 2560 f.8*). I am grateful to Jeremy Naydler for bringing this treatise to my attention and to Bronac Holden for kind permission to include her English translation. See also Gabriele (*Don de Dieu*, 28–9), for a later varaint of this text. Cf. also the *Turba philosophorum*'s reference to the transformation of copper (Sermon 32) being associated with the death and regeneration of humans, and the appearance of the colours (Ruska, *Turba*, 139–40, 217–18). Ruska called it a rare reference to metal transformation and the powers ascribed to it (*ibid*. 218, n.5), though it is paralleled here in *Donum Dei*.

36) For copper being symbolic of 'life' in the Middle Ages, see Pastoureau, 'Couleur verte', 156.

37) For this correlation with the zodiac, see Vereno, *Studien*, 142[42–45].

38) Mertens, *Zosime de Panopolis*, 39–40[5], from the first lesson *On Excellence*.

39) British Library *Harley Ms. 3469*. For the grouping of *Splendor Solis* with *Aurora consurgens*, see Obrist, *Débuts*, 256, with n.9. For its 22 illustrations and their correspondences with the ancient Egyptian night journey of death and regeneration, see Roberts, *My Heart My Mother*, 205–17. Like Zosimus's death and regeneration 'visions' *(see chapter 15, n.15)*, it seems to be based on a 'copper to silver to gold' transformational sequence. Thus, the 'copper' (or 'Venus') phase is symbolized by the fruit-bearing tree of life in illustration 6, which includes a scene of naked women bathing beneath it; the 'silver' phase culminates in the depiction of the moon (illustration 19); and the final 'gold' phase is shown in the concluding scene of the rising sun.

40) Cf. medieval illustrated *Books of Hours*, which sometimes show the Infancy cycle of Christ (his conception, birth and rearing) accompanied by episodes from his Passion, synchronized also with the monastic Divine Office through the day; see Wieck, *Time Sanctified*, 60–72. I am grateful to Jeremy Naydler for this reference.

41) See Fowden, *Egyptian Hermes*, 66–7. For the difficulties of recognizing transformed Egyptian material in early Christian sources, see also Roberts, *Golden Shrine*, 94, 101–103, 107–108, 142–3, n.21 (with further references).

42) See Fowden, *op.cit*. 66. Kingsley (*Ancient Philosophy*, 208–10) observed how Syriac, Islamic and medieval writers were far more comfortable with a mythical vision of the world than Greek writers steeped in Platonic and Aristotelian philosophy.

43) For Syrian elements in the text, see Sezgin, *ZGAW* 12 (1998), 363. For Islamic knowledge of Dendara, see El Daly, *Egyptology*, 31, 40, 118–19.

44) Vereno's view that the *Epistle of the Secret* is essentially a reworked Islamic version of much earlier Greek texts (*op.cit*. 332, 339) needs to be treated with caution, as Sezgin rightly pointed out (*op.cit*. 355, 366). Whilst she agreed its origins predated the Islamic period (*ibid*. 366), she thought it too limiting to cite only Greek (and possibly Syriac) compositions as sources (*ibid*. 355), and, judging by her brief comments, evidently thought ancient Egyptian beliefs still known to Copts could have contributed to the transmission (*ibid*. 365). She also noted (with further references) that ancient Egyptian elements are much more detectable in Coptic Hermetic treatises from the Nag Hammadi cache than in the Greek *Corpus Hermeticum* (*ibid*. 355–6). For the Copts as keepers of ancient wisdom, see El Daly, *op.cit*. 26, 62–3. Frankfurter (*Religion in Roman Egypt*, 216–17) noted the exchange of knowledge between Jewish, Coptic and Islamic ritual specialists in Egypt.

45) Vereno, *op.cit*. 339.

46) Sūra 22:5–7, Sūra 23:12–16 (Jones edition, 306–307, 314–15). For the importance of this generative cycle in Islam, see Haleem, 'Early *kalām*', 76. Cf. also the Ismā'īlī view of creation, which uses the biological metaphor of a seed out of which the cosmos develops. In this process of generation and development (called *inbi'āth*), the Intellect participates as a 'vital' principle in the cosmos, progressively manifesting itself in both material and spiritual forms; see Nanji, 'Ismā'īlī Philosophy', 151.

47) On the compatibility of the study of nature and the Hermetic 'cosmological sciences' with the *Qur'ān*, see Nasr, *Cosmological Doctrines*, 5.

48) For a translation of both parts of the text, see Ferguson (trans.), *Aurora consurgens (Morning Rising)*. The first part was published by M.-L. von Franz, who annotated the many biblical and alchemical quotations in the text; see *Aurora consurgens* (trans. R.F.C. Hull and A.S.B. Glover). See also von Franz, *Alchemy*, 177–272, for an extended commentary from a Jungian perspective, including translations of many passages. For a list of the manuscripts, the latest dating to the 17th century, see Ferguson, *op.cit*. 1; Obrist, *op.cit*. 275–84. A. Haaning has identified a complete version in a 14th-century manuscript in Hanover (*Niedersächsischen Landesbibliotek IV 339*), which is the earliest so far known (see *Journal of Analytical Psychology* 59 [2014], 20). However, he adheres to the critical consensus, interpreting the text as 'a blend

296

of two different traditions: on the one hand the Gnostic paradigm of antiquity, the pagan spirituality, and on the other the Christian paradigm' (*ibid*. 22), without reference to ancient Egypt, though he entertains the possibility of influence by 'Arabian texts' (*ibid*. 24).

49) Ferguson (trans.), *op.cit*. 14–15; von Franz, *Aurora*, 50–51.

50) Cf. Wisdom's revelations in the *Horologium Sapientiae* (translated as *Clock of Wisdom* or *Wisdom's Watch upon the Hours*), composed by the German mystic Henry Suso in 1334. Suso based his work on the 24-hour cycle, and here the biblical texts are chosen for him by Wisdom, who, as in *Aurora*, manifests as queen of the universe (see Newman, *God and the Goddesses*, 206–22 for a detailed discussion). Newman noted briefly the similarity with *Aurora consurgens*, see *ibid*. 236, though her unfavourable interpretation of *Aurora* as a 'chemical' work with no real spiritual content is regrettable (*see also below chapter 23, n.71, for her view*). The 13th-century poet and mystic Hadewijch of Brabant also refers to the 12 nameless hours of Minne, or Lady Love; see *ibid*. 377, n.50.

51) Quoted in Vereno, *Studien*, 158[174].

52) *Ibid*. 158[176]. This gift has the potential for infinite multiplication beyond itself. Cf. also Maat's inexhaustible gift associated with regeneration mentioned in Ihy's chamber in the Crypt South 1 at Dendara (*chapter 7*).

53) Ferguson (trans.), *op.cit*. 12; von Franz, *Aurora*, 46–7.

54) Vereno, *op.cit*. 140[30]. The advice to remain silent is usual in alchemical texts, e.g. in Zosimus's treatise *On Excellence*, he says, 'Silence teaches excellence'; Mertens, *Zosime de Panopolis*, 41[7], 225, n.47.

CHAPTER 21

1) Vereno, *Studien*, 142[53–5].

2) Quoted *ibid*. 160[1].

3) See *ibid*. 282–3, for similarities with Horus of Edfu's epithets. For the secrecy surrounding the beloved's identity in Arabic mystical poetry, see Schimmel, *As Through a Veil*, 27–8. As Schimmel noted, it is an old Sufi rule not to reveal the beloved's name. Cf. for example, 'Umar ibn al-Fāriḍ's statement in his Sufi *Poem of the Way*, composed in Egypt in the early 12th century: 'A name is a sign, therefore, if you would allude to me, use metaphors or epithets'; see Nicholson, *Studies*, 230 (verse 325). Cf. also the ancient Egyptian sun god's 'great, secret name', which Isis mercilessly extracts from him to gain power over him; Borghouts, *Magical Texts*, 51–5 (no.84); Roberts, *Hathor Rising*, 102.

4) Vereno, *op.cit*. 144[64].

5) Vereno, *op.cit*. 144[71]. For Hermes's reference to the seven, see *ibid*. 150[106]. Vereno noted the compatibility of a marriage with four wives with Islamic tradition (*ibid*. 240, n.232).

6) For the four women as pearls, see *ibid*. 144[72]. For pearls being associated with the alchemical secrets of colouring, cf. Komarios's teaching to Cleopatra: 'Look, from a pearl, and another pearl you receive, O Cleopatra, the whole colouring' (Berthelot, *Alchimistes grecs* 3, 279[4]).

7) Vereno, *op.cit*. 146[75]. For the 'division of Hermes' and the astrological division of the planet Mercury, being associated with the sea and everything living in the watery realm, see *ibid*. 186, with n.10.

8) Vereno identified the *Epistle's* four women (and also Theosebia) solely with mercury/quicksilver (*op.cit*. 186), thus overlooking that 'four' is the alchemical number of copper. In his view, Hermes's 'technical' instructions are based on the distillation of quicksilver/mercury from cinnabar (see *ibid*. 230–40), though he noted that mercury is only specifically mentioned once in the text (*ibid*. 235). Cf. however, the well of 'copper and quicksilver', and its 'foetid water', associated with the beginning of the work in the *Rosarium*; see Telle (ed.), *Rosarium* 1, 41–2, and 2, 41–2. In the *Rosinus ad Sarratantam episcopum*, Venus's womb is referred to as 'corrupt, mixed quicksilver' even though her metal is more usually copper (see Jung, *Mysterium*, 305 [para 417]). The *Book of the Holy Trinity* calls Venus the 'lunar, mercurial mother'; see Ganzenmüller, *Buch der heiligen Dreifaltigkeit*, 242, n.43. Cf. also the transformational battle between 'copper' and 'mercury' as a union between male and female in Sermon 42 of the *Turba philosophorum* (Ruska, *Turba*, 149, 229), and Sermon 60's emphasis on the womb's moisture to preserve the 'seed' developing within it (*ibid*. 162–3, 247–8). The importance of this copper-mercury relationship in 'marriage' alchemy has remained unnoticed (*see also chapter 19, n.11*).

9) *See chapter 11, nn.11–13*.

10) Sethe, *Amun*, 84[173].

11) Utterance 205, *PT* §123a–c. Cf. also the four companions of Horus who are invoked in a magical spell to assist a woman in a difficult childbirth. Evidently aspects of Hathor, they place her 'amulet of health' on the parturient mother (Borghouts, *op.cit*. 39, no.61).

12) Vereno, *Studien*, 146[80]. Vereno (*ibid*. 280) recognizes this 'house' symbolism is ancient Egyptian. For 'pillar' and 'bull' as interchangeable symbols of fertility and regeneration in ancient Egypt; see Frankfort, *Kingship*, 169.

13) For the four elements as a square, see Abt *et al.*, *Explanation of the Symbols* 2, 95, *fig.30*; as four faces, cf. the first two woodcuts in Reusner's *Pandora* series of the royal marriage (1582). The first shows the couple holding hands in a vessel and surrounded by four faces which have 'one father'. In the second woodcut the 'faces' are identified as the four elements; see Fabricius, *Alchemy*, 70–71 (*figs.117–18*).

14) Abt *et al.*, *op.cit*. 1, 157 (line 3). For a goddess enclosing the Egyptian king as his temple, *see chapter 7, n.17*.

15) Vereno, *op.cit*. 146–8[85–91]. For blindfolding rites in the ancient mysteries, see *ibid*. 267–8.

16) Olympiodorus describes the 'place of pulverization' as a consecrated 'house' surrounded by water and gardens; see Berthelot, *Alchimistes grecs* 3, 109[50].

17) For these colour changes, see Vereno, *op.cit*. 148[101–104]. Cf. also the heating of the king and queen in a generative bath so they change from black to white to red in Sermon 29 of the *Turba philosophorum* (Ruska, *op.cit*. 219). For the 'Elephant's Head' vessel, cf. a curious illustration of an alchemist standing beside an oven with a distillation apparatus on top (Abt [ed.], *Book of Pictures: Edition of the Pictures*, 67, *fig.11*). Both the protuberance from the distillation vessel and the alchemist's nose are pictured like an elephant's trunk. For this 'Jupiter' transformation, cf. the *Rosarium's* description of 'tin' (the

metal of Jupiter) changing from black to a 'white gleam'. It then goes on to say that when water mixes with copper, it makes this white inwardly, and this white colour is called 'pregnancy' by some people 'because the earth is coloured white' (see Telle [ed.], *op.cit.* 1, 78, and 2, 70[78]).

18) Rūmī compares love to a mortar in which lovers are pounded and antimony beaten to become 'material for vision'; see Schimmel (*As Through a Veil*, 122), who noted that antimony was believed to enhance eyesight.

19) For this first parable, see Ferguson (trans.), *Aurora*, 18–20; von Franz, *Aurora*, 56–65.

20) Austrian National Library, *Codex Vindobonensis 2372*. For an English summary of the text, which is written in Middle Dutch, see Birkhan, 'Alchemical Tracts', 162–70. Birkhan noted that Gratheus was turning his back on technological alchemy, 'perhaps also on the commonly accepted mercury–sulphur theory' (*ibid.* 163).

21) For the four flasks, see Birkhan, *Lehrdichtung* 2, 9 (lines 63–70). For the couple in the flask (*Miniature 15*), see *ibid.* 2, 66; Obrist, *Débuts*, ill.7. Called Ylarius and Virgo, they are shown wearing crowns with a male figure called Multipos between them. In *Miniature 16* Multipos is outside the flask, wielding his staff to silence the couple. They are human-headed and the rest of their bodies animal-like (see Birkhan, *op.cit.* 2, 70), though, unlike in later 'marriage' treatises, their coition is not depicted. For the child in the flask, see *ibid.* 2, 78–9.

22) Cf. the depiction of Christ coming forth from the grave in *Miniature 12*; Birkhan, *ibid.* 2, 54; Obrist, *op.cit.* ill.6.

23) The *Rosarium* was first printed in Frankfurt in 1550 and attributed to various authors, including Arnald of Villanova (*c.*1240–1311) and Petrus Toletanus, but the authorship remains uncertain; see Telle (ed.), *Rosarium* 2, 170. The illustrations here in this book come from an earlier version dating from around 1530, now in the Stadtbibliothek, St. Gallen, which has 20 coloured illustrations (for the manuscript's date, see Gamper and Hofmeier [eds.], *Alchemische Vereinigung*, 13, 213). For the *Rosarium*'s putative link with the Sulphur–Mercury theory of metals in modern-day literature, *see chapter 19, n.11*.

24) *See also above, n.13*, for these four faces in the *Pandora* series of woodcuts. For the treatise's attribution to various authors, see Gabriele (ed.), *Don de Dieu*, 15–18. In the British Library's manuscript only two female heads are visible above the mating couple. Gabriele (*ibid.* 8) compared the flasks with medieval medical illustrations of the womb as a flask containing the foetus.

25) The imagery here visually replicates Hermes's statement in Ibn Umail's the *Silvery Water and the Starry Earth*: 'Know that this Marriage and Conception takes place by Putrefaction in the lower part of the vessel. The Birth of this "Child" which will be born to them, will be in the Air, viz., in the head of their vessel. The head of the vessel is the top of the Dome and the Dome is the Anbīq' (quoted in Stapleton *et al.*, *Ambix* 3 [1949], 76[X]).

26) Illustrated in Newman, *God and the Goddesses*, 161 (*fig.4.1*).

27) *Ibid.* 164 (*fig.4.4*). See also *figs.4.2* and *4.3* for other examples.

28) For their presence at the Annunciation in a 15th-century *Book of Hours* illustration, see *ibid.* 44–6, with *fig.1.8*.

29) For the problems surrounding the origin of these 'four daughters', see *ibid.* 160. Some of the controversy surrounding alchemy in the 13th century is discussed by Principe, *Secrets of Alchemy*, 58–62.

30) For this second parable, see Ferguson (trans.), *Aurora*, 21–2; von Franz, *Aurora*, 66–71.

31) Von Franz noted the 'Egyptianness' of the parable's life-giving and destructive female, referring to Hathor's role in the *Myth of the Destruction of Humanity* (*ibid.* 246, n.14). Cf. also Frau Minne's fierce manifestation in 13th-century German mystical texts (see Newman, *op.cit.* 157–9) and Mary Magdalene's role as the bearer of life and death in the Middle Ages (Haskins, *Mary Magdalen*, 217–18).

32) For *Aurora* as an 'incomprehensible text', see Halleux, *Textes alchimiques*, 104.

33) Von Franz (*op.cit.* 256–7) noted the identification with Eve here. For Eve as the 'separating-uniting' power in the early *Gospel of Philip*, see Roberts, *Golden Shrine*, 117.

34) The only complete edition currently available is a revised version of the original text prepared in 1433 for John, Margrave of Brandenburg, which only survives, however, in later copies; see the edition of Junker, *Das "Buch der heiligen Dreifaltigkeit"*. Ganzenmüller (*Buch der heiligen Dreifaltigkeit*, 231–71) commented on an early 15th-century version now in the German National Museum in Nuremberg (*Ms. 80061*). See also Obrist (*Débuts*, 117–82) for a detailed discussion of the Nuremberg manuscript's illustrations.

35) Illustrated in Obrist (*ibid.* ill.10), who describes her as serpent-tailed (*ibid.* 161–2); also in Roob, *Alchemy*, 209. The scene is replicated in Reusner's *Pandora* (see Jung, 'Paracelsus as a Spiritual Phenomenon', 144, with *fig.B4*). Jung (*Mysterium*, 30–31 [paras 23–4]) compared it with Honorius of Autun's 12th-century commentary on a passage in the *Song of Songs*: 'You have wounded my heart, my sister, my spouse, you have wounded my heart with one of your eyes…' Cf. also a melusine shown upside down on a pillar and embracing a man rising forth from water in a manuscript now in the British Library (*Ms. Sloane 5025*); see Jung, *op.cit.* *fig.B5*. Jung noted Paracelsus associated the melusine with the 'characters of Venus' (*ibid.* 174), but had to conclude that what he meant here 'remains obscure' (*ibid.* 187). The relevant Arabic alchemy would have been unavailable to him at the time of writing.

36) The 'calcine copper' text is mentioned by Obrist (*op.cit.* 150–51, with n.163, with reference to Nuremberg *Ms. 80061*, fol.24v). For Christ as the 'black sun' associated with lead and Saturn, see *ibid.* 149; and identified with the transformed *tetrasomia*, see *ibid.* 170, with n.246. Ganzenmüller (*op.cit.* 244) noted the mermaid's spearing of Adam reflected the 'calcination' of metals.

37) It is beyond the scope of this book to compare *Scivias* and *Aurora consurgens* in detail, but wherever possible, the parallels will be referred to. For these themes in the second vision, see *Scivias* (trans. Hart and Bishop), 88–9 (Section 32), and in the fourth vision, *ibid.* 118–19 (sections 14, 16). *See also below, n.41*, for parallels with *Aurora*'s third parable.

38) Vereno, *Studien*, 150[108–109].

39) For the third parable, see Ferguson (trans.), *op.cit.* 23–4; von Franz, *op.cit.* 72–9. Both authors translate porta aerea as 'gate of brass', which obscures copper's importance here, especially as aereus also means 'copper' or 'bronze'. The reference surely is to copper (Venus) and iron (Mars) as members of the transforming *tetrasomia* during ascent; lead (Saturn) and tin (Jupiter) already having been transformed in the previous phases. *See below, n.43,* for their association in alchemy.

40) For the noonday demon during the time of Sext in the monastic day, see Steindl-Rast and Lebell, *Music of Silence*, 68–78. Cf. also Olympiodorus's warning about the dangers of the demon Ophiuchus, who seeks to lure alchemists away from the work (Berthelot, *Alchimistes grecs* 3, 92–3[28]).

41) Cf. also Hildegard of Bingen's description of the musical daughters of Zion in *Scivias*, who have laid aside 'stiffnecked wantonness' (trans. Hart and Bishop, 206, seventh section), which once again offers a direct parallel with *Aurora consurgens*, though her focus in this vision is mainly on the virtuous daughters, shown in the bosom of Mater Ecclesia (or Mother Church), who, interestingly, is depicted with fish scales forming the lower half of her body; see *Illuminations*, 70; Newman, *God and the Goddesses*, 42 *(fig.1.7).*

42) Illustrated in Obrist, *op.cit. ills.23 (colour),* 35; Roob, *op.cit.* 463.

43) Cf. the depiction of Venus (copper) and Mars (iron) embracing in a flask in Reusner's *Pandora* (illustrated in Ploss *et al.*, *Alchimia*, 131; Fabricius, *Alchemy*, 132). The flask is placed next to a winged and a wingless bird, which Ibn Umail associates with ascent and the first work of 'Whitening' in his temple vision; *see chapter 16.*

44) See Telle (ed.), *Rosarium*, 1, 95–6, and 2, 83 (my translation from the German). For this female lifecycle in the ancient Egyptian ascent to the zenith, see Roberts, *Hathor Rising*, 66–7. Following Roob's translation (*op.cit.* 452), I have translated unnd ohn alles mahl as 'without nourishment', but equally it might mean 'without anything bad', cf. the St Gallen earlier version, where it is written und on alles mal (Gamper and Hofmeier, *Alchemische Vereinigung*, 46).

45) For this nocturnal festival, *ibid.* 24–36.

46) The inscription accompanies the offering of Hathor's sacred drink; see Sternberg-el Hotabi and Kammerzell, *Hymnus*, 22–3(d).

CHAPTER 22

1) Vereno, *Studien*, 150[112, 116].

2) For Hermes's instruction, see Vereno, *ibid.* 150[116]. Vereno is perplexed by what this 'reduction by a third' means (*ibid.* 199–200). Maria's statement is quoted in a Greek treatise attributed to Zosimus entitled *On the Fundamental Materials(?) and the Four Bodies according to Democritus* (see Berthelot, *Alchimistes grecs* 3, 151[3]; Patai, *Jewish Alchemists*, 82). It is not classified as a genuine Zosimus work by Mertens (*Zosime de Panopolis*, lviii[2], lx). The association between 'refining copper' and 'three' may well have ancient Egyptian antecedents. Cf. Nibbi's discussion of the three basic processes for working copper ore (roasting, smelting and refining), and the way three crucibles symbolized the third and final refining

stage, as well as the refined copper alloy, in ancient Egypt (*JARCE* 14 [1977], 63).

3) Immediately after mentioning copper's reduction by a third, Maria refers to changes occurring 'when one whitens and when one makes yellow', evidently in a copper process designed to remove sulphurous substances. For these colours being associated with the alchemical 'egg' and second birth, see Abt *et al.*, *Explanation of the Symbols* 2, 226–7. Abt traces the egg's origins back to ancient Egypt.

4) Vereno, *op.cit.* 152[124].

5) Cf. Ficino's reference to Saturn as the 'highest of planets' in chapter 22 of his third *Book on Life*; see Kaske and Clark (trans.), *Three Books on Life*, 367.

6) See Stricker, *Geboorte* 3, 251 (nos.828, 838), 254 (nos.857, 859). For citations from Jewish writers, see *ibid.* 254–6.

7) Abt *et al.*, *op.cit.* 1, 163 (lines 3–5). Cf. the *Turba philosophorum*'s statement that seed is formed in the womb after 40 days and nights (Ruska, *Turba*, 163, 247; Roberts, *My Heart My Mother*, 205). For this time period in Gratheus's untitled treatise being associated with the soul's manifestation in the body, see Birkhan, *Lehrdichtung* 2, 57, (lines 885–89). The *Rosarium* mentions 'whitening' and 'purifying the stone' for a period of 40 days (Telle [ed.], *Rosarium* 1, 169, and 2, 143[169]).

8) Abt (ed.), *Book of Pictures* 1, fol.175b, and 2, 485–6.

9) See Berthelot, *op.cit.* 3, 197–8[22–3]; Festugière, *Hermétisme*, 235. The text, called *On the Philosophers' Stone*, is attributed to Zosimus, but is probably a later work. Earlier it refers to a 40-day period for the 'work of fire' (Berthelot, *op.cit.* 195[10]) and also mentions the 'etesian stone', thus paralleling the *Epistle*'s transformations of the 'stone'.

10) See Berthelot, *op.cit.* 3, 209[5]. From the section on the 'etesian stone' in the *Chapters of Theodore* (attributed to Zosimus, but a later text). Cf. Stephanus's description of copper becoming the 'etesian stone' and being treated for 41 days in the gentle heat of a vapour-tight vessel, indicative of an embryological process, see Taylor, *Ambix* 2 (1938), 43. See also Festugière, *op.cit.* 235–7, for various references to the foetus being conceived and nourished in the mother's womb.

11) Abt *et al.*, *Explanation of the Symbols* 1, 163 (lines 5–7). Some Classical authors, e.g. Galen, favoured a four-phase embryological period (see Stricker, *op.cit.* 246[818], 253[852], 274[935]). For the three-phase period, which Stricker considered 'metaphysical' in origin, see *ibid.* 276. A *Qur'ānic* verse refers to the foetal three phases as a 'triple darkness' (Sūra 39:6).

12) See Festugière, *op.cit.* 244, nn.63–5.

13) For this combination of 'light', 'air' and 'water' in Ramessid hymns to Amun-Re, see Assmann, 'Primat und Transzendenz', 11, with n.20, also 13–15, 28. Cf. also Plato's explanation in the *Timaeus* (56b–57c) that the three primary elements of fire, air and water— corresponding respectively to the tetrahedron, the octahedron and the icosahedron, the faces of which form an equilateral triangle—are transformed one into another without earth as the fourth element intervening.

14) In chapter 11 of his third *Book on Life* Ficino writes: 'This universal life indeed flourishes much more above the earth in the subtler bodies, which are nearer to Soul. Through its inward power, water, air, and fire possess

living things proper to them and partake of motion';
see Kaske and Clark (trans.), *op.cit.* 289.

15) Book Five [5–6]; Copenhaver, *Hermetica*, 19[5–6]. Cf.
Ficino's statement in the third chapter of his third *Book on
Life*: 'Assuredly, the world's body is living in every part, as is
evident from motion and generation' (Kaske and Clark
[trans.], *Three Books on Life*, 255).

16) According to the Arabic *Book of the Lover (Kitāb al-Ḥabīb)*,
the foetus acquires its complete form after 40 nights, and
then, after 80 nights, begins to feed, sustained by a 'gentle
heat' in the maternal womb; see Festugière, *op.cit.* 236–7.

17) Vereno (*Studien*, 272) recognized this time period's
significance for bringing Theosebia's work into the sign
of Leo. He also associated it with the new year period in
ancient Egypt, inaugurated by the heliacal rising of Sothis.
For 'cooking the stone', linked with 'the one born every
year' and embryonic development, in the Arabic *Book
of Pictures*, see Abt (ed.), *Book of Pictures* 2, 481.

18) Vereno, *op.cit.* 152[126–7]. For black and white being
compared to the cosmic Eye in Olympiodorus's work,
see Berthelot, *Alchimistes grecs* 3, 100.

19) For Ibn Umail's association of 'second whiteness' with
alabaster, see Abt *et al.*, *Explanation of the Symbols* 1, 9
(lines 10–12), with n.42. For the 'alabaster stone' as a code
name for the alchemical egg, see Mertens, *Zosime de
Panopolis*, 48, n.1.

20) Rūmī, *Masnavi* 1. 3193; Williams (trans.),
Spiritual Verses, 294.

21) Vereno, *op.cit.* 152[131–2].

22) Ferguson (trans.), *Aurora*, 34; Von Franz, *Aurora*, 116–17,
with n.67. In the fourth chapter of his third *Book on Life*,
Ficino warns against the 'third degree of heat and dryness'
when the sun is in the sign of Leo, which can create
'martian' rather than the sought-after solar and jovial
qualities; see Kaske and Clark (trans.), *op.cit.* 261.

23) Vereno, *op.cit.* 152[133, 136].

24) *Ibid.* 152[134].

25) *Ibid.* 152[136].

26) *Ibid.* 154[142–3].

27) *Ibid.* 154[146–7].

28) See Alliot, *Culte d'Horus à Edfou*, 353–4. Vereno noted
the similarities with the Edfu rites (*op.cit.* 281–2).

29) Ibn Umail calls heavenly pyrite water 'sulphur water';
see Abt *et al.*, *Explanation of the Symbols* 1, 69 (lines 1–2).

30) Vereno, *op.cit.* 156[156]. He connects Theosebia's vessels
with Vitruvius's statement that Nile water was brought
into the temples on the day of flood (*ibid.* 279).

31) For an English translation of Ibn Umail's version of the
lion hunt, see Abt *et al.*, *Explanation of the Symbols* 2,
186–8. For Jung's commentary, based on a Latin version,
see *Mysterium*, 298–9 (para 409).

32) Cf. also the fable 'The Lion in Search of Man' in the
Demotic *Goddess in the Distance* myth, in which a proud
lion falls into a pit dug by a huntsman and is entangled
in a net. He is eventually rescued by a tiny mouse, whom
he had previously captured but decided not to kill,
and who now gnaws through his fetters to release him;
see Lichtheim, *Ancient Egyptian Literature* 3, 157–9
(with further references).

33) The lion's fate in the vessel is graphically depicted in
Reusner's *Pandora*; see Fabricius, *Alchemy*, 66 *(fig.109)*.
The 'lion hunt' theme also runs through Book 1 of Rūmī's

Masnavi, creatively retold and interrupted by different
sub-stories, in order to teach about the egotistic attitudes
hindering the soul's development. Initially, the self-
regarding lion repeatedly harasses the other animals, and,
tricked by the hare, falls into a well (*Masnavi* 1. 1060–209,
1272–398; Williams [trans.], *Spiritual Verses*, 103–116,
122–33). His different bodily parts, identified with the
astrological sign of Leo, are then 'amputated' in the
tattoo parlour (*ibid.* 2994–3025; 276–8), and ultimately
his royal nature manifests, represented by the 'lion of
God', the Prophet Muḥammad's son-in-law, 'Alī, who
magnanimously spares an infidel warrior he has
overcome in battle (*ibid.* 3735ff; 341ff).

34) For the scene's caption, see Abt (ed.), *Book of Pictures* 2,
477.

35) Cf. Sūra 13 of the *Qur'ān* (verse 17): 'He makes water
descend from the sky, so there are torrents that flow
according to their measure.' For the association between
this life-giving water and the animation of corpses, see
Corbin, 'Realism and Symbolism of Colors', 90–91.

36) Abt (ed.), *op.cit.* 2, 481. The scene belongs to the
'Eighth Book about the Operation', in which Zosimus
explains various purifying copper procedures to obtain
the 'purple' dyes.

37) *Masnavi* 1. 3287; Williams (trans.), *op.cit.* 301.

38) Cf. also Zosimus's first lesson *On Excellence*, in which
Ion is both the priest at the phial-shaped altar and the one
who suffers intolerable violence, the transformer and the
transformed, the sacrifice and the sacrificed; see Mertens,
Zosime de Panopolis, 35–6.

39) See Abt *et al.*, *Explanation of the Symbols* 1, 11 (line 2).
For the divine water as 'the Mother of Gods', see *ibid.* 81
(line 2).

40) For the fourth parable, see Ferguson (trans.), *Aurora*,
25–30; von Franz, *Aurora*, 80–99. Von Franz is puzzled by
the seemingly sudden introduction of the Trinity here
(*ibid.* 280), but it is entirely consistent with the shift to the
'three' during this 'purifying copper' phase.

41) 'The Philosophers have commented on the first, second
and third forms of baptism, pointing out that in the first
trimester of pregnancy it is water that nourishes the foetus
in the womb, and in the second trimester air, while in the
third trimester it is fire that nourishes and guards it, and
that the burden will never pass to the child until these
months have passed, when he or she shall be born and
be given life by the Sun, because the Sun gives life to
everything' (Ferguson [trans.], *op.cit.* 26–7).

42) Later, this 'womb' alchemy reappears in Goethe's *Faust*
(Part 2, Act 2, scene 2), being associated with the creation
of the human-like *homunculus* in a phial, and with the
question of how 'body and soul really hang together'.
For Paracelsian notions that living beings can be produced
in the laboratory; see Principe, *Secrets of Alchemy*, 131–2,
with further references.

43) For the association with distillation, see Ferguson
(trans.), *op.cit.* 30; von Franz, *op.cit.* 98–9; and with the
Holy Spirit's fiery sevenfold gifts, see Ferguson (trans.),
op.cit. 25–6; von Franz, *op.cit.* 86–7. For Hermes's
instruction, see Vereno, *op.cit.* 150[113–14]. Ibn Umail
refers to the ninth month as the time from 'the first
work until the end of the distillation', see Abt *et al.*,
Explanation of the Symbols 1, 61 (lines 7–8).

44) Abt (*Book of Pictures: Edition of the Pictures*, 64–7) compares this scene with the distillation oven shown in the later *Mutus Liber*. Over the centuries, alchemists associated distillation with numerous procedures, including the preparation of healing remedies and essential oils, as well as the extraction of mercury from cinnabar, already mentioned by Pliny and other Greek writers. In the European Middle Ages it was also applied to the art of distilling alcohol.

45) Birkhan, *op.cit.* 2, 76–7 (lines 1245–52).

46) *Masnavi* 1. 3789–91; Williams (trans.), *op.cit.* 345.

47) *Ibid.* 3795–7.

48) Taylor, *Ambix* 1 (1937), 131–3.

49) Abt *et al.*, *Explanation of the Symbols* 1, 53 (lines 6–7), and 2, 228. For the distillation process being analogous to the meteorological phenomenon of rain, see Martelli, *Four Books*, 240–41[6].

50) Cited in Jung, *Psychology and Alchemy*, 128, n.44 (para 167).

51) For the fifth parable, see Ferguson (trans.), *Aurora*, 31–5; von Franz, *Aurora*, 100–119.

52) For the 14 virtues, see von Franz, *op.cit.* 104–19. Their unusual number may relate to the fourth parable's promise of enthronement on David's throne, the numerical value of King David's name being 14 in ancient Jewish numerology, or *gematria*. It also resonates with the 'artificial' genealogy of Christ as the 'son of David' set out in the Gospel of Matthew, in which, through a triadic structure of 14 generations, including 'four mothers', he becomes the descendant of the royal Davidic line as the Messiah (Matthew 1: 1–17). For the link with Matthew's gospel, see von Franz (*ibid.* 331, with n.89), who also suggests an ancient Egyptian influence, referring to the sun god's 14 *Kas*, the physical and moral attributes the pharaoh also radiates as 'son of Re'.

53) The 'plea of justification' is in the fourth parable, 'sowing and reaping' in the fifth.

54) Ferguson (trans.), *op.cit.* 34; von Franz, *op.cit.* 114–15. Cf. Ficino's statement in chapter 19 of his third *Book on Life*: 'For we see composite things acquire life only then, when a perfect commixture of qualities seems to have broken up the initial contrariety, as in plants' (Kaske and Clark [trans.], *Three Books on Life*, 347).

55) For the timing with the ninth hour, see Dante, *Divina Commedia*, (ed. Oelsner), 320, nn.1–3 (referring to *Purgatorio* XXV, 1–3).

56) *Ibid.* XXV, 37–108.

57) For entry into the fiery womb, see *ibid.* XXVII, 25–7; for Virgil's crowning of Dante, see *ibid.* XXVII, 142.

58) *Ibid.* XXXIII, 142–5.

59) Anderson, *Dante the Maker*, 352.

60) See chapter 21 of his third *Book on Life* (Kaske and Clark [trans.], *op.cit.* 355). Ficino also associates these seven steps with the planets.

61) The connection with Dante's poetic life is implied by the 'sandwiching' of canto XXV's embryological rebirth account between cantos XXIV and XXVI. Previously, among the emaciated souls in canto XXIV, Dante had met a poet of the 'old school', who had questioned him about the 'new style' of poetry. Then, in canto XXVI, he encounters Guido Guinizelli, founder of the 'new style' of poetry inspiring Dante's Florentine circle, the implication

being this new poetic form is a rebirth of creativity.

62) *See chapter 20, n.48*, for the dating of the known manuscripts.

63) For this embryological vision, see *Scivias* (trans. Hart and Bishop), 109–29.

64) Hildegard compares the activity of human seed in the womb to the process of making cheese from milk (*ibid.* 118), and people carrying vessels containing milk turning into cheese are depicted in the egg-like shape containing the pregnant mother and child *(fig.186)*. For Jewish and Classical comparisons of male seed in the womb and cheese-making, see Stol, 'Embryology in Babylonia and the Bible', 144. For Ibn Umail's description of the fertilizing 'water' in the womb as cheese, see Abt *et al.*, *Explanation of the Symbols* 1, 163 (line 1).

CHAPTER 23

1) For the final part of the *Epistle*, see Vereno, *Studien*, 156–8[162–178].

2) This statement about the non-repetition of the work probably reflects the irreversibility of the alchemical process at a certain stage. Cf. Nasr's comments (*Cosmological Doctrines*, 266) about the irreversibility of the process of illumination and gnosis when the angelic realm is attained, which he compared to alchemy: 'Once the gold is made it cannot be unmade into a base metal.'

3) Vereno, *op.cit.* 156–7[163–170].

4) For the arrival of the ten 'powers' or 'virtues' presaging rebirth and divinization in Book 13 of the *Corpus Hermeticum*, see Copenhaver, *Hermetica*, 51[10], 188. Cf. also the *anthropos* associated with 'ten' in an alchemical text attributed to Albert the Great (Jung, *Psychology and Alchemy*, 233, *fig.117*). Here the completed human being, bearing the four elements within a circle between his outstretched arms, is placed in a circle with 24 dots around the rim for the hours of day and night. Above him are the numerals one to ten placed beneath a human head.

5) Abt *et al.*, *Explanation of the Symbols* 1, 43 (lines 7–8). For red mercury being identified with the purified red slave in Ibn Umail's treatise the *Pure Pearl*, see *ibid.* 2, 99, with n.244.

6) Finlay (*Colour*, 180–82) gives a fascinating account of visiting a mercury mine in Spain and holding a two-litre bottle of mercury which was so heavy she needed two hands to lift it, as 'mercury is seventeen times as dense as water'. She also describes placing her hand in a tub 'full of this strange liquid-solid … I submerged my arm and swirled this pool of pure mercury around: it was a wonderful sensation. When I went with it, it felt like water. When I went against it, it was an almost unstoppable force. An elephant of elements.'

7) In chapter 26 of his third *Book on Life*; see Kaske and Clark (trans.), *Three Books on Life*, 387. For the two-headed hermaphrodite as a symbol of Mercurius and the 'two in one', see Jung, *Mysterium*, 14–15 (para 11).

8) Ferguson (trans.), *Aurora*, 36; von Franz, *Aurora*, 120–21. Cf. also *Aurora*'s illustration of a man working a fermenting golden substance, kneading it like dough in a container placed on a table (Ferguson, *op.cit.* 89). Obrist (*Débuts*, 226–7) notes the dough here symbolizes the transmutation of the first material.

9) Ferguson (trans.), *op.cit.* 36; von Franz, *op.cit.* 122–3.

10) Von Franz (*ibid.* 121 n.3) refers to Morienus's statement, 'But this earth is a body and a ferment.' Cf. also Ibn Umail's description of the fermenting body: 'Water is the ferment of the body and the body is the ferment of the water, and each one of them is a ferment to its companion … thus at that time they became one single dyeing dye' (Abt *et al.*, *op.cit.* 1, 45 [lines 5–7], and 2, 209). Cf. also Mary the Copt's statement about the 'second body' quoted by al-'Irāqī: 'The second body is derived from the first and is not a strange body, but is that very one from which the spirits have been extracted. It is, indeed, the soul's body in essence and in species, although it is not the same body from which the soul went out' (see Holmyard [ed.], *Kitāb al-'Ilm*, 51).

11) Ferguson (trans.), *op.cit.* 37; von Franz, *op.cit.* 124–5.

12) Al-'Irāqī also refers to the transformed earth providing stability for 'water, air and fire' (Holmyard [ed.], *op.cit.* 28). According to Ficino, the influence of the celestial rays in the earth is essential for the creation of images. The intensity of the rays in the dry earth sets it on fire, he says, and when kindled, the earth is 'vaporized' and then 'dispersed through channels in all directions and blows out both flames and sulfur'. The rays of the stars also penetrate the metal and precious stones used to create the images, bestowing on them wonderful gifts and making them living bodies; see Kaske and Clark (trans.), *op.cit.* 321–3.

13) Ferguson (trans.), *op.cit.* 38; von Franz, *op.cit.* 128–9. In his 12th-century poem *Anticlaudianus*, Alan of Lille describes the Christ like 'Perfect Man' created by Nature and the Virtues as 'human on the earth and divine among the stars' (see Newman, *God and the Goddesses*, 75), his role being to 'go safely in the middle way' (*ibid.* 81). Cf. Hermes's description of the marvellous powers of the human being as mediator between heaven and earth in the *Asclepius* (*Asclepius* 6; Copenhaver, *Hermetica*, 69–70[6]), a treatise already circulating in the Latin West in the 12th century.

14) *Paradiso* VII, 67–72.

15) Mertens, *Zosime de Panopolis*, 4–6[8–10]. See Festugière, *Révélation* 1, 268–9, cited also in Jung, *Psychology and Alchemy*, 363 (para 456). Jung notes how the union of opposites 'symbolizes the production of the hermaphroditic second Adam, namely Christ and the *corpus mysticum* of the Church', the alchemical earth being 'equated with the body of Christ and with *adamah*, the red earth of paradise' (*Mysterium*, 440, paras 631–2). Cf. also Olympiodorus's description of Adam as 'virgin earth and igneous earth, carnal earth and blood earth' (Berthelot, *Alchimistes grecs* 3, 95[32]).

16) See Abt *et al.*, *Explanation of the Symbols* 1, 71 (lines 9–10), and 258.

17) Ferguson (trans.), *Aurora*, 38; von Franz, *Aurora*, 130–31.

18) The six figures perhaps symbolize the six soakings during the 'Reddening' process for the second body; see Abt *et al.*, *op.cit.* 2, 150–51.

19) For the figure as Hermes, see Obrist, *Débuts*, 236–7. She noted the 'distant echo' with the myth of Osiris. Cf. the figure named 'the Headless One' depicted in the *Book of Day*'s sunset phase *(see chapter 2)*. For the restoration of Osiris's head in the Khoiak rites, *see chapter 11*.

20) For the caption to this picture, see Abt (ed.), *Book of Pictures* 2, 555.

21) Quoted from verses 528 and 532; see Nicholson, *Studies*, 248.

22) Verse 299; see *ibid.* 227. According to Ibn Sīnā (or Avicenna, as he was known in medieval Europe), miracle-working power depends on the extent to which the soul can purify itself and concentrate its energies, enabling it to influence other human beings of a weaker nature, and even the elements; see Nasr, *Cosmological Doctrines*, 260–61; Abt *et al.*, *op.cit.*, 2, 211.

23) Verse 237; see Nicholson, *op.cit.* 222.

24) Cited by Jung, *Mysterium*, 482 (para 685); *Psychology and Alchemy*, 255 (para 358), with n.29.

25) Dorn wrote: 'Within the human body is concealed a certain metaphysical substance, known to very few, which needs no medicament, being itself an incorrupt medicament… As faith works miracles in man, so this power, the *veritas efficaciae*, brings them about in matter. This truth is the highest power and an impregnable fortress wherein the stone of the philosophers lies hid.' See Jung, *ibid.* 269 (para 377).

26) The Arabic Zosimus says the same: 'This stone is the gift of God, which He bestows graciously upon whom He wants' (Abt [ed.], *Book of Pictures* 2, 172). For alchemy being regarded as the 'gift of God' (*donum dei*), see Principe, *Secrets of Alchemy*, 192–206.

27) For this passage, see Bonus, *New Pearl of Great Price*, 123–8. Quoted also in Jung, *Psychology and Alchemy*, 374–5 (para 462). Hermes teaches this same wisdom in Book 13 of the *Corpus Hermeticum*: 'The sensible body of nature is far removed from essential generation. One can be dissolved, but the other is indissoluble; one is mortal, the other immortal.' He also explains that to experience this regeneration requires 'no longer picturing things in three bodily dimensions'; see Copenhaver, *Hermetica*, 52[13–14]. For the mysteries of 'essential regeneration' as a gift which cannot be taught, 'but god reminds you of it when he wishes', see *ibid.* 49[2].

28) *Masnavi* 1. 3889–91; Williams (trans.), *Spiritual Verses*, 354.

29) *Ibid.* 3894. Cf. also the philosophy of the medieval 'faithful of love' that as long as external sensory perception and memory are relied on, no inward renewal is possible. Only by silencing memory-imagination, and impressing the mind with the image of the beloved is it possible to enter the new life and be truly regenerated. For a succinct summary of their philosophy, see Zolla, *Seshat* 6 (2003), 64–72.

30) Cited by Holmyard, *Isis* 8 (1926), 412; Obrist, *Débuts*, 206–207. Cf. also Zosimus's advice to Theosebia in the Arabic *Book of Pictures* to reflect on the pictures he has made for her (Abt [ed.], *op.cit.* 2, 85).

31) See Telle (ed.), *Rosarium* 1, 173–4, and 2, 146.

32) For the association of this copper text with the green lion, *see chapter 19, n.11*. Earlier in the *Rosarium*, the green lion is referred to as 'the copper of Hermes' (*ibid.* 1, 33, and 2, 34[33]).

33) For the original colour scheme in the Zurich manuscript (*Ms. Rhenoviensis 172*, f.29v), see Obrist, *op.cit.* 238. Newman mistakenly identifies the woman's flesh colour as black, interpreting her as the only representation of a Black Madonna in painting, known otherwise exclusively as a sculptural type; see *God and the Goddesses*, 238–40.

More convincing is her association of *Aurora*'s woman with the biblical Book of Revelation's pregnant female 'clothed with the sun, with the moon beneath her feet', a text particularly favoured by *Aurora*'s author (*ibid*. 238–9).

34) For a depiction of Thoth holding a snake-staff in Seti I's temple at Abydos, see Roberts, *Golden Shrine*, 107, fig.66. Ibn Umail refers to the unifying power of 'the stick' in his *Poem Rhyming on the Letter Nūn*: 'Thus the hot one escaped from the cold one. But when I brought the stick, which is the mediating spirit, the hot one mixed with the cold one, and the moist one with the dry one, and they became one thing, which is the three: the spirit, the soul, and the body' (quoted in Abt *et al.*, *Explanation of the Symbols* 2, 242). Cf. Ficino's description of Mercury's caduceus, with which he can 'put minds to sleep and wake them up.' With it he can also 'blunt the wits or sharpen them, weaken or strengthen them, agitate or calm them'; see Kaske and Clark (trans.), *Three Books on Life*, 293. For alchemical transmutation as a process of 'turning the inside out', see Dufault, *Ambix* 62 (2015), 215–44.

35) For Adam and Eve's relationship interpreted in terms of 'separation' and 'union', rather than a fall into sin, in the early alchemical *Gospel of Philip*, see Roberts, *op.cit.* 117–18.

36) For similarities between *Aurora*'s 'mediatrix of the elements' and the medieval iconic type of the Virgin of the Sign, shown with a medallion on her breast within which is Christ yet-to-be-born, see Obrist, *op.cit.* 239, and Newman, *op.cit.* 202 (*fig.5.2*), 239. Von Franz (*Aurora*, 257) notes that *Aurora*'s author understands the second parable's woman 'as a unified embodiment of the figure that was otherwise personified separately in Eve and Mary'.

37) Obrist (*op.cit.* 239) noted the allusion to Venus and copper here.

38) See Abt (ed.), *Book of Pictures* 2, 225. For reddened copper as the goal of the work, *see chapter 20, n.34.*

39) Emblem 38. See de Jong (ed.), *Atalanta Fugiens*, 251–5, 414, *fig.38*. The motto states: 'Like the Hermaphrodite, the Rebis is born out of two mountains, of Mercury and Venus', though de Jong associated this with the Sulphur–Mercury theory (*ibid*. 255), not copper alchemy.

40) Sermon 29 (Ruska, *Turba*, 137, 215).

41) Ferguson (trans.), *Aurora*, 39; von Franz, *op.cit.* 132–3.

42) Ferguson (trans.), *op.cit.* 41; von Franz, *op.cit.* 140–43.

43) Ferguson (trans.), *op.cit.* 41–2; von Franz, *op.cit.* 142–3.

44) Here I have followed von Franz's translation of '12th hour' in this passage (*Aurora*, 140–41), rather than Ferguson's rendering of 'second hour' (*op.cit.* 41).

45) Ferguson (trans.), *ibid*. 41; von Franz, *op.cit.* 138–9.

46) Ferguson (trans.), *op.cit.* 40, 42; von Franz, *op.cit.* 134–7. For other examples of the adorned king's apotheosis in alchemy, see Jung, *Mysterium*, 327–9 (paras 460–61).

47) For this passage about entering through the ear, see von Franz (*op.cit.* 134–5). She suggested that 'entering through the ear' alluded to the medieval belief that the Virgin Mary conceived Christ through her ear (*conception per aurem*), thus conveying the supernatural nature of this final union (*ibid*. 371). Cf. also 'hearing' being the only faculty allowed to Wisdom when she ascends beyond the sphere of the fixed stars to seek the divine soul created by God in Alan of Lille's *Anticlaudianus* (see Newman, *God and the Goddesses*, 76). Hermes calls the illuminated Tat 'the hearer' at the close of his teaching about 'essential

regeneration' in Book 13 of the *Corpus Hermeticum* (see Copenhaver, *Hermetica*, 54[22]).

48) Cf. Hildegard of Bingen's description of descent and return in *Scivias*: 'The Word went forth from the Father a Spirit and returned again to the Father in fruitful flesh' (trans. Hart and Bishop, 417 [third part, vision seven, section eight]).

49) For love as the unifying power in the *Dialogue of the Philosophers and Cleopatra*, see Roberts, *Golden Shrine*, 118–19, and in the early *Gospel of Philip*, *ibid*. 112–13, 117–18.

50) For the blue eagle being associated with Hermes and quicksilver here, see Obrist (*Débuts*, 210), who compared the scene with Adam and Eve's iconography in an early 15th-century Bohemian manuscript (*ibid*. 279–80).

51) For the 12th vision and musical celebration, see *Scivias* (*op.cit.* 514–36).

52) Dante, *Divina Commedia*, (ed. Oelsner), 4–6 (*Paradiso* I, 53–63).

53) *Ibid*. X, 139–48.

54) For the redemption of Adam and Eve, see *ibid*. VII, 25–148; for unifying love and the 'two in one', see *ibid*. IX, 81. See Newman, (*op.cit.* 186–7) for the ninth canto's 'coinherence' (with further references), though, in light of the many parallels in Islamic metaphysics and alchemy, her claim this 'two in one' was 'Dante's most original contribution to the metaphysics of love' (*ibid*. 186) is doubtful.

55) For love eternally multiplying, see Dante, *Paradiso* X, 82–5.

56) Vereno, *Studien*, 158[170].

57) Ferguson (trans.), *Aurora*, 7–8; von Franz, *Aurora*, 36–7.

58) For Aquinas, Albert the Great and alchemy, see Principe, *Secrets of Alchemy*, 59–60, 63, 78–9 (with further references).

59) For the dating of *Aurora*'s known manuscripts, *see chapter 20, n.48*. Von Franz accepted the attribution to Aquinas (*op.cit.* 407–31; *Alchemy*, 179–81, 193), concluding the text probably originated in notes made by those present at the discourse on the *Song of Songs* that he gave on his deathbed. Birkhan (*Lehrdichtung* 1, 68, n.49) noted Aquinas believed in the possibility of transmutation and cautiously refrained from rejecting his authorship outright. Principe (*op.cit.* 63) called *Aurora* a 'pseudonymous' work. Jung suggested an anonymous 14th-century cleric as the author (*Psychology and Alchemy*, 376–7 [para 465]).

60) Dante undoubtedly placed alchemists among the 'Falsifiers' in hell (*Inferno* XXIX, 73–139; XXX, 46–129), but those he was criticizing were 'false' alchemists (of whom there were many in his day) who conned gullible patrons with their alleged powers to imitate nature and also create false coinage (see Principe, *op.cit.* 59–62, for the medieval controversies surrounding 'false' alchemy). This says little, though, about Dante's attitude to 'true' alchemy, since alchemists themselves often criticized the 'puffers'.

61) Whilst Renaissance alchemy is beyond the scope of this book, it should be noted that the *Chemical Wedding of Christian Rosencreutz* (1616) is divided, like *Aurora consurgens*, into seven sections, or days, and is full of alchemical symbolism, opening with a vision of a celestial Lady inviting Christian Rosencreutz to a marriage; see Yates, *The Rosicrucian Enlightenment*, 82–96. Prominent

in the early 17th-century Rosicrucian movement was the copper-loving alchemist Michael Maier (see *ibid.* 109–25), and though Yates did not mention earlier 'rising dawn' alchemy, she recognized alchemy's influence on the development of Rosicrucianism throughout her book.

62) Here I have followed von Franz's translation of the second earth as 'leaves' (*op.cit.* 142–3), not 'sons', as in Ferguson's translation (*op.cit.* 41), though it probably should be identified with 'silver'. Cf. the 'three earths' in Ibn Umail's work (Abt *et al.*, *Explanation of Symbols* 2, 94–5), where Ibn Umail identified the second earth as 'silver'. Abt noted how the Arabic word *ard al-waraq* was wrongly translated into Latin, being confused with the Arabic word *waraq*, meaning 'leaf' or 'sheet' of paper, rather than referring, as in Arabic alchemical texts, to the 'silver of the sages' (*ibid.* 94, n.223).

63) Ferguson (trans.), *op.cit.* 43; von Franz, *op.cit.* 146–9. For a three-chambered mysticism in the early *Gospel of Philip*, see Bourgeault, *Mary Magdalene*, 131–3.

64) See Voragine, *Golden Legend*, 374–5. Haskins (*Mary Magdalen*, 226) observed these three aspects were 'curiously reminiscent of Gnostic ideas about her', without further elaboration.

65) Cf. her role in the *Gospel of Philip* (see Roberts, *Golden Shrine*, 96, 99, 102, 123–4). For the three transformations of the black Shulamite, the beloved of Solomon, in an 18th-century alchemical treatise by Abraham Eleazar, see Jung, *Mysterium*, 451–4 (paras 645–8). Jung compared these with the three stages of mystical ascent in the work of Pseudo-Dionysius (or Denis) the Areopagite. For the three transformations of the Feminine in Goethe's *Faust*, see Raphael, *Goethe and the Philosophers' Stone*, 231.

66) Cf. also the *Rosarium*'s mention of 'foetid water' associated with the well of copper and mercury, *see chapter 21, n.8.*

67) Jung (*Mysterium*, 504–505 [para 718]) compared the three stages of the alchemical 'conjunction' in Dorn's work with the exercises for spiritual contemplation in the Middle Ages, especially the three stages of illumination of St Bonaventure, who, interestingly, associated them with 'the one day'. Dorn, though, was an alchemist, and alchemy already had a long tradition on which he could have drawn. For the three stages of the conjunction in his work, see also *ibid.* 469–77 (paras 669–80).

68) Quoted in Jung, *ibid.* 99–100 (para 120).

69) Quoted from the Kadolzburg version of the text; see Junker, Das *"Buch der heiligen Dreifaltigkeit"*, 40, 224, fol.100r, v, with *fig.25* (my translation from the German). The sword of justice Christ holds contrasts with the sword held by the Antichrist and his mother, embodying the sick *'tetrasomia'* associated with the love-denying passions of 'greed', 'pride' and 'envy', headed by 'anger' and contrary to 'justice'. For Christ's identification with the transformed *tetrasomia*, see Obrist, *Débuts*, 170. Cf. also Suso's visualization of Christ on the cross as a three-coloured rainbow, manifesting peace and a new covenant, in his *Wisdom's Watch upon the Hours*: 'Look, how Love has reddened, greened and yellowed him!' (quoted in Dronke, 'Tradition and Innovation', 72). For alchemical references to the rainbow, see Jung, *Mysterium*, 286, 288.

70) A detailed study of 'two suns' alchemy in the *Book of the Holy Trinity* is beyond the scope of this book, but cf. the statement that 'the Sun of the heavenly holy Trinity' has placed a 'natural Sun' and 'strong king' as ruler of his people and their sensual natures (Junker, *op.cit.* 129, fol. 32r–32v), which is extended then to the German rulers of the time. For the text's political context more generally, see Obrist, *op.cit.* 119–34. Junker discussed the treatise solely in the context of Aristotle and St Augustine's philosophy (*op.cit.* 49–54), and the Sulphur–Mercury theory of metals (*ibid.* 53–4). Similarly, Obrist (*op.cit.* 174–5, with n.273), in citing a passage about the supernatural and natural suns, seemed unaware of the long tradition of 'two suns' copper alchemy influencing the text.

71) Whilst a great debt is owed to von Franz for her erudite publication of *Aurora consurgens*, her conclusion that the work was a 'product of the unconscious' (*Aurora*, 153–4), is untenable. She compared it unfavourably to St Bernard's sermons, where 'the associations follow in a known, i.e., conscious pattern', whereas in the case of *Aurora* 'the context can be reconstructed only if one considers the unconscious background' (*ibid.* 407, n.1). Cf. also Jung's statement that 'the misfortune of the alchemists was that they themselves did not know what they were talking about' (*Mysterium*, 125 [para 147]). Cf. also his description of alchemical symbolism as 'the nearest relatives of those serial fantasies which underlie the delusions of paranoid schizophrenia' (*ibid.* 105 [para 125]). Junker noted several 'schizophrenic' features in the *Book of the Holy Trinity* (*op.cit.* 72–6); and Obrist (*op.cit.* 117–18) uncritically cited R. Minzloff's opinion, written in 1874, that anyone meditating too deeply on the *Book of the Holy Trinity* could go mad 'and that is no doubt what happened to him [the Franciscan author]'. Regrettably, such views have coloured perceptions of *Aurora*. Cf. Newman's description of the illustrations as 'grotesque and sexually titillating imagery' (*God and the Goddesses*, 237). She also observed that more than one scholar had questioned 'the sanity' of the *Book of the Holy Trinity*'s author (*ibid.* 240).

CHAPTER 24

1) For the *Gospel of Philip* and alchemy, see Roberts, *Golden Shrine*, 112–19.

2) For this 'Morienus' tradition, see Principe, *Secrets of Alchemy*, 29–30. He reappears as Doctor Marianus in Goethe's *Faust*; see Raphael, *Goethe and the Philosophers' Stone*, 243–6.

3) For Suhrawardī's tradition of the transmission, see Corbin, *En Islam iranien*, 36; Kingsley, *Ancient Philosophy*, 388–90 (with further references).

4) For Ibn Masarrah's connection with Dhū'l-Nūn, see Goodman, 'Ibn Masarrah', 279; Stroumsa and Sviri, *JSAI* 36 (2009), 202 n.4, though they reject the notion that he was a Sufi (*ibid.* 209–11, 214). Yet his quest, like Dhū'l-Nūn's, was to behold 'the hidden' with the eye of the heart, and 'to know the science of the Book' (*ibid.* 218–19).

5) For this episode in Medina, see Goodman, *op.cit.* 278.

6) For this ancient Egyptian 'building' tradition, see Roberts, *My Heart My Mother*, 196–7. The Ptolemaic *Book of Traversing Eternity* makes numerous allusions to the Osirian cult rites as secret 'work', in the sense of 'building'; see Herbin, *Parcourir l'éternité*, 317.

7) For these 'building' references in the *Pyramid Texts*, see Jacq, *Tradition primordiale*, 166. For Osiris being

8) identified with the pyramid as 'the building work [*k3t*] of the pharaoh' enclosing the resurrection mysteries, see *ibid.* 239.

8) See Roberts, *My Heart My Mother*, 157–8.

9) For brief bibliographic details of his life, see Deladrière (trans.), *La vie merveilleuse*, 15–16, 57–8, 62–3. Though he is usually known by the name Dhū'l-Nūn, this was his surname, and his disciples addressed him as Abū-l-Fayḍ (*ibid.* 15). For his birthplace as Adwa in Nubia, see Edwards, *A Thousand Miles up the Nile*, 312, n.1. Edwards wondered if this was the same place as Kalat Adda on the east bank of the Nile, not far from Abu Simbel.

10) The Min temple at Akhmim was almost intact when visited in 1183 by the Spanish traveller Ibn Jubayr, who described it as 'one of the wonders of the world'; see Broadhurst (trans.), *Travels of Ibn Jubayr*, 53–5; El Daly, *Egyptology*, 51–2. For the temple's association with the Islamic tradition of the three Hermes, see van Bladel, *The Arabic Hermes*, 125–6; Roberts, *My Heart My Mother*, 246–7 n.13. For Dhū'l-Nūn and the temple, see El Daly (*op.cit.* 51), who mentions an unpublished long poem called *Poem on the Noble Craft* attributed to Dhū'l-Nūn, now in the British Library (*Add. Ms. 7590*), in which he affirms his study of Egyptian priestly knowledge still preserved on temple walls (*ibid.* 164). The Min temple was finally dismantled in the 14th century, the materials being used to build a *madrasa*, see van Bladel, *op.cit.* 126 n.21.

11) See, for example, Kákosy ('Das Ende des Heidentums in Ägypten', 76), who relegated Dhū'l-Nūn's discovery of knowledge in the Akhmim temple to the realm of fable.

12) For this episode, see Deladrière, *op.cit.* 63–4.

13) Kingsley tellingly observed that Dhū'l-Nūn's connection with alchemy and Hermetism in medieval sources has 'proved a major source of embarrassment for those interested in maintaining the purely Islamic nature of Sufism and denying its links with previous, non-Arab traditions' (*op.cit.* 389). For a positive assessment of his alchemical links, see *ibid.* 388–90; also El Daly, *op.cit.* 164.

14) See Sezgin, *Geschichte*, 4, 273; Ullmann, *Natur*, 196. For a treatise preserving a conversation between Dhū'l-Nūn and his pupil Ya'qūb about the 'stone of the wise', see *ibid.* 197. Schimmel refers to a brief poem about a black cat with a tail which is attributed to Dhū'l-Nūn and 'is supposed to contain alchemistic information' (*As Through a Veil*, 167–8). Deladrière (*op.cit.* 17) gives little credence to the idea of Dhū'l-Nūn as an alchemist, arguing that the treatises attributed to him were written by alchemists using his name as a cover.

15) For Isrāfīl as an alchemist, see *ibid.* 17.

16) *Ibid.* 20 (Isrāfīl as teacher); *ibid.* 236–7 (for Fātima). Another important teacher was Chuqrān al-'Abid from Kairouan; see *ibid.* 241–2. See also Deladrière's discussion, *ibid.* 19–24.

17) See Sezgin, *op.cit.* 273. Cf. also the Akhmim alchemist Buṭrus al-Ḥakīm al-Iḥmīmī, who apparently cited Hermes and Zosimos in his now-lost work (*ibid.* 274).

18) See Deladrière (trans.), *La vie merveilleuse*, 69–72.

19) For Ibn Umail's references to the *Epistle of the Secret*, see Stapleton *et al.*, *Ambix* 3 (1949), 78–9[XV]. The quotation from Dhū'l-Nūn states: 'And regarding this, Dhū'l-Nūn al-Miṣrī said: "When you have completed it [the Salting] thrice, do not fear the permanency of its operations."' See *ibid.* 77–8[XIV]; *al-Mā'al-waraqī* 45 (line 15).

20) When the *Epistle of the Secret* achieved the form in which it is known today is unclear. All the known manuscripts come from a much later date, raising the question as to whether the teaching was re-expressed at some point to bring it into wider circulation among Sufis. What matters, though, is that in the early tenth century Ibn Umail was associating its marriage with Dhū'l-Nūn.

21) In his unpublished doctoral thesis Hallum carefully traced the history of the *Book of Pictures* (or *Tome of Images* as he calls it) from the various stamps and owners' names included in the manuscript ('Zosimus Arabus', 261–7). The Bursa manuscript collection is uncatalogued, so how many other alchemical treatises it might have possessed is unclear, but Hallum (*ibid.* 267, n.42) noted it did contain the oldest known copy of a work by Jābir ibn-Ḥayyān (*Kitab al-sab'in*), which again indicates an interest in alchemy among these Turkish Sufis.

22) Rūmī's knowledge of Egyptian alchemy in the *Masnavi* has gone unnoticed, partly because scholars of the Persianate world are unfamiliar with its importance in Islam, but I have referred to passages where relevant.

23) Cf. also an illustration in the same manuscript of a great fish rising up towards a three-branched tree bearing nine fish-like heads, each one with sharp teeth and a single eye. To the left of the tree is a red-clothed man joyfully drumming; see Abt *et al.*, *Explanation of the Symbols* 2, 302, *fig.94*.

24) The 'Man of the Fish' reference is in verse 493: 'Therefore the best of God's creatures forbade us to prefer him to the Man of the Fish, although he is worthy of preference.' See Nicholson, *Studies*, 245. Nicholson (*ibid.* 245, n.493) thought the reference was to the Prophet Muḥammad's reported statement about Jonah, but equally it could be a coded reference to Dhū'l-Nūn.

25) For the 'two in one' in Ibn al-Fāriḍ's work, *see above, chapter 23, nn.22, 23*. For the 'two in one' in Jābirian alchemy and Ismā'īlī thought, associated with becoming the 'Glorious One' who accomplishes the resurrection in a miraculous birth whereby the Father becomes the Son, see Corbin, 'Le livre du glorieux', 102, also *Creative Imagination*, 147–52. For the 'two in one' in Ibn Umail's alchemy, see Abt, *op.cit.* 2, 153–4.

26) Verse 527; see Nicholson, *op.cit.* 247–8.

27) Verse 759; *ibid.* 265.

28) For the role of death-dealing and life-giving Wisdom as the 'Creative Feminine' holding the 'secret of the Godhead' in Ibn 'Arabī's spiritual life, see Corbin, *Creative Imagination*, 136–45, 326, n.16.

29) For a French translation, together with an introduction and notes, see Deladrière (trans.), *La vie merveilleuse*. Ibn 'Arabī's homage is best preserved in a manuscript copied in Cairo in 1312, and now in the Topkapi Library, Istanbul (no.1378). There is also a 19th-century copy preserved in a manuscript in the University Library, Leiden; see *ibid.* 37–8.

30) For this praise-poem, see *ibid.* 89–105. Whether this is an original composition by Dhū'l-Nūn or whether its arrangement into 22 verses reflects Ibn 'Arabī's own interpretation of Dhū'l-Nūn's spiritual path, as drawn from various sources, is unclear. For the different ways of reading Ibn 'Arabī's homage, see Deladrière, *ibid.* 36.

31) Hirtenstein, (trans.), *Alchemy of Human Happiness*, 50,

32) For Ibn 'Arabī's distinction between Muḥammad's ascent and the ascent of seekers on the path of sainthood, who share a covenant of mutual responsibilities with their Lord, see Uždavinys, *Ascent to Heaven*, 106–109. The aim of the 'saintly' path is to actualize and rediscover the divine Attributes and Names, particularly the divine form or eternal 'Name' hidden uniquely within each person; see also Corbin, *Creative Imagination*, 132, 160–61. This same aim informs Ibn Masarrah's spiritual journey in his *Epistle on Contemplation*, *see below n.41*. For a similar divine–human relationship defining Amun-Re's cult in ancient Egypt, marked by the reciprocal relationship between the god and his faithful worshippers, see Roberts, *Hathor Rising*, 72–5.

33) Verse 1 (Deladrière, *op.cit.* 89–90).

34) Verse 5 (*ibid.* 91, my translation from the French). In this verse Dhū'l-Nūn also mentions 'three darknesses' into which he has been born, bound to flesh and blood, and born in a masculine form, which, as Deladrière noted, echoes the 'three darknesses' in Sūra 39:6 in the *Qur'ān*. For these three darknesses defining foetal life in the womb, *see above chapter 22, n.11*.

35) Verses 6 to 9 particularly mention the 'trials', including 'straying from the path', 'lack of certainty', 'forgetfulness of God' and 'carelessness'. Verse 13 also refers to the 'covetousness' and 'concupiscence' that can afflict the soul. For Dhū'l-Nūn's participation in the 'royalty of your heavens', see verse 7.

36) Verse 12. Cf. also the importance of finding the Lord and Creator in the kingdom of *Malakūt* in Ibn Masarrah's *Epistle on Contemplation*, see Stroumsa and Sviri, *op.cit.* 223–4[36–8].

37) Verse 13.

38) Verses 14 and 15.

39) For the adept's ascent to the mysterious realm of 'Silence' and the revelation of the Face in Jābirian alchemy, see Corbin, 'Le livre du glorieux', 114.

40) Verse 17 (Deladrière [trans.], *La vie merveilleuse*, 102).

41) Verse 19. Cf. also Ibn Masarrah's contemplation of the signs of creation in his *Epistle on Contemplation*, which is closely associated with the contemplation of God's beautiful Names and Attributes; see Stroumsa and Sviri, *op.cit.* 225[43–44], 243 n.44.

42) These 'worlds' are named in verse 14: 'Place us, O my God, among those who have drunk the cup of purity, creating patience in them during the long trial, until their hearts will be transported in *Malakūt*, evolving in the midst of the mysterious veils of *Jabarūt*, their spirits seeking to reach the shade where the breeze blows, refreshing 'the people of desire…' (Deladrière, *op.cit.* 101). Previously, Dhū'l-Nūn included the 'corporeal' world when describing his birth and growth. For Dhū'l-Nūn as the first to give a systematic teaching concerning the 'states and stations' of the Sufi mystic way and the true nature of gnosis, see *Encyclopaedia of Islam* 2, 242; Deladrière, *op.cit.* 13.

43) Cf. also Muḥammad Wafā's naming of the three worlds as: 'Corporeal' (*Mulk*), 'Sovereignty' (*Malakūt*) and 'Omnipotence' (*Jabarūt*); see McGregor, *Sanctity and Mysticism*, 79, 98. He founded a Sufi order on Roda Island in Cairo in the 14th century, having previously established a flourishing Sufi community in Akhmim, which perhaps suggests a wish to connect with Dhū'l-Nūn (see *ibid.* 50, 57–8, 184, n.55), though McGregor makes no mention of Dhū'l-Nūn in his book. Muḥammad Wafā's son, Ali Wafā', reveals a distinctly alchemical outlook when stating how God's self-disclosure must be discovered among the lesser forms of creation, being hidden among base material existence as a 'black veil covering the happy moon-lit face' (*ibid.* 123–4), which McGregor noted was 'reminiscent of a Gnostic theurgy'. Cf. also his unusual embryological comparison, saying that without fire, a soul is like a cold, barren womb, unreceptive to the knowledge a teacher imparts, incapable of bringing it to birth (*ibid.* 128). McGregor compared Wafā' mysticism with Neoplatonism (see *ibid.* 5, 85), particularly emphasizing Ibn 'Arabī's influence (*ibid.* 89–91, 157–8), and without reference to alchemy and Dhū'l-Nūn's 'three worlds'. Similarly, Ibn Masarrah's contemplative path is interpreted as Neoplatonic, see Stroumsa and Sviri, *op.cit.* 207–11, even though they recognize it does not follow a typical Neoplatonic pattern (see *ibid.* 212).

44) The description comes from Ibn Jubayr, who visited the mosque in 1183, and also the mosque of Daud (David); see Broadhurst (trans.), *Travels of Ibn Jubayr*, 53; *also above, n.10*.

45) *Masnavi* 2. 933, 939; Nicholson (trans.), *The Mathnawī*, 270.

46) For the passage about Jonah, *ibid.* 3135–144; Nicholson (trans.), 384. For resurrection from the tomb of the heart, *ibid.* 3132. It echoes one of Dhū'l-Nūn's most famous sayings: 'The breasts of free people are the tombs of secrets'; see Deladrière, *op.cit.* 158, with n.180.

47) *Masnavi* 2. 1389; Nicholson (trans.), 292. As Nicholson noted (*ibid.* 292, n.3), this phrase meant that in his ecstasy Dhū'l-Nūn had no regard for his formal religion.

48) *Ibid.* 2. 1394–6; Nicholson, 293.

49) *Ibid.* 2. 1461; Nicholson, 296.

50) For these lines, see Schimmel, *As Through a Veil*, 124.

51) *Ibid.* 106. However, for Rūmī's deprecatory reference to copper, *see chapter 18, n.31*.

52) This criticism has been made against Corbin's studies of the eastern stream of Suhrawardī's genealogy, which he rooted in the Zoroastrian religion of ancient Persia (*see above, n.3*). Ziai ('Shihāb al-Dīn Suhrawardī', 443) maintained Suhrawardī's mention of Persian kings and heroes simply reflected the wish to invoke ancient authority rather than to recover some lost systematic philosophy. Similarly, Walbridge argued Suhrawardī was primarily a reviver not of Iranian but of Platonic mysticism (*Wisdom of the Mystic East*, 13).

53) Deladrière (trans.), *La vie merveilleuse*, 12, 106–107.

54) See Homerin, *From Arab Poet to Muslim Saint*, 52.

55) Abt *et al.*, *Explanation of the Symbols* 2, 253. Cf. also the 'green bird' in the *Book of Western Mercury* from the Jābirian corpus, where it states the 70 books are symbolic, and that 'myrtle branches' correspond to what Maria called the 'steps of gold' and Democritus the 'green bird'; see Corbin, 'Le livre du glorieux', 50.

BIBLIOGRAPHY

Abou Simbel. Desroches-Noblecourt, C. and Kuentz, C. *Le petit temple d'Abou Simbel*. 2 vols. Centre de documentation et d'études sur l'ancienne Égypte. Cairo, 1968.

Abram, D. *The Spell of the Sensuous: Perception and Language in a More-Than-Human-World*. Paperback edn. New York, 1997.

Abt, Th. *The Great Vision of Muḥammad Ibn Umail*. Los Angeles, 2003.

—*The Book of Pictures Muṣḥaf aṣ-ṣuwar by Zosimos of Panopolis: Edition of the Pictures and Introduction*. CALA (Supplement). Zurich, 2007.

— (ed.). *The Book of Pictures Muṣḥaf aṣ-ṣuwar by Zosimos of Panopolis 1: Facsimile*. CALA 2.1. Zurich, 2007.

— *The Book of Pictures Muṣḥaf aṣ-ṣuwar by Zosimos of Panopolis 2: Translation*. Trans. S. Fuad and Th. Abt. CALA 2.2. Zurich, 2011.

— Madelung, W. and Hofmeier, Th (eds.). *Book of the Explanation of the Symbols: Kitāb Ḥall ar-Rumūz by Muḥammad ibn Umail* 1. Text in Arabic and English. Trans. S. Fuad and Th. Abt. CALA 1. Zurich, 2003.

— *Book of the Explanation of the Symbols: Kitāb Ḥall ar-Rumūz by Muḥammad ibn Umail* 2. Psychological Commentary by Th. Abt. CALA 1B. Zurich, 2009.

Al-ʻIrāqī, Abū al-Qāsim Muḥammad ibn Aḥmad. *Kitāb al-aqālīm al-ṣabʻah (Book of the Seven Climes)*. British Library, London, Add. Ms. 25724.

Allen, J.P. 'The Cosmology of the Pyramid Texts' in J.P. Allen *et al.*, *Religion and Philosophy in Ancient Egypt*. Yale Egyptological Studies 3. New Haven, CT, 1989, 1–28.

Alliot, M. *Le culte d'Horus à Edfou au temps des Ptolémées*. Cairo, 1949–54.

Anderson, W. *Dante the Maker*. London, Boston and Henley, 1980.

Assmann, J. *Liturgische Lieder an den Sonnengott: Untersuchungen zur altägyptischen Hymnik* 1. MÄS 19. Berlin, 1969.

— *Ägyptische Hymnen und Gebete*. Zurich and Munich, 1975.

— *Das Grab der Mutirdis: Grabung im Asasif 1963–70*. AV 13. Mainz am Rhein, 1977.

— 'Primat und Transzendenz: Struktur und Genese der ägyptischen Vorstellung eines "Höchsten Wesens"' in W. Westendorf (ed.), *Aspekte der spätägyptischen Religion*. GOF IV, 9. Wiesbaden, 1979.

— *Sonnenhymnen in thebanischen Gräbern*. Theben 1. Mainz am Rhein, 1983.

— *Egyptian Solar Religion in the New Kingdom: Re, Amun and the Crisis of Polytheism*. Trans. A. Alcock. London and New York, 1995.

Aubourg, É. 'La date de conception du zodiaque du temple d'Hathor à Dendera'. *BIFAO* 95 (1995), 1–10.

— and Cauville, S. 'En ce matin du 28 Décembre 47…' in Clarysse *et al.* (eds.), *Egyptian Religion: The Last Thousand Years* 2, 767–72.

Aufrère, S. *L'univers minéral dans la pensée égyptienne*. 2 vols. IFAO. Cairo, 1991.

Bachelard, G. *Psychoanalysis of Fire*. Trans. A.C.M. Ross. Paperback edn. Boston, 1968.

Barb, A.A. 'Der Heilige und die Schlangen'. *MAGW* 82 (1952), 1–21.

Barguet, P. 'Remarques sur quelques scènes de la salle du sarcophage de Ramsès VI'. *RdÉ* 30 (1978), 51–6.

Barton, T. *Ancient Astrology*. Sciences of Antiquity series. London and New York, 1994.

Beck, R. *A Brief History of Ancient Astrology*. Paperback edn. Oxford, 2007.

Bell, L. 'Luxor Temple and the Cult of the Royal *Ka*'. *JNES* 44 (1985), 251–94.

Belmonte Avilés, J.A. *Pirámides, templos y estrellas: Astronomía y arqueología en el Egipto antiguo*. Barcelona, 2012.

Berlandini, J. 'L'"acéphale" et le rituel de revirilisation'. *OMRO* 73 (1993), 29–41.

Berlekamp, P. 'Painting as Persuasion: A Visual Defence of Alchemy in an Islamic Manuscript of the Mongol Period'. *Muqarnas* 20 (2003), 35–59.

Berthelot, M. *La chimie au moyen âge 2: L'alchimie syriaque* (with R. Duval). Paris, 1893.

— *La chimie au moyen âge 3: L'alchimie arabe* (with O. Houdas). Paris, 1893.

— and Ruelle, C-É. *Collection des anciens alchimistes grecs*. 3 vols. Paris, 1888.

Betrò, M. Review of G. Roulin, *Le livre de la nuit*. *Or* 67 (1998), 509–22.

Betz, H.D (ed.). *The Greek Magical Papyri in Translation, including the Demotic Spells 1: Texts*. Second edn. Chicago and London, 1992.

Birkhan, H. *Die alchemistische Lehrdichtung des Gratheus Filius Philosophi in Cod.Vind.2372: Zugleich ein Beitrag zur okkulten Wissenschaft im Spätmittelalter*. 2 vols. Vienna, 1992.

— 'The Alchemical Tracts of Gratheus Filius Philosophi in Codex Vindobonensis 2372' in von Martels (ed.), *Alchemy Revisited*, 162–70.

Bissing, F.W. von. 'Tombeaux d'époque romaine à Akhmîm'. *ASAE* 50 (1950), 547–76, with *pls.1–4*.

Bladel, K. van. *The Arabic Hermes: From Pagan Sage to Prophet of Science*. Oxford, 2009.

Bleeker, C.J. *Egyptian Festivals: Enactments of Religious Renewal*. Leiden, 1967.

Bomhard, A.-S. von. *The Egyptian Calendar: A Work for Eternity*. London, 1999.

Bonus, P. *The New Pearl of Great Price: A Treatise concerning the Treasure and most Precious Stone of the Philosophers*. Trans. A.E. Waite. Second impression. London, 1963.

Borghouts, J.F (trans.). *Ancient Egyptian Magical Texts*. Nisaba 9. Leiden, 1978.

Bourgeault, C. *The Meaning of Mary Magdalene: Discovering the Woman at the Heart of Christianity*. Boston and London, 2010.

Broadhurst, R.J.C (trans.). *The Travels of Ibn Jubayr*. London, 1952.

Bruyère, B. *Rapport sur les fouilles de Deir el Médineh (1930)*. FIFAO. Cairo, 1933.

Budge, E.A.W. *The Book of the Dead. The Chapters of Coming Forth by Day. The Egyptian Text according to the Theban Recension in Hieroglyphic edited from Numerous Papyri, with a Translation, Vocabulary, etc.* London, 1898.

— *Hieroglyphic Texts from Egyptian Stelae, etc., in the British Museum* 4. London, 1913.

Bulté, J. 'Une "Thouéris" rare et couronnée'. *RdÉ* 54 (2003), 1–20.

Cameron, A. *Daughters of Copper Woman*. Revised edn. Madeira Park, 2002.

Cauville, S. *Le temple de Dendera: Guide archéologique*. IFAO. Cairo, 1990.
Le temple de Dendera: Les chapelles osiriennes 1: *Transcription et traduction*. 2: *Commentaire*. 3: *Index*. IFAO. Cairo, 1997.
— *Le temple de Dendera* 10: *Les chapelles osiriennes*. 2 vols. IFAO. Cairo, 1997.
— *Le zodiaque d'Osiris*. Leuven, 1997.
— *Dendara III: Traduction*. OLA 95. Leuven, 2000.
— *Dendara: Les fêtes d'Hathor*. OLA 105. Leuven, 2002.
— 'Dendara: du sanatorium au tinctorium'. *BSFE* 161 (2004), 28–40.
— and Ibrahim Ali, M. *Dendara: Itinéraire du visiteur*. Leuven, Paris and Bristol, CT, 2015.

Chassinat, É. *Le temple de Dendera* 5. 2 vols. FIFAO. Cairo, 1947, 1952, reprinted 2004.
— *Le mystère d'Osiris au mois de Khoiak*. 2 vols. IFAO. Cairo, 1966, 1968.

Clarysse, W., Schoors, A. and Willems, H (eds.). *Egyptian Religion: The Last Thousand Years: Studies Dedicated to the Memory of Jan Quaegebeur* 2. OLA 85. Leuven, 1998.

Cleary, T (trans.). *The Secret of the Golden Flower: The Classic Chinese Book of Life*. San Francisco, 1991.

Conman, J. 'The Egyptian Origins of Planetary Hypsomata'. *DE* 64 (2006–2009), 7–20.

Copenhaver, B.P. *Hermetica: The Greek Corpus Hermeticum and the Latin Asclepius in a New English Translation, with Notes and Introduction*. Third reprinted paperback edition. Cambridge, 1998.

Corbin, H. 'Le livre du glorieux de Jâbir ibn Ḥayyân (Alchimie et Archétypes)'. *Eranos-Jahrbuch* 18 (1950), 47–114.
— *Creative Imagination in the Ṣūfism of Ibn 'Arabī*. Trans. R. Manheim. London, 1969.
— *En Islam iranien: Aspects spirituels et philosophiques 2: Sohrawardî et les Platoniciens de Perse*. Paris, 1971.
— 'The Realism and Symbolism of Colors in Shiite Cosmology according to the "Book of the Red Hyacynth" by Shaykh Muḥammad Karīm-Khān Kirmānī (d.1870)' in K. Ottmann (ed.), *Color Symbolism: The Eranos Lectures*. Second revised edition. Putnam, CT, 2005.

Dalrymple, W. *Nine Lives: In Search of the Sacred in Modern India*. London, 2009.

Dante Alighieri. *La Divina Commedia: Inferno, Purgatorio, Paradiso*. H. Oelsner (ed.), English translations by J.A. Carlyle, T. Okey and P.H. Wicksteed. London and Toronto, 1933.

Darnell, J.C. *The Enigmatic Netherworld Books of the Solar-Osirian Unity: Cryptographic Compositions in the Tombs of Tutankhamun, Ramesses VI and Ramesses IX*. OBO 198. Freiburg and Göttingen, 2004.
— 'Ancient Egyptian Cryptography: Graphic Hermeneutics' in D. Klotz and A. Stauder (eds.), *Enigmatic Writing in the Egyptian New Kingdom*. ZÄS Beihefte 1. Berlin, (forthcoming).

Daumas, F. 'Sur trois représentations de Nout à Dendara'. *ASAE* 51 (1951), 373–400.
— 'La valeur de l'or dans la pensée égyptienne'. *RHR* 149 (1956), 1–17.

— 'Les objets sacrés de la déesse Hathor à Dendara'. *RdÉ* 22 (1970), 63–78.
— 'L'alchimie a-t-elle une origine égyptienne?' in *Das römisch-byzantinische Ägypten. Akten des internationalen Symposions 26–30 September 1978 in Trier*. Mainz am Rhein, 1983, 109–18.
— *Le temple de Dendara* 9. 2 vols. IFAO. Cairo, 1987.

Davies, N. de G. *The Tomb of Puyemrê at Thebes*. Robb de Peyster Tytus Memorial Series 2–3. New York, 1922–3.

Deladrière, R (trans.). Ibn 'Arabî: *La vie merveilleuse de Dhû-l-Nûn l'Égyptien d'après le Traité hagiographique al-Kawkab al-durrî fî manâqib Dhî-l-Nûn al-Misrî 'L'astre éclatant des titres de gloire de Dhû-l-Nûn l'Égyptien'*. Paris, 1988.

Derchain, P. *Hathor Quadrifrons: Recherches sur la syntaxe d'un mythe égyptien*. Uitgaven van het nederlands historisch archaeologisch Instituut te Istanbul 28. Istanbul, 1972.
— 'L'*Atelier des Orfèvres* à Dendara et les origines de l'alchimie'. *CdÉ* 65 (130) (1990), 219–42.

Desroches-Noblecourt, C. 'Un "Lac de Turquoise": Godets à onguents et destinées d'outre-tombe dans l'Égypte ancienne'. *Monuments et mémoires de la Fondation Eugène Piot* 47 (1953), 1–34.
— *Ramsès le Grand, Exhibition Catalogue*. Galeries nationale du Grand Palais, Paris, 1976.
— 'Le zodiaque de Pharaon'. *Archéologia* 292 (1993), 21–45.
— and Kuentz, C. *Le petit temple d'Abou Simbel*. 2 vols. Centre de documentation et d'études sur l'ancienne Égypte. Cairo, 1968.

Dijk, J. van (ed.). *Essays on Ancient Egypt in Honour of Herman te Velde*. Egyptological Memoirs 1. Groningen, 1997.

Diodorus Siculus, *Library of History* 1. Trans. E. Murphy, *The Antiquities of Egypt: A Translation with Notes of Book 1 of the Library of History of Diodorus Siculus*. Revised edn. New Brunswick and London, 1990.

Doresse, J. *The Secret Books of the Egyptian Gnostics: An Introduction to the Gnostic Coptic Manuscripts discovered at Chenoboskion*. London, 1960.

Dronke, P. 'Tradition and Innovation in Medieval Western Colour-Imagery' in P. Dronke (ed.), *The Medieval Poet and his World*. Storia e letteratura raccolta di Studi e Testi 164. Rome, 1984, 55–103.

Dufault, O. 'Transmutation Theory in the Greek Alchemical Corpus'. *Ambix* 62 (2015), 215–44.

Ebeling, F. *Das Geheimnis des Hermes Trismegistos: Geschichte des Hermetismus von der Antike bis zur Neuzeit*. Munich, 2005.

Eckmann, C. and Schafik, S. *"Leben dem Horus Pepi": Restaurierung und technologische Untersuchung der Metallskulpturen des Pharao Pepi I. aus Hierakonpolis*. Römisch-Germanisches Zentralmuseum Mainz, Forschungsinstitut für Vor-und Frühgeschichte, Monographien 59. Mainz, 2005.

Edwards, A.B. *A Thousand Miles up the Nile*. Reprinted paperback edn. London, 1984.

El-Daly, O. *Egyptology: The Missing Millennium. Ancient Egypt in Medieval Arabic Writings*. London, 2005.

El-Raziq, M. Abd., Castel, G., Tallet, P. and Ghica, V. *Les inscriptions d'Ayn Soukhna*. MIFAO 122. Cairo, 2002.

El-Sawi, A. 'The Deification of Sety Ist in his Temple of Abydos'. *MDAIK* 43 (1987), 225–7.

Eliade, M. *The Forge and the Crucible*. Trans. S. Corrin. Second paperback edn. Chicago and London, 1978.

The Encyclopaedia of Islam. New edition. 12 vols. Leiden,1960–2004.

Erman, A. *Hymnen an das Diadem der Pharaonen*. Berlin, 1911.

Evans, J. 'The Astrologer's Apparatus: A Picture of Professional Practice in Greco-Roman Egypt'. *JHA* 35 (2004), 1–44.

Fabricius, J. *Alchemy: The Medieval Alchemists and their Royal Art*. Revised edn. Wellingborough, 1989.

Falchetti, A.M. 'The Seed of Life: The Symbolic Power of Gold-Copper Alloys and Metallurgical Transformations' in J. Quilter and J. W. Hoopes (eds.), *Gold and Power in Ancient Costa Rica, Panama, and Colombia: A Symposium at Dumbarton Oaks, 9 and 10 October 1999*. Dumbarton Oaks, 2003, 345–81.

Faulkner, R.O (trans.). *The Ancient Egyptian Book of the Dead*, ed. C. Andrews. Revised paperback fourth impression. London, 1993.

— *The Ancient Egyptian Pyramid Texts Translated into English*. Paperback edition. Warminster and Oak Park, Illinois [No date].

Ferguson, P (trans.). *Aurora Consurgens (Morning Rising): Books I and II, attributed to St. Thomas Aquinas*. Magnum Opus Hermetic Sourceworks 40. Glasgow, 2011.

Festugière, A.-J. *Hermétisme et mystique païenne*. Paris, 1967.

— *Proclus: Commentaire sur le "Timée". Traduction et notes*. 5 vols. Paris, 1966–68.

— *La révélation d'Hermès Trismégiste 1: L'astrologie et les sciences occultes*. Reprinted second edition. Paris, 1989.

Finlay, V. *Colour: Travels through the Paintbox*. London, 2002.

Fowden, G. *The Egyptian Hermes: A Historical Approach to the Late Pagan Mind*. Paperback edn. Princeton, NJ, 1993.

Frankfort, H. *Kingship and the Gods: A Study of Ancient Near Eastern Religion as the Integration of Society and Nature*. Paperback edn. Chicago, 1978.

— de Buck, A. and Gunn, B. *The Cenotaph of Seti I at Abydos*. 2 vols. The Egypt Exploration Society Memoir 39. London, 1933.

Frankfurter, D. *Religion in Roman Egypt: Assimilation and Resistance*. Princeton, NJ, 1998.

Franz, M.-L. von (ed.). *Aurora Consurgens: A Document Attributed to Thomas Aquinas on the Problem of Opposites in Alchemy*. Trans. R.F.C. Hull and A.S.B. Glover. London and New York, 1966.

— *Alchemy: An Introduction to the Symbolism and the Psychology*. Toronto, 1980.

Gabriele, M (ed.). *Le don de Dieu: Introduction, transcription et notes par Mino Gabriele; suivi d'une autre version commentée par Albert Poisson*. (Cover title: *Le précieux don de Dieu*). Trans. F. Isidori. Paris, 1988.

Gage, J. *Colour and Culture: Practice and Meaning from Antiquity to Abstraction*. London, 1993.

Gamper, R. and Hofmeier, T. *Alchemische Vereinigung: Das Rosarium Philosophorum und sein Besitzer Bartlome Schobinger*, mit einem Beitrag von D. Oltrogge and R. Fuchs. Zurich, 2014.

Ganzenmüller, W. 'Das Buch der heiligen Dreifaltigkeit: Eine deutsche Alchemie aus dem Anfang des 15. Jahrhunderts' in W. Ganzenmüller, *Beiträge zur Geschichte der Technologie und der Alchemie*. Weinheim, 1956, 231–71.

Gardiner, A. *Egyptian Grammar: Being an Introduction to the Study of Hieroglyphs*. Third revised edn. London, 1973.

Gardiner, A.H., Peet, T.E. and Černý, J. *The Inscriptions of Sinai*. 2 vols. The Egypt Exploration Society Memoir 45. London, 1952, 1955.

George, B. *Zu den altägyptischen Vorstellungen vom Schatten als Seelen*. Bonn, 1970.

Germer, R. *Die Textilfärberei und die Verwendung gefärbter Textilien im alten Ägypten*. ÄA 53. Wiesbaden, 1992.

Germond, P. *Sekhmet et la protection du monde*. AH 9. Basle and Geneva, 1981.

Giumlia-Mair, A. and Quirke, S. 'Black Copper in Bronze Age Egypt'. *RdÉ* 48 (1997), 95–108.

Goddio, F. and Masson-Berghoff, A. *The BP Exhibition: Sunken Cities. Egypt's Lost World*. London, 2016.

Goodman, L.E. 'Ibn Masarrah' in Nasr and Leaman (eds.), *History of Islamic Philosophy*, 277–93.

Goyon, J.C. 'Répandre l'or et éparpiller la verdure: Les fêtes de Mout et d'Hathor à la néoménie d'Epiphi et les prémices des moissons' in van Dijk (ed.), *Essays on Ancient Egypt in Honour of Herman te Velde*, 85–100.

Graefe, E. *Untersuchungen zur Wortfamilie bi3*. Cologne, 1971.

Graindorge-Héreil, C. *Le dieu Sokar à Thèbes au Nouvel Empire 1: Textes*. GOF IV, 28,1. Wiesbaden, 1994.

Griffiths, J. Gwyn (ed.). *Plutarch's de Iside et Osiride*. [Cardiff], 1970.

Griggs, E.L (ed.). *The Collected Letters of Samuel Taylor Coleridge 4: 1815–1819*. Revised edn. Oxford, 2000.

Guilmant, F. *Le tombeau de Ramsès IX*. MIFAO 15. Cairo, 1907.

Haaning, A. 'Jung's quest for the "Aurora consurgens"'. *Journal of Analytical Psychology* 59 (2014), 8–30.

Haarmann, U.W. 'Islam and Ancient Egypt' in D.B. Redford (ed.), *The Oxford Encyclopedia of Ancient Egypt* 2. Oxford, 2001.

Haleem, M. Abdel. 'Early *kalām*' in Nasr and Leaman (eds.), *History of Islamic Philosophy*, 71–88.

Halleux, R. *Les textes alchimiques*. Turnhout, 1979.

— *Les alchimistes grecs 1: Papyrus de Leyde, Papyrus de Stockholm, fragments de recettes*. Paris, 1981.

Hallum, B.C. 'Zosimus Arabus: The Reception of Zosimos of Panopolis in the Arabic/Islamic World'. Unpublished Ph.D. thesis, Warburg Institute, London, 2008.

— 'The *Tome of Images*: An Arabic Compilation of Texts by Zosimos of Panopolis and a Source of the *Turba Philosophorum*'. *Ambix* 56 (2009), 76–88.

Harris, J.R. *Lexicographical Studies in Ancient Egyptian Minerals*. Deutsche Akademie der Wissenschaften zu Berlin. Institut für Orientforschung, Veröffentlichung 54. Berlin, 1961.

Haskins, S. *Mary Magdalen: Myth and Metaphor*. Paperback edn. London, 1994.

Herbert, E.W. *Red Gold of Africa: Copper in Precolonial History and Culture*. Madison, WI, 1984.

Herbin, F.R. *Le livre de parcourir l'éternité*. OLA 58. Leuven, 1994.

Hershbell, J.P. 'Democritus and the Beginnings of Greek Alchemy'. *Ambix* 34 (1987), 5–20.

Hildegard of Bingen. *Scivias*. Trans. Mother Columba Hart and J. Bishop. New York and Mahwah, NJ, 1990.

— *Illuminations of Hildegard of Bingen*. Text by Hildegard of Bingen with commentary by Matthew Fox. Santa Fe, 1985.

Hillman, J. *A Blue Fire: Selected Writings by James Hillman* (ed. T. Moore). Paperback edn. New York, 1991.

Hirtenstein, S (trans.). Muḥyiddīn Ibn 'Arabī: *The Alchemy of Human Happiness (fī ma'rifat kīmiyā' al-sa'āda).* Oxford, 2017.

Holmyard, E.J (ed.). Abu'l-Qāsim Muḥammad ibn Aḥmad al-'Irāqī, *Kitāb al-'Ilm al-muktasab fi zirā'at adh-ḏhahab: Book of Knowledge acquired concerning the Cultivation of Gold.* Paris, 1923.

—'Abu'l-Qāsim al-'Irāqī'. *Isis* 8 (1926), 403–26.

Homerin, Th. E. *From Arab Poet to Muslim Saint: Ibn al-Fāriḍ, His Verse, and His Shrine.* Paperback edn. Cairo and New York, 2001.

Hopkins, A.J. *Alchemy: Child of Greek Philosophy.* New York, 1934.

Hornung, E. *Das Amduat: Die Schrift des verborgenen Raumes.* 3 vols. ÄA 7, 13. Wiesbaden, 1963–67.

— *Das Buch der Anbetung des Re im Westen (Sonnenlitanei): Nach den Versionen des Neuen Reiches.* 2 vols. AH 2–3. Geneva, 1975–6.

— *Das Buch von den Pforten des Jenseits: Nach den Versionen des Neuen Reiches.* 2 vols. AH 7–8. Basle and Geneva, 1979–80.

— *The Valley of the Kings: Horizon of Eternity.* Trans. D. Warburton. New York, 1990.

— *Zwei ramessidische Königsgräber: Ramses IV. und Ramses VII.* Theben 11. Mainz am Rhein, 1990.

— *Das esoterische Ägypten: Das geheime Wissen der Ägypter und sein Einfluß auf das Abendland.* Munich, 1999.

Iamblichus. *De mysteriis.* Trans. E.C. Clarke, J.M. Dillon and J.P. Hershbell. Atlanta, 2003.

Jacq, C. *Egyptian Magic.* Trans. J.M. Davis. Warminster and Chicago, IL, 1985.

— *La tradition primordiale de l'Égypte ancienne selon les Textes des Pyramides.* Paris, 1998.

Jong, H.M.E de. *Michael Maier's Atalanta Fugiens: Sources of an Alchemical Book of Emblems.* Leiden, 1969.

Jung, C.G. 'The Psychology of the Transference Interpreted in Conjunction with a Set of Alchemical Pictures' in *The Practice of Psychotherapy: Essays on the Psychology of the Transference and other Subjects.* Collected Works 16. Trans. R.F.C. Hull. Second revised edn. New York 1966, 163–323.

— 'Paracelsus as a Spiritual Phenomenon' in *Alchemical Studies.* Collected Works 13. Trans. R.F.C. Hull. Princeton, NJ, 1967, 109–89.

— 'The Visions of Zosimos' in *Alchemical Studies.* Collected Works 13. Trans. R.F.C. Hull. Princeton, NJ, 1967, 57–108.

— *Psychology and Alchemy.* Collected Works 12. Trans. R.F.C. Hull. First paperback printing. London, 1980.

— *Mysterium Coniunctionis: An Inquiry into the Separation and Synthesis of Psychic Opposites in Alchemy.* Collected Works 14. Trans. R.F.C. Hull. Paperback edn. Princeton, NJ, and London, 1977.

Junker, H. 'Poesie aus der Spätzeit'. *ZÄS* 43 (1906), 101–27.

Junker, U. *Das "Buch der heiligen Dreifaltigkeit" in seiner zweiten, alchemistischen Fassung (Kadolzburg 1433).* Cologne, 1986.

Kahn, D. *Hermès Trismégiste: La Table d'Émeraude et sa tradition alchimique.* Paris, 1994.

— and Matton, S (eds.). *Alchimie: Art, histoire et mythes. Actes du 1er colloque international de la Société d'Étude*

de l'Histoire de l'Alchimie (Paris, Collège de France, 14–15–16 mars 1991). Textes et Travaux de Chrysopœia 1. Paris and Milan, 1995.

Kákosy, L. 'Decans in Late-Egyptian Religion'. *Oikumene* 3 (1982), 163–91.

— 'Das Ende des Heidentums in Ägypten' in P. Nagel (ed.), *Graeco-Coptica: Griechen und Kopten im byzantinischen Ägypten.* Halle, 1984, 61–76.

Kaper, O.E. 'The Astronomical Ceiling of Deir el-Haggar in the Dakhleh Oasis'. *JEA* 81 (1995), 175–95.

— 'A Fragment from the Osiris Chapels at Dendera in Bristol'. *JEOL* 41 (2008–9), 31–45.

Karageorghis, V. *Kition: Mycenaean and Phoenician Discoveries in Cyprus.* London, 1976.

Kaske, C.V. and Clark, J.R (eds.). Marsilio Ficino: *Three Books on Life. A Critical Edition and Translation with Introduction and Notes.* Binghamton, NY, 1989.

Kees, H. 'Die Schlangensteine und ihre Beziehungen zu den Reichsheiligtümern'. *ZÄS* 57 (1922), 120–36.

— 'Göttinger Totenbuchstudien: Ein Mythus vom Königtum des Osiris in Herakleopolis aus dem Totenbuch Kap.175'. *ZÄS* 65 (1930), 65–83.

Kingsley, P. *Ancient Philosophy, Mystery and Magic: Empedocles and Pythagorean Tradition.* Oxford, 1995.

Kitchen, K.A. *Ramesside Inscriptions: Historical and Biographical.* 8 vols. Oxford, 1968–90.

— *Pharaoh Triumphant: The Life and Times of Ramesses II, King of Egypt.* Warminster, 1982.

Klibansky, R., Panofsky, E. and Saxl, F. *Saturn and Melancholy: Studies in the History of Natural Philosophy, Religion and Art.* London, 1964.

Krupp, E.C. 'Astronomers, Pyramids, and Priests' in E.C. Krupp (ed.), *In Search of Ancient Astronomies.* Paperback edn. Harmondsworth, Middlesex, 1984, 186–218.

Kurth, D. 'Der kosmische Hintergrund des grossen Horusmythos von Edfu'. *RdÉ* 34 (1982–3), 71–5.

— '"Same des Stieres" und "Same". Zwei Bezeichnungen der Maat' in *Studien zu Sprache und Religion Ägyptens 1: Sprache. Zu Ehren von Wolfhart Westendorf überreicht von seinen Freunden und Schülern.* Göttingen, 1984, 273–81.

— 'Zur Nord–Süd–Fahrt des Sonnengottes'. *GM* 83 (1984), 39–41.

— 'Zu den Darstellungen Pepi I. im Hathortempel von Dendera' in W. Helck (ed.), *Tempel und Kult.* ÄA 46. Wiesbaden, 1987, 1–23.

Lalouette, C. 'Le "firmament de cuivre": contribution à l'étude du mot bi3'. *BIFAO* 79 (1979), 333–53.

Leclant, J. 'Sur un contrepoids de menat au nom de Taharqa: allaitement et "apparition" royale' in *Mélanges Mariette.* BdÉ 32. Cairo, 1961, 251–84.

Leitz, Ch. *Studien zur ägyptischen Astronomie.* ÄA 49. Wiesbaden, 1989.

— 'Die Sternbilder auf dem rechteckigen und runden Tierkreis von Dendara'. *SAK* 34 (2006), 285–318.

— Mendel, D. and El-Masry, Y. *Athribis 2. Der Tempel Ptolemaios XII: Die Inschriften und Reliefs der Opfersäle, des Umgangs und der Sanktuarräume.* 3 vols. IFAO. Cairo, 2010.

Letrouit, J. 'Chronologie des alchimistes grecs' in Kahn and Matton (eds.), *Alchimie: Art, histoire et mythes,* 11–93.

Lexikon der Ägyptologie. 7 vols. Wiesbaden, 1972–92.

Lichtheim, M. *Ancient Egyptian Literature: A Book of*

Readings. 3 vols. Berkeley, Los Angeles and London, 1973–80.

Lieven, A. von. *Der Himmel über Esna: Eine Fallstudie zur religiösen Astronomie in Ägypten am Beispiel der kosmologischem Decken- und Architravinschriften im Tempel von Esna.* ÄA 64. Wiesbaden, 2000.

— *Grundriss des Laufes der Sterne: Das sogenannte Nutbuch 1: Text.* Carsten Niebuhr Institute publications 31. Copenhagen, 2007.

— 'Im Schatten des Goldhauses: Berufsgeheimnis und Handwerkerinitiation im Alten Ägypten'. *SAK* 36 (2007), 147–55.

Lindsay, J. *The Origins of Alchemy in Graeco-Roman Egypt.* London, 1970.

Lippmann, E.O. von. *Entstehung und Ausbreitung der Alchemie mit einem Anhange: Zur älteren Geschichte der Metalle. Ein Beitrag zur Kulturgeschichte.* 3 vols. Berlin, 1919–54.

Lory, P. *Alchimie et mystique en terre d'Islam.* Lagrasse, 1989.

Maehler, H. and Strocka, V.M (eds.). *Das ptolemäische Ägypten: Akten des internationalen Symposions 27.–29. September 1976 in Berlin.* Mainz am Rhein, 1978.

Mahé, J.-P. *Hermès en Haute-Égypte 1: Les textes hermétiques de Nag Hammadi et leurs parallèles Grecs et Latins.* Bibliothèque copte de Nag Hammadi 3. Quebec, 1978.

Manilius. *Astronomica.* Trans. G.P. Goold. Cambridge (Mass.), 1977.

Manniche, L. 'The Complexion of Queen Ahmosi Nefertere'. *AcOr* 40 (1979), 11–19.

Martelli, M. *The Four Books of Pseudo-Democritus.* Ambix 60 (Supplement 1). Leeds, 2013.

Martels, Z.R.W.M. von (ed.). *Alchemy Revisited: Proceedings of the International Conference on the History of Alchemy at the University of Groningen, 17–19 April 1989.* Collection de Travaux de l'Académie Internationale d'Histoire des Sciences 33. Leiden, New York, Copenhagen, Cologne, 1990.

McGregor, R.J.A. *Sanctity and Mysticism in Medieval Egypt: The Wafāʾ Sufi Order and the Legacy of Ibn ʿArabī.* Albany, NY, 2004.

McLean, A. URL:http://www.alchemywebsite.com/Emblems manuscripts.html.

Medinet Habu. The Epigraphic Survey, Medinet Habu 6: The Temple Proper Part 2. The Re Chapel, the Royal Mortuary Complex, and Adjacent Rooms, with Miscellaneous Material from the Pylons, the Forecourts, and the First Hypostyle Hall. University of Chicago Oriental Institute Publications 84. Chicago, 1963.

Meeks, D. 'Les "*quatre Ka*" du démiurge memphite'. *RdÉ* 15 (1963), 35–47.

Mercer, S.A.B. *The Pyramid Texts in Translation and Commentary.* 4 vols. New York, London and Toronto, 1952.

Mertens, M. 'Une scène d'initiation alchimique: La "Lettre d'Isis à Horus"'. *RHR* 205(1) (1988), 3–23.

— *Les alchimistes grecs 4(1): Zosime de Panopolis. Mémoires authentiques.* Paris, 1995.

— 'Alchemy, Hermetism and Gnosticism at Panopolis c.300 A.D.: The Evidence of Zosimus' in A. Egberts, B.P. Muhs and J. van der Vliet (eds.), *Perspectives on Panopolis: An Egyptian Town from Alexander the Great to the Arab*

Conquest. Acts from an International Symposium held in Leiden on 16, 17 and 18 December 1998. Leiden, 2002, 165–75.

— 'Graeco-Egyptian Alchemy in Byzantium' in P. Magdalino and M. Mavroudi (eds.), *The Occult Sciences in Byzantium.* Geneva, 2006, 205–30.

Mistree, K.P. *Zoroastrianism: An Ethnic Perspective.* Bombay, 1982.

Moran, W.L. *The Amarna Letters.* Baltimore and London, 1992.

Moret, A. *Le rituel du culte divin journalier en Égypte d'après les Papyrus de Berlin et les textes du temple de Séti Ier à Abydos.* Annales du Musée Guimet, Bibliothèque d'Études 14. Paris, 1902.

Motte, L. 'La vache multicolore et les trois pierres de la régéneration' in *Études coptes III. Troisième journée d'études, Musée du Louvre, 23 Mai 1986.* Cahiers de la Bibliothèque Copte 4. Louvain and Paris, 1989, 130–49.

Moussa, A.M. and Altenmüller, H. *Das Grab des Nianchchnum und Chnumhotep.* AV 21. Mainz am Rhein, 1977.

Müller-Roth, M. *Das Buch vom Tage.* OBO 236. Freiburg and Göttingen, 2008.

Nanji, A. 'Ismāʿīlī Philosophy' in Nasr and Leaman (eds.), *History of Islamic Philosophy*, 144–54.

Nasr, S.H. *An Introduction to Islamic Cosmological Doctrines: Conceptions of Nature and Methods Used for its Study by the Ikhwān al-Ṣafāʾ, al-Bīrūnī, and Ibn Sīnā.* Cambridge (Mass.), 1964.

— and Leaman, O (eds.). *History of Islamic Philosophy.* Reprinted paperback edn. London and New York, 2003.

Naville, É (ed.). *Das aegyptische Todtenbuch der XVIII. bis XX. Dynastie aus verschiedenen Urkunden 1: Text und Vignetten.* Berlin, 1886.

Naydler, J. *Temple of the Cosmos: The Ancient Egyptian Experience of the Sacred.* Rochester, VT, 1996.

— *Shamanic Wisdom in the Pyramid Texts: The Mystical Tradition of Ancient Egypt.* Rochester, VT, 2005.

— 'Plato, Shamanism and Ancient Egypt' in *Temenos Academy Review* (2006), 67–92.

Neugebauer, O. 'Demotic Horoscopes'. *JAOS* 63 (1943), 115–27.

— *A History of Ancient Mathematical Astronomy* 2. Studies in the History of Mathematics and Physical Sciences 1. Berlin and New York, 1975.

— and Parker, R.A. *Egyptian Astronomical Texts.* 3 vols. Brown Egyptological Studies 3, 5, 6. Providence, Rhode Island, and London, 1960–69.

Newman, B. *God and the Goddesses: Vision, Poetry, and Belief in the Middle Ages.* Paperback edn. Philadelphia, PA, 2005.

Nibbi, A. 'Some Remarks on *Copper*'. *JARCE* 14 (1977), 59–66.

Nicholson, P.T. and Shaw, I (eds.). *Ancient Egyptian Materials and Technology.* Cambridge, 2000.

Nicholson, R.A. *Studies in Islamic Mysticism.* Cambridge, 1921.

Nock, A.D. and Festugière, A.-J (eds.). *Corpus Hermeticum 4: Fragments extraits de Stobée (XXIII–XXIX), fragments divers.* Paris, 1954.

Obrist, B. *Les débuts de l'imagerie alchimique (XIVe–Xve siècles).* Paris, 1982.

Ogden, J. 'Metals' in P.T. Nicholson and I. Shaw (eds.), *Ancient Egyptian Materials and Technology*. Cambridge, 2000, 148–76.

Osing, J. *et al. Denkmäler der Oase Dachla aus dem Nachlass von Ahmed Fakhry*. AV 28. Mainz am Rhein, 1982.

Papathanassiou, M.K. 'L'œuvre alchimique de Stéphanos d'Alexandrie: structure et transformations de la matière, unité et pluralité, l'énigme des philosophes'. *Chrysopœia 7* (2000–2003), 11–31.

Parkinson, R.B (trans.). *The Tale of Sinuhe and other Ancient Egyptian Poems, 1940–1640 BC*. Oxford World's Classics. Paperback edn. Oxford, 1998.

Pastoureau, M. 'Couleur verte et sensibilité médiévale: de l'héraldique à l'alchimie et retour' in Kahn and Matton (eds.), *Alchimie: Art, histoire et mythes*, 151–6.

Patai, R. *The Jewish Alchemists: A History and Source Book*. Princeton, NJ, 1994.

Pereira, M. *The Alchemical Corpus Attributed to Raymond Lull*. Warburg Institute Surveys and Texts 18. London, 1989.

Piankoff, A. *Le livre du jour et de la nuit*. BdÉ 13. Cairo, 1942.

— *La création du disque soleil*. BdÉ 19. Cairo, 1953.

— *The Tomb of Ramesses VI*. 2 vols. Bollingen Series 40. New York, 1954.

Pinch, G. *Votive Offerings to Hathor*, Oxford, 1993.

Plato, *Phaedrus*. Trans. C.J. Rowe. Warminster, 1986.

— *Politicus (The Statesman)*. Trans. J.B. Skemp. Paperback edn. London, 1961.

— *Plato's Cosmology: The Timaeus of Plato translated with a Running Commentary*. Trans. F.M. Cornford. London, 1937.

Pliny the Elder. *Natural History 9: Libri XXXIII–XXXV*. Trans. H. Rackham. London and Cambridge (Mass.), 1952.

Ploss, E.E. *et al. Alchimia: Ideologie und Technologie*. Munich, 1970.

Plutarch. 'The Dialogue on Love' in *Moralia* 9. Trans. E.L. Minar, F.H. Sandbach and W.C. Helmbold. London and Cambridge (Mass.), 1961.

— *De Iside et Osiride*: See J. Gwyn Griffiths (ed.), *Plutarch's de Iside et Osiride*. [Cardiff], 1970.

Powell, R.A. *Hermetic Astrology: Towards a New Wisdom of the Stars 1: Astrology and Reincarnation*. Kinsau, 1987.

Principe, L. *The Secrets of Alchemy*. Chicago, 2013.

Quack, J.F. Review of G. Roulin, *Le livre de la nuit. Die Welt des Orients* 28 (1997), 177–81.

— 'Les mages égyptianisés? Remarks on Some Surprising Points in supposedly Magusean Texts'. *JNES* 65 (2006), 267–82.

Quaegebeur, J. 'Reines ptolémaïques et traditions égyptiennes' in Maehler and Strocka (eds.), *Das ptolemäische Ägypten*, 245–62.

— 'Cléopâtre VII et le temple de Dendara'. *GM* 120 (1991), 49–72.

Quirke, S. *The Cult of Ra: Sun-worship in Ancient Egypt*. London, 2001.

— *Going Out in Daylight: prt m hrw. The Ancient Egyptian Book of the Dead: Translation, Sources, Meanings*. London, 2013.

The Qur'ān. Trans. A. Jones. [Cambridge], 2007.

Ramsay, J. *Alchemy: The Art of Transformation*. Reprinted edn. Shaftesbury, 2017.

Raphael, A. *Goethe and the Philosophers' Stone: Symbolical Patterns in 'The Parable' and the Second Part of 'Faust'*. London, 1965.

Ratié, S. 'Un Osiris alchimique au musée de Chambéry'. *La revue du Louvre et des musées de France* (1980), 219–21.

Richter, B.A. *The Theology of Hathor of Dendera: Aural and Visual Scribal Techniques in the Per-Wer Sanctuary*. Wilbour Studies in Egyptology and Assyriology 4. Atlanta, 2016.

Richter, T.S. 'Miscellanea magica, III: ein vertauschter Kopf? Konjekturvorschlag für P. Berlin P 8313 ro, col.II, 19–20'. *JEA* 93 (2007), 259–63.

Riggs, C. 'Roman mummy masks from Deir el-Bahri'. *JEA* 86 (2000), 121–44.

Ritner, R.K. *The Mechanics of Ancient Egyptian Magical Practice*. Studies in Ancient Oriental Civilization 54. Chicago, 1993.

Roberson, J.A. *The Ancient Egyptian Books of the Earth*. Wilbour Studies in Egypt and Ancient Western Asia 1. Atlanta, GA, 2012.

— *The Awakening of Osiris and the Transit of the Solar Barques: Royal Apotheosis in a Most Concise Book of the Underworld and Sky*. OBO 262. Freiburg and Göttingen, 2013.

Roberts, A. 'Cult Objects of Hathor: An Iconographic Study'. 2 vols. Unpublished D. Phil. thesis, Oxford, 1984.

— *Hathor Rising: The Serpent Power of Ancient Egypt*. Totnes, 1995.

— *My Heart My Mother: Death and Rebirth in Ancient Egypt*. Rottingdean, 2000.

— *Golden Shrine, Goddess Queen: Egypt's Anointing Mysteries*. Rottingdean, 2008.

— 'Invisible Hathor: Rising Dawn in the Book of Day' in E. Frood and A. McDonald (eds.), *Decorum and Experience: Essays in Ancient Culture for John Baines*. Oxford, 2013, 163–9.

Robinson, J.M (ed.). *The Nag Hammadi Library in English*. Third revised edn. Leiden, New York, Copenhagen and Cologne, 1988.

Rochberg-Halton, F. 'Elements of the Babylonian Contribution to Hellenistic Astrology'. *JAOS* 108 (1988), 51–62.

Roob, A. *Alchemy and Mysticism: The Hermetic Museum*. Cologne, 2005.

Roth, A.M. and Roehrig, C. 'Magical Bricks and the Bricks of Birth'. *JEA* 88 (2002), 121–39.

Rothenberg, B. *Timna: Valley of the Biblical Copper Mines*. London, 1972.

— *et al.* (eds.). *Sinai: Pharaohs, Miners, Pilgrims and Soldiers*. Bern, 1979.

— and Bachmann, H.G. *The Egyptian Mining Temple at Timna*. London, 1988.

Roulin, G. *Le livre de la nuit: Une composition égyptienne de l'au delà*. 2 vols. OBO 147. Freiburg and Göttingen, 1996.

Rūmī, J. *Masnavi*, Book I. Trans. A. Williams, *Spiritual Verses: The First Book of the Masnavi-ye Ma'navi*. London, 2006.

— *Masnavi*, Book 2. Trans. R.A. Nicholson, *The Mathnawi of Jalálu'ddin Rúmi edited from the Oldest Manuscripts Available: With Critical Notes, Translation & Commentary 2: Containing the Translation of the First & Second Books*. E.J.W. Gibb Memorial Publications, New Series 4. Reprinted edn. London, 1972.

— *Selected Poems*. Trans. C. Barks with J. Moyne, A.J. Arberry and R. Nicholson. London, 2004.

Rundle Clark, R.T. *Myth and Symbol in Ancient Egypt*. Second paperback edn. London, 1978.

Ruska, J. *Turba Philosophorum: Ein Beitrag zur Geschichte der Alchemie*. Quellen und Studien zur Geschichte der Naturwissenschaften und der Medizin 1. Berlin, 1931.

Salazar, E. Garcés. *Astronomia Inka: Arqueoastronomía & Etnoastronomía*. Second edn. Lima, 2014.

Scharff, A. 'Ein Denkstein der römischen Kaiserzeit aus Achmim'. *ZÄS* 62 (1927), 86–107.

Schenkel, W. 'Die Farben in ägyptischer Kunst und Sprache'. *ZÄS* 88 (1963), 131–47.

Schimmel, A. *As Through a Veil: Mystical Poetry in Islam*. Paperback edn. Oxford, 2001.

Schmidt, H.C. and Willeitner, J. *Nefertari: Gemahlin Ramses' II*. Mainz am Rhein, 1994.

Schorsch, D. 'Precious-Metal Polychromy in Egypt in the Time of Tutankhamun'. *JEA* 87 (2001), 55–71.

Schott, S. *Urkunden mythologischen Inhalts 2: Bücher und Sprüche gegen den Gott Seth*. Leipzig, 1939.

Schwaller de Lubicz, R.A. *Sacred Science: The King of Pharaonic Theocracy*. Trans. A. and G. VandenBroeck. Rochester, VT, 1982.

Sethe, K. *Amun und die acht Urgötter von Hermopolis: Eine Untersuchung über Ursprung und Wesen des ägyptischen Götterkönigs*. Berlin, 1929.

Sezgin, F. *Geschichte des arabischen Schrifttums 4: Alchimie–Chemie–Botanik–Agrikultur bis ca. 430H*. Leiden, 1971.

Sezgin, U. Review of I. Vereno, *Studien zum ältesten alchemistischen Schrifttum*. *ZGAW* 12 (1998), 350–67.

— 'Ein arabischer Text (4./10. Jahrhundert) über Könige von Ägypten gewährt Einblicke in das spätantike Ägypten zugleich V. und letzter Teil von *Pharaonische Wunderwerke* bei Ibn Waṣīf aṣ-Ṣābi' und al-Mas'ūdī'. *ZGAW* 16 (2004/5), 149–223.

Shaltout, M. and Belmonte, J.A. 'On the Orientation of Ancient Egyptian Temples: (1) Upper Egypt and Lower Nubia'. *JHA* 36 (2005), 273–98.

Sherwood, D.N. 'Alchemical Images, Implicit Communication, and Transitional States'. *Spring: A Journal of Archetype and Culture* 74 (2006), 233–62.

Siorvanes, L. *Proclus: Neo-Platonic Philosophy and Science*. Edinburgh, 1996.

Smith, M. *The Mortuary Texts of Papyrus BM 10507*. Catalogue of Demotic Papyri in the British Museum 3. London, 1987.

— *Following Osiris: Perspectives on the Osirian Afterlife from Four Millennia*. Oxford and New York, 2017.

Sophocleous, S. *Atlas des représentations chypro-archaïques des divinités*. Gothenburg, 1985.

Spalinger, A.J. 'Historical Observations on the Military Reliefs of Abu Simbel and other Ramesside Temples in Nubia'. *JEA* 66 (1980), 83–99.

Stapleton, H.E., Lewis, G.L. and Sherwood Taylor, F. 'The Sayings of Hermes quoted in the *Mā'-al-waraqī* of Ibn Umail'. *Ambix* 3 (1949), 69–90.

Steindl-Rast, D. and Lebell, S. *Music of Silence: A Sacred Journey through the Hours of the Day*. Berkeley, CA, 2002.

Sternberg el-Hotabi, H. and Kammerzell, F. *Ein Hymnus an die Göttin Hathor und das Ritual 'Hathor das Trankopfer*

Darbringen' nach den Tempeltexten der griechisch-römischen Zeit. Rites égyptiens 7. Brussels, 1992.

Stewart, H.M. 'A Crossword Hymn to Mut'. *JEA* 57 (1971), 87–104.

Stol, M. 'Embryology in Babylonia and the Bible' in V.R. Sasson and J.M. Law (eds.), *Imagining the Fetus: The Unborn in Religion, Myth and Culture*, Oxford and New York, 2009, 137–75.

Stolzenberg, D. 'Unpropitious Tinctures. Alchemy, Astrology & Gnosis according to Zosimos of Panopolis'. *Archives Internationales d'Histoire des Sciences* 49 (1999), 3–31.

Strabo. *The Geography of Strabo*. 8 vols. Trans. H.L. Jones. Cambridge (Mass.) and London, 1932.

Stricker, B.H. 'La prison de Joseph'. *AcOr* 19 (1943), 101–37.

— *De Geboorte van Horus*. 5 vols. Mededelingen en Verhandelingen van het Vooraziatisch-Egyptisch Genootschap "Ex Oriente Lux" 14, 17–18, 22, 26. Leiden, 1963–89.

Stroumsa, S. and Sviri, S. 'The Beginnings of Mystical Philosophy in al-Andalus: Ibn Masarra and his *Epistle on Contemplation*'. *JSAI* 36 (2009), 201–53.

Taylor, F. Sherwood. 'The Alchemical Works of Stephanos of Alexandria, part 1'. *Ambix* 1 (1937), 116–39.

— 'The Alchemical Works of Stephanos of Alexandria, part 2'. *Ambix* 2 (1938), 38–49.

Taylor, J.H (ed.). *Journey through the Afterlife: Ancient Egyptian Book of the Dead*. London, 2010.

Telle, J (ed.). *Rosarium Philosophorum: Ein alchemisches Florilegium des Spätmittelalters*. 2 vols. Weinheim, 1992.

Thompson, D.J. *Memphis under the Ptolemies*, Princeton, NJ, 1988.

Trismosin, S. *Splendor Solis*. Trans. J. Godwin. Introduction and Commentary, A. McLean. Grand Rapids, MN, 1991.

Troy, L. *Patterns of Queenship in Ancient Egyptian Myth and History*. Uppsala Studies in Ancient Mediterranean and Near Eastern Civilizations 14. Uppsala, 1986.

— 'Mut Enthroned' in van Dijk (ed.), *Essays on Ancient Egypt in Honour of Herman te Velde*, 301–15.

Ullmann, M. *Die Natur- und Geheimwissenschaften im Islam*. Handbuch der Orientalistik 1, Der nahe und der mittlere Osten. Ergänzungsband 6.2. Leiden, 1972.

Uždavinys, A. *Philosophy as a Rite of Rebirth: From Ancient Egypt to Neoplatonism*. Westbury, 2008.

— *Ascent to Heaven in Islamic and Jewish Mysticism*. London, 2011.

Velde, H. te. *Seth, God of Confusion: A Study of His Role in Egyptian Mythology and Religion*. Leiden, 1977.

Vereno, I. *Studien zum ältesten alchemistischen Schrifttum: Auf der Grundlage zweier erstmals edierter arabischer Hermetica*. Islamkundliche Untersuchungen 155. Berlin, 1992.

Viano, C. 'Olympiodore l'alchimiste et les présocratiques: une doxographie de l'unité (De arte sacra, §18–27)' in Kahn and Matton (eds.), *Alchimie: Art, histoire et mythes*, 95–150.

Vittmann, G. 'Beobachtungen und Überlegungen zu fremden und hellenisierten Ägyptern im Dienste einheimischer Kulte' in Clarysse et al. (eds.), *Egyptian Religion: The Last Thousand Years* 2, 1231–50.

Voragine, J. de. *The Golden Legend: Readings on the Saints* 1. Trans. W.G. Ryan. Princeton, NJ, 1993.

Waitkus, W. *Die Texte in den unteren Krypten des Hathortempels von Dendera: Ihre Aussagen zur Funktion und Bedeutung dieser Räume.* MÄS 47. Mainz am Rhein, 1997.

Walbridge, J. *The Wisdom of the Mystic East: Suhrawardī and Platonic Orientalism.* Albany, 2001.

Wegner, J. 'A Decorated Birth-Brick from South Abydos: New Evidence on Childbirth and Birth Magic in the Middle Kingdom' in D.P. Silverman, W.K.Simpson and J.Wegner (eds.), *Archaism and Innovation: Studies in the Culture of Middle Kingdom Egypt.* New Haven and Philadelphia, 2009, 447–96.

Wente, E.F. 'Mysticism in Pharaonic Egypt?'. *JNES* 41 (1982), 161–79.

Whitehouse, H. 'Roman in Life, Egyptian in Death: The Painted Tomb of Petosiris in the Dakhleh Oasis' in O.E. Kaper (ed.), *Life on the Fringe: Living in the Southern Egyptian Deserts during the Roman and early-Byzantine Periods. Proceedings of a Colloquium held on the Occasion of the 25th Anniversary of the Netherlands Institute for Archaeology and Arabic Studies in Cairo, 9–12 December 1996.* CNWS Publications 71. Leiden, 1998, 253–70.

Wieck, R.S. *Time Sanctified: The Book of Hours in Medieval Art and Life.* New York and Baltimore, 1988.

Winter, E. 'Der Herrscherkult in den ägyptischen Ptolemäertempeln' in Maehler and Strocka (eds.), *Das ptolemäische Ägypten,* 147–60.

Winter, J.G (ed.). *Papyri in the University of Michigan Collection: Miscellaneous Papyri.* Michigan Papyri 3. University of Michigan Studies, Humanistic Series 40. Ann Arbor, MI, 1936.

Yates, F.A. *The Rosicrucian Enlightenment.* Paperback edn. London and New York, 2002.

Youtie, H.C. 'The Heidelberg Festival Papyrus: A Reinterpretation' in P.R. Coleman-Norton (ed.), *Studies in Roman Economic and Social History in Honor of Allan Chester Johnson.* Princeton, NJ, 1951, 178–208.

Zandee, J. *De Hymnen aan Amon van Papyrus Leiden I 350.* OMRO 28. Leiden, 1947.

Ziai, H. 'Shihāb al-Dīn Suhrawardī: Founder of the Illuminationist School' in Nasr and Leaman (eds.), *History of Islamic Philosophy,* 434–64.

Zolla, E. 'The Archetype of the Supernatural Lady from ancient Arabia to Edwardian England'. *Seshat: Cross-Cultural Perspectives in Poetry and Philosophy* 6 (2003), 60–85.

INDEX

A

SOURCES OF THE ILLUSTRATIONS